Mastering

Biology

D0363230

Macmillan Master Series

Accounting
Advanced English Language
Advanced Pure Mathematics
Arabic
Banking
Basic Management
Biology
British Politics
Business Administration
Business Communication
Business Law
C Programming
C++ Programming
Catering Theory
Chemistry
COBOL Programming
Communication
Databases
Economic and Social History
Economics
Electrical Engineering
Electronic and Electrical Calculations
Electronics
English Grammar
English Language
English Literature
French
French 2
German
Global Information Systems
Human Biology

Internet
Intranets
Italian
Italian 2
Java
Manufacturing
Marketing
Mathematics
Mathematics for Electrical and Electronic
   Engineering
Microsoft Office
Modern British History
Modern European History
Modern World History
Network Operating Systems
Pascal and Delphi Programming
Philosophy
Photography
Physics
Psychology
Science
Shakespeare
Social Welfare
Sociology
Spanish
Spanish 2
Statistics
Study Skills
Systems Analysis and Design
Visual Basic
World Religions

Macmillan Master Series
Series Standing Order ISBN 0–333–69343–4

You can receive future titles in this series as they are published by placing a standing order.
Please contact your bookseller or, in case of difficulty, write to us at the address below with
your name and address, the title of the series and the ISBN quoted above.

Customer Services Department, Macmillan Distribution Ltd
Houndmills, Basingstoke, Hampshire RG21 6XS, England

Mastering

Biology

Third Edition

O F G Kilgour and P D Riley

MACMILLAN

Dedication

To Irene Rachelle and Janet Katherine

First published 1999 by
MACMILLAN PRESS LTD
Houndmills, Basingstoke, Hampshire RG21 6XS
and London
Companies and representatives throughout the world

ISBN 0–333–66058–7 paperback

A catalogue record for this book is available from the British Library.

This book is printed on paper suitable for recycling and made from
fully managed and sustained forest sources.

10   9   8   7   6   5   4   3   2   1
08 07 06 05 04 03 02 01 00 99

Typeset by EXPO Holdings, Malaysia

Printed in Hong Kong

Contents

Preface to third edition

This book is about living things and the genetic material they contain. Life is concerned with the process of duplication of genetic material; death discontinues the process.

In this book the biological principles of life support, life continuation and organism interrelationships are considered as they relate to the vital component of living things – genetic material.

The book covers the needs of all GCSE Biology examination syllabuses in the UK, and meets the requirements of first year college courses worldwide.

The practical work of scientific investigation required for the GCSE Science One syllabus receives specific coverage in chapter one, together with instructions for practical work relevant to the topic under consideration in all chapters.

Terminology, classification of organisms, units and symbols used throughout the book conform to the recommendations of the Institute of Biology 1989.

O.F.G. Kilgour
Bae Colwyn, Cymru, Wales, 1998

Acknowledgement

Fred Kilgour would like to acknowledge the following people.

Peter Riley, who contributed a substantial number of revisions to this edition, as well as objectives, summaries and an abundance of questions in every chapter.

Suzannah Tipple, the senior commissioning editor, who has brought all our diverse contributions together in an attractive new format for this third edition.

Meinwen Parry, Vera Stirling and David Kilgour who contributed to the original edition which continues to provide the main central structure of the book and to whom Fred would like to give his sincere thanks.

Introduction

What is biology?

Everything on the Earth is either a living thing or a non-living thing. Living things have a very distinctive feature, or **characteristic**, in that they **grow** or increase in amount; compared to non-living things which do not grow. The non-living surface of the first formed Earth is crust. It has not grown or changed since its formation – it is still rock, but the living things have grown and covered the Earth's surface.

Genetic material

All living things have the vital substance of life, called **genetic material**; non-living things are without this vital substance. The genetic material is a very complex chemical substance with the outstanding characteristic of being able to double or **duplicate** itself. This process of duplication is the basis of life, or the 'life process'.

ONE part of genetic material	+	nutrient materials and energy	→	TWO parts of genetic material

Once this process of duplication of genetic material stops, the living organism **dies**.

What is the job of genetic material?

Apart from duplicating itself during growth, the genetic material has two important jobs concerned with:
• **structure** – how the living organism is built up, and
• **function** – how the living organism works.

Living organisms have form

The consequence of building up a structure gives the living organism a characteristic **form** to what is called its **body**; this makes it possible to recognise a buttercup plant and distinguish it from a fish, for example.

Living organisms have functions

A number of important and different jobs or **functions** take place within the body of a living organism. All are concerned with keeping the duplication of genetic material or 'life process' going – keeping the organism alive, protecting it from harm, and making sure the genetic material is continued on Earth in the future.

Life support functions

The duplication process needs raw materials (nutrients) and energy in order to make new genetic material. The following are distinctive life support jobs in living things:
- **Nutrition** or 'feeding' brings the raw materials (nutrients) to make genetic material.
- **Respiration** supplies the energy to make the new genetic material.
- **Transport** circulates materials through the organism's body.
- **Homeostasis** maintains relatively steady conditions inside the organism's body; it involves controlling temperature, water levels, sugar levels and removing waste chemical materials. Without this the organism could overheat, become too cool, swell up, shrink or become poisoned by excess sugar and chemical waste.

Life protection functions

Genetic material is an extremely delicate material in need of continuous protection. The main protective function is provided by a characteristic known as irritability. This gives a living organism:
- **Sensitivity** or the ability to detect changes in and around its body;
- the ability to **respond** or act against the changes.

The most important responses are **movement** of part or all of the body, together with **secretion** of various fluids, both of which try to compensate for the changes which are taking place.

Both processes must happen smoothly and a further process of **coordination** makes sure that response and sensitivity work together.

Life continuation functions

The continuation of life on Earth happens through:
- **growth** or the increase of an organism's body, seen particularly well in the growth or spread of a tree body;
- **reproduction** by various means, e.g. rabbit numbers increase rapidly by reproduction; grass seeds into a lawn or meadow by reproduction. In both processes the genetic material has either been passed on into new body parts or into completely new individuals.

Interaction

This is a feature of all living things – they must interact with others of the same kind, or with others of a different kind, or with non-living things which make up the soil, seas, lakes and air around them, in other words their **environment**.

Summary

The following list summarises the essential characteristics of living organisms.
- **Body form and structure:** all living things have a distinct body form and structure.
- **Life support functions:** nutrition, respiration, transport and homeostasis.
- **Life protection functions:** irritability, which encompasses sensitivity, response (movement or secretion) and coordination.
- **Life continuation functions**: growth and reproduction.
- **Interaction** of living things with each other and with non-living things.

This is what biology is all about. This first introduction may seem rather formidable, but it will soon fall into place as you proceed through the main themes of this book, where the structure and function of living organisms are explained chapter by chapter, finally closing with an explanation of how living and non-living things interact on Earth.

Introducing biology and the scientific investigation process

Objectives

When you have completed this chapter you should be able to:
- identify parts of the scientific process
- identify safety isssues in experimental work
- operate biological instruments
- make accurate measurements
- display your work clearly
- perform calculations on your results
- evaluate your work

Biology is a science, and in this first chapter the investigation processes of science are introduced. It is through them that we have built up our knowledge and understanding of living things. In this book the study of biology is presented in five themes. They take you from considering how to identify a living organism to examining ways of maintaining life on Earth for future generations. The themes are listed in Table 1.1 on page 5, together with a brief description of their contents, so you can see at a glance where the topic you are studying relates to the whole of the subject. (Use the contents list and the index to find the information you need.) The table also shows how biology can be studied from considering living organisms to examining how they work and finally learning about the conditions they need for survival.

1.1 The scientific investigation process

Science is knowledge concerning things both living and non-living. **Scientific investigation process** is the *way* in which this knowledge is collected together.

The scientific investigation process can be divided into five stages:
1 Planning experimental procedures;
2 Using apparatus and materials;
3 Obtaining evidence;
4 Analysing evidence and drawing conclusions;
5 Evaluating evidence.

Each stage is divided further. For example, planning experimental procedures involves turning an idea into one that can be investigated, making a prediction and considering how factors can be isolated and observed.

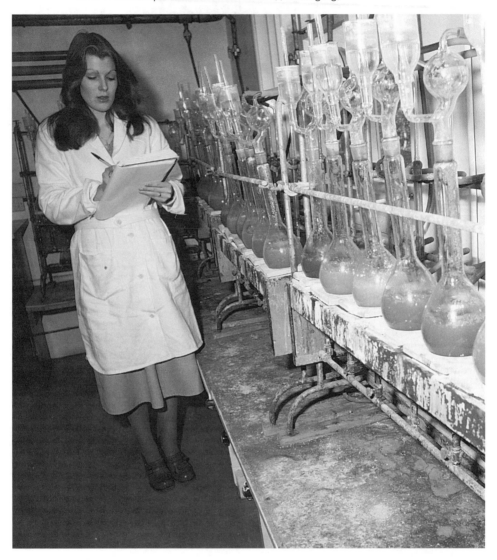

Figure 1.1 Biologists enjoy very wide-ranging work from studying molecules to the whole living environment of the Earth (the biosphere) and even searching for life in other places in the universe (f/8 Imaging)

Preliminary work

Some experiments which have a large number of stages or things to manipulate may be rehearsed first to see if there are any problems that may not have been thought of, such as having enough space or difficulties with the arrangement of apparatus.

Predictions

Where possible you should make a prediction about what you think might happen in the experiment. This helps to keep your mind focused through the whole experiment.

Table 1.1   The themes of the book.

Theme	Chapter
I Body form and structure	2
II Life support processes	
• Green plants – the food producers	3
• Nutrition and the food consumers	4
• How living things transport materials	5
• Respiration	6
• Homeostasis – the steady state	7
III Life protection processes	
• Irritability	8
• Response, support and movement	9
IV Life continuation processes	
• Reproduction	10
• Growth and development	11
• Genetics	12
• Evolution	13
V Interrelationships	
• Natural ecosystems	14
• The effect of modern humans on the ecosystem	15

However, you should not let it influence your observations and measurements. You must make these as accurately as you can and compare your results with your prediction at the end of the investigation.

Variable factors

There are variable factors in every experiment. These are components that have different values. They may not have a number value, for example the different colours of flowers visited by bees. The number value may be a definite number, such as the number of seeds you use in an experiment, or it may be one of a range of values, such as 35.5°C on a thermometer scale.

In the planning of an experiment you must identify all those variable factors which may affect the result. For example, in an experiment on photosynthesis, both light and temperature are important but in an experiment on enzymes, light may not be important while temperature is extremely important. You may find that revising your work on the topic under investigation may help you think of more variables that may be important.

When you have identified all the variable factors, decide which one you are going to change (the **independent variable factor**) and which one you are going to measure (the **dependent variable factor**). Follow this by identifying all the variable factors you are going to control so that they do not affect the relationship between the independent and dependent variable factors. For example, if you are measuring the effect of temperature on the speed of enzyme action you would keep the concentrations of the enzyme and the substrate (the substance the enzyme acts on) the same throughout the tests. You would change only the temperature (the independent variable) and measure the rate of the reaction (the dependent variable).

In the laboratory, factors can be carefully controlled but in fieldwork, factors such as temperature, humidity and light intensity cannot be controlled. They can be measured during the experiment and their readings compared with the result of the experiment. For example, in an investigation into the movements of arthropods using harmless pitfall traps, the numbers of animals captured can be compared with the air temperature and its humidity. Such a comparison may reveal a relationship between animal activity and environmental factors which can be taken into consideration in planning further experiments.

Safety

In the course of planning an investigation you must take account of safety requirements. Here are some general points that you must consider. Your teacher or lecturer will tell you of others which are specific to your school or college or to the investigation you are carrying out.

Personal protection and hygiene

For your own personal protection you must wear a cotton laboratory coat correctly buttoned up, or a disposable polythene apron. Also you will need to know where protective eyeshields, goggles, gloves, dust face-masks and safety screens are located. Personal hygiene means washing your hands before and after experimental work. You *must not* eat or drink anything in the laboratory.

Laboratory protection and hygiene

1 Fire-drill instructions must be known and understood and the fire exit doors located.
2 Unsupervised work is forbidden and a teacher must be present at all times.
3 The first aid kit position must be known.
4 Spillages and breakages must be reported immediately and the location of a special refuse disposal container must be known. There must be an understanding of approved first aid emergency treatments.
5 Water and gas taps and electric switches must be turned off after use.
6 Bench tops should be wiped down after use with disinfectant solution.
7 Pieces of equipment must be returned clean and washed to their correct locations unless otherwise advised by the teacher/lecturer. Do not interfere with experimental work that does not concern you, and leave the laboratory as you found it!
8 Electrical equipment must never be operated with wet hands or near spilt water.

General laboratory equipment

Glassware is of two main kinds:
• soft or soda glass, which is cheap and melts easily; it may crack if heated suddenly;
• hard, or boro-silicate 'Pyrex' glass, which is more expensive and does not melt easily or crack on heating.
Glassware includes glass rods, tubing, test tubes, beakers, flasks, filter funnels, watch glasses, specimen bottles and Petri dishes. Wash glassware after use in a warm detergent solution, rinse and dry or leave to air dry. If your teacher or lecturer advises picking up broken glassware, gloves must be worn for protection. If your teacher or lecturer advises connecting glass tubing to rubber tubing and rubber stoppers, the tubing and rubber stopper holes should be moistened with soap solution before connection.

Hardware is mainly metal or wooden apparatus: iron tripod supports, retort stands, clamps and boss heads, wire gauzes. Wood is used for filter-funnel stands and test-tube racks.

- Always allow hot ironware to cool before handling.
- Recognise and know the names of all the general laboratory equipment.

Equipment for providing heat

Gas can come from three main sources: natural gas, bottled LPG (liquid petroleum gas) and coal gas. Check the source used in the laboratory. The most common type of burner used in the laboratory is the **Bunsen burner**. Instructions for correct use are as follows:

1 Connect the burner correctly to the gas tap, pressing the connecting gas tube firmly over the burner inlet and gas-tap outlet to prevent leakage of gas.
2 Close the air hole on the burner. Have a match, taper or lighter ready and turn on the gas but not fully. Place the flame or sparking device next to the top of the burner's chimney and light the gas quickly.
3 Regulate the height of the gas flame by turning the gas tap. Make sure that you do not lean over the burner at any time.
4 Regulate the heat of the flame by turning the collar or air regulator at the base of the chimney as shown in Figure 1.2. When the burner is on but not in use, turn the air regulator to make the cool or luminous flame so that it can easily be seen.

Figure 1.2 The three flames of a Bunsen burner

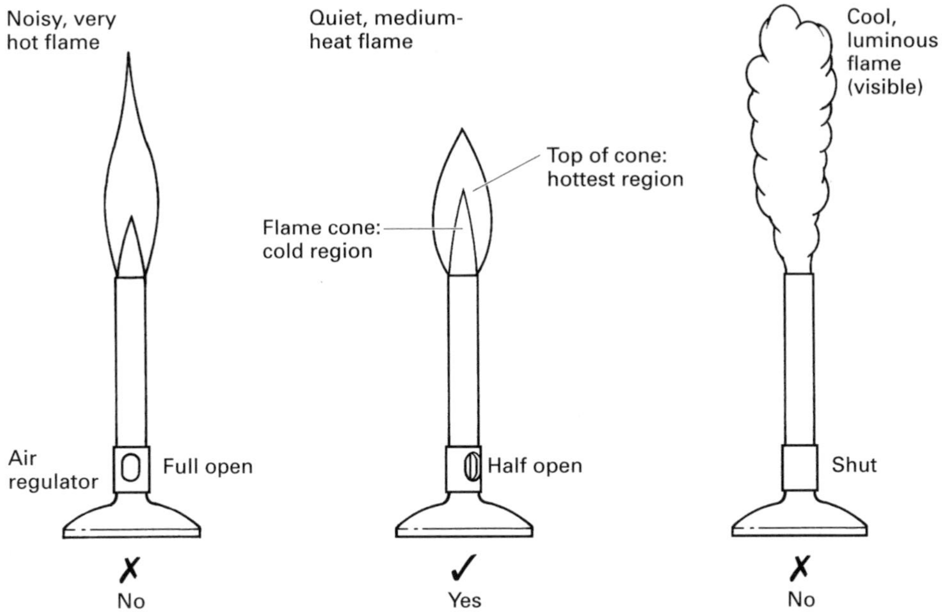

NB: Small ethanol spirit burners consisting of a glass container and wick can be used where there is no gas supply.

Test tube heating technique

Gas burners are sources of direct heat which can be applied directly to hard glass ('Pyrex') boiling tubes. These must be supported by a test tube holder. The boiling tube should not be more that one-fifth full of water or solution and its mouth must be directed away from yourself and other people. The boiling tube should be shaken *gently* throughout the heating process (see Figure 1.3).

A wire gauze must be positioned between a glass beaker or flat-bottomed flask and the iron supporting tripod stand; beakers and flasks must not be heated with a direct gas burner flame. Contents of beakers – water and aqueous solutions – should be stirred gently with a glass rod during heating. Wear protective eye shields in case hot liquid splashes outwards from the beaker while it is heating and boiling.

Only water and aqueous solutions of substances (i.e. substances dissolved in water) can be heated by these methods. Ethanol, which is flammable, cannot be heated in these ways and an indirect form of heating must be used.

Water-bath heating technique

Small amounts of flammable liquids such as ethanol must always be heated indirectly in a water-bath. Test tubes of water and aqueous solutions can be heated more safely in a water-bath than by direct heating.

Figure 1.3 Heating liquids in a boiling tube

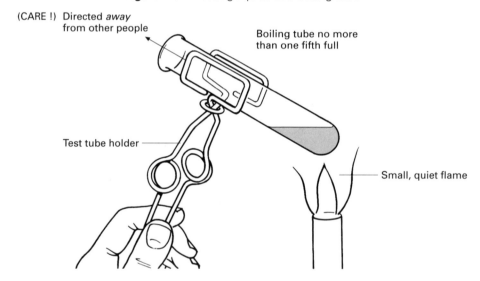

Note: If ethanol is being heated in a water-bath using gas, the gas flame *must* be extinguished before the tube of ethanol is put in the water-bath.

Biological processes require heat energy and function best at between 35–40°C. This is called the **optimal temperature**. It is achieved by using a bath of warm water, the temperature of which is carefully noted with a 0–110°C thermometer. The water-bath can be a beaker, a metal pan or a small saucepan, heated by gas or electricity (see Figure 1.4).

If the water-bath temperature rises too rapidly, it can be cooled down by adding small amounts of cold water. You should be able to maintain the water-bath tempera-

Figure 1.4 A simple water-bath

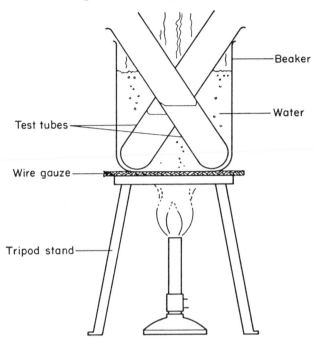

Beaker

Water

Test tubes

Wire gauze

Tripod stand

ture within 2°C of the working temperature. Thermostats are present in some water-baths to control the bath temperature automatically.

Chemical reagents

Chemical reagents must be used with care and under supervision. Some are potentially hazardous and can cause fire, explosion or death by poisoning. Others are harmful to the body as irritants, or are corrosive. Hazard symbols will be seen on the reagent bottle label, as indicated in Figure 1.5. Table 1.2 lists some chemical reagents you may expect to use, together with their hazards.

Solid chemical reagents must always be handled by means of special metal spatulas or spoons, placing the reagent on clean glass dishes or watch glasses. After use the spatula must be washed and dried to prevent contamination of the reagent-bottle contents. NEVER touch, taste or smell any chemical reagent. Droppers are used to transfer small amounts of liquid chemical reagents. THOROUGH RINSING of all containers, test tubes and beakers must always be carried out in order to remove all traces of chemical reagents, and prevent cross-contamination.

Thermometers

Thermometers are made of glass and are easily broken because of their long thin design. Great care should be taken at all times in handling thermometers and putting them safely on the bench. Many thermometers contain mercury but some thermometers are mercury-free. Mercury is toxic and the breakage of a mercury-in-glass thermometer must be reported to your teacher or lecturer so the hazard may be removed. Clinical thermometers (see Figure 1.14) must be disinfected after use.

Figure 1.5 Chemical reagent hazard symbols

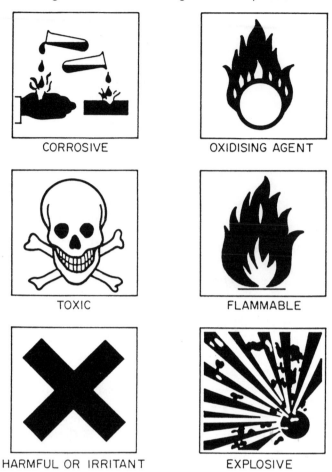

CORROSIVE

OXIDISING AGENT

TOXIC

FLAMMABLE

HARMFUL OR IRRITANT

EXPLOSIVE

Working with microorganisms

Cultures of microorganisms such as bacteria and fungi can be dangerous and a potential source of infectious disease. Some microorganisms such as fungal spores can bring on an asthma attack. All microorganisms must be treated as a potential hazard. Your teacher or lecturer will give you the latest information you need for working with microorganisms.

1.2 Obtaining evidence using biological instruments

During the course of your biological investigations you will use a range of equipment and instruments. They are used for examining and transfering organisms or parts of organisms (see Figure 1.6 for examples).

Table 1.2 Chemical reagents and their hazards

Chemical reagent	Hazards in use
Acetic acid (ethanoic acid)	*Corrosive*, vapour *irritates* eyes and lungs; burns skin and eyes
Alkaline pyrogallol	*Corrosive*, causes skin and eye burns *Poisonous* if swallowed and absorbed by the skin
Ammonium hydroxide or ammonia solution	Vapour *irritates* lungs; solution *corrosive* to skin and eyes, *poisonous* if swallowed
Copper sulphate, Benedict's and Fehling's solution	*Irritates* eyes; if swallowed causes vomiting and diarrhoea
Calcium chloride	*Irritant* dust and highly hygroscopic
Ethanol	*Flammable*; *poisonous* if swallowed; may cause blindness
Formaldehyde (methanal)	*Irritant* vapour affects lungs and eyes; *poisonous* if swallowed
Hydrochloric acid	Vapour *irritates* eyes, skin and lungs; liquid *corrosive* and *poisonous*
Hydrogen peroxide	*Oxidiser* supports burning; liquid *irritates* eyes and skin; causes vomiting and internal bleeding if swallowed
Iodine solution	Vapour *irritates* eyes and lungs; highly *poisonous* if swallowed
Soda lime	Causes skin and eye burns; *irritant* dust affects lungs; harmful if swallowed
Sodium and potassium hydroxide	*Corrosive*; solutions burn the skin and eyes; severe damage caused if swallowed
Propanone (acetone: nail-varnish remover)	*Flammable*; poisonous if inhaled or swallowed

These reagents are only available for use under supervision and issue by a qualified instructor.

Always wash instruments in hot soapy water and disinfectant after use. Keep metal instruments clean and dry in a cloth instrument roll.

Following instructions

At an early stage in your practical course or where a new technique or piece of apparatus is introduced you may be asked to perform an experiment following a set of instructions. They may be in the form of a **flow diagram** (see Figure 1.8) or as **written instructions**, for example as in Microscope use. The experiments featured in this book are presented as flow diagrams and instructions. You can build on them by planning and carrying out your own investigations under the supervision of your teacher or lecturer.

Magnifiers and microscopes

Magnifiers and microscopes magnify the size of an object's image. The degree of magnification is shown by the symbol × followed by the number of times the image is magnified.

Magnifiers come in a range of magnifications such as ×5, ×10 and ×20. Most small objects, such as certain insects and flower parts, are examined with a magnifier,

Figure 1.6 Biological instruments

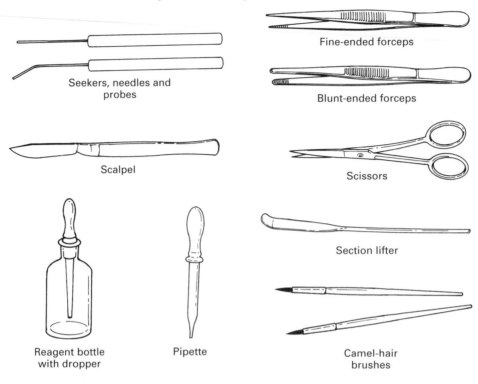

Seekers, needles and probes

Fine-ended forceps

Blunt-ended forceps

Scalpel

Scissors

Section lifter

Reagent bottle with dropper

Pipette

Camel-hair brushes

- **Pipettes** are used to transfer and collect small biological specimens in water.
- **Camel-hair brushes** and **section lifters** are used to collect and transfer thin sections of biological material.
- **Sharp needles** are used to pierce materials; blunt **seekers** and **probes** are used for parting tissues.
- **Forceps** are used to hold or transfer material; blunt forceps for large items, fine forceps for small items.
- **Scissors** and **scalpels** are used to cut through biological material.
- **Pooters** are used for picking up small organisms (see Figure 1.7).

also called a hand lens. You may need to draw objects using a magnifier and should practice this skill before carrying out an investigation where it is needed.

Microscopes (see Figure 1.9) are a means of magnifying an object using two lenses: the **eyepiece lens**, and the **objective lens**.

The simplest microscope for GCSE examination use has an eyepiece with a ×10 magnification, and objective lenses giving ×4, ×10 or ×20 magnification. The magnifications that can be made by combining the eyepiece with each objective lens in turn are:

Eyepiece	Objective	Magnification
×10	×4 →	×40
×10	×10 →	×100
×10	×20 →	×200

Figure 1.7   A pooter collecting tube

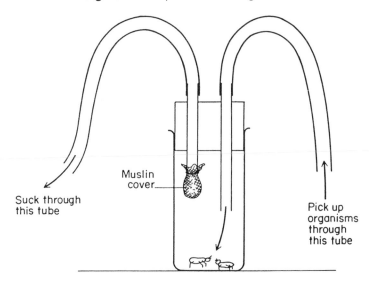

Suck through
this tube

Muslin
cover

Pick up
organisms
through
this tube

Figure 1.8   Flow diagram for sugar testing in leaf material

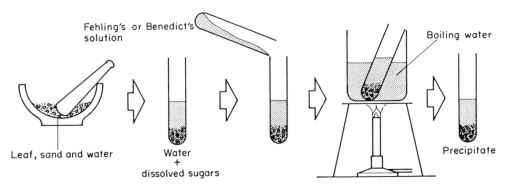

Fehling's or Benedict's
solution

Boiling water

Leaf, sand and water

Water
+
dissolved sugars

Precipitate

Microscope care

Microscopes are valuable biological instruments and expensive to buy. Lenses should not be touched with the fingers, and should be dusted with a clean camel-hair brush or a lens tissue. Always keep the microscope under its cover when not in use. Soft paper tissues can be used to dry a wet lens.

Microscope use

1   Direct light from a table lamp (not sunlight) onto the mirror and adjust the diaphragm to vary the illumination to suit you and the object you are looking at. (Some microscopes have a built-in light source and no mirror.)
2   Select the low-power ×4 objective lens and, with your eye level with the microscope stage, slowly bring the objective downwards by turning the focusing wheel until you see the objective nearly touch the microscope slide.

3 Looking through the eyepiece, slowly bring the microscope tube upwards by turning the focusing wheel until the image comes into focus.
4 Change to the high power ×10 or ×20 objective lenses and adjust the illumination to brighten the object.

Figure 1.9 The parts of a light or optical microscope

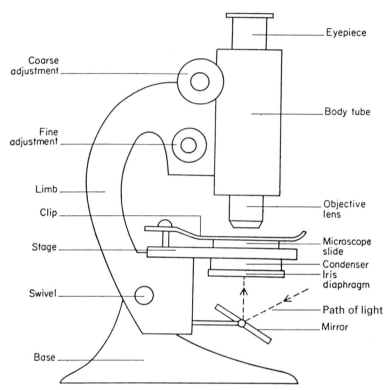

Preparation of specimens for microscopic examination

Cavity microscope slides

These slides have a small hollow into which a drop of water containing small organisms can be placed from a dropper pipette. The slide is then placed on the microscope stage, which must be in the horizontal and not the tilted position. After viewing, the specimen can be returned by gently rinsing it off into its original container.

Plain flat microscope slides

These slides can be used for preparing temporary mounts of sectioned or sliced specimens of, for example, potato sliced with a razor (see Figure 1.10). Alternatively, thin strips of lining tissue, or epidermis removed from onion bulbs, can be mounted in the following way:
1 Pick up the thin specimen tissue by means of a camel-hair brush and transfer it to the centre of the slide. It may be necessary to spread it out gently with needles.
2 Carefully add one drop of iodine solution to stain the tissue.

Figure 1.10 Specimen preparation for microscopic examination

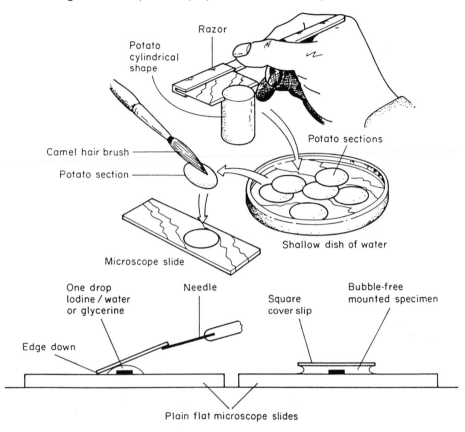

3 Carefully lower a square cover slip by one edge, gently guiding it into position with a needle. Continue lowering the cover slip until it lies flat over the specimen without any air bubbles being trapped between the glass.

4 If air bubbles are present, discard the specimen and repeat the procedure until no air bubbles are present. Any surplus liquid on the slide surface can be mopped up with blotting paper. This procedure is also illustrated in Figure 1.10.

Photographs through the microscope

Photographs of specimens taken through a light microscope are called **photo-micrographs** (see Figures 1.11 and 1.12) and are available as prints or 35 mm transparencies. The electron microscope provides photographic prints called **electron micrographs** which can show cell structures magnified as much as 500 000 times.

Making observations and measurements

There are two main kinds of observation: qualitative and quantitative.
- **Qualitative observations** involve descriptions and comparisons;
- **Quantitative observations** involve measurements of amounts.

Figure 1.11 A photomicrograph showing the structure of a maize, *Zea* sp. stem seen in transverse section (Griffin Biological Laboratories)

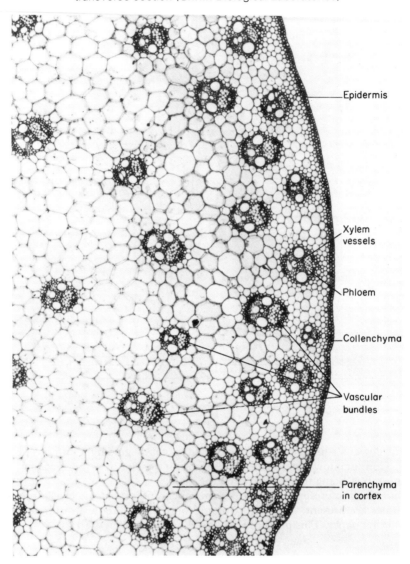

Epidermis

Xylem vessels

Phloem

Collenchyma

Vascular bundles

Parenchyma in cortex

Qualitative observations

Qualitative variables include colour, smell, taste, temperature and moisture (humidity). Qualitative observations include tests for chemicals which are of biological importance (see Table 1.3).

pH determination

Qualitative observations on the pH of a liquid can be made by comparing the colour of **universal indicator paper** with a colour chart after the paper has been dipped in the liquid. The range of colours span a range of pH values. The pH or hydrogen ion concentration scale ranges from pH 1 to pH 14. Liquids with a pH value of less than 7 are

Figure 1.12   A photomicrograph showing the structure of a maize, *Zea* sp. root seen in transverse section (Griffin Biological Laboratories)

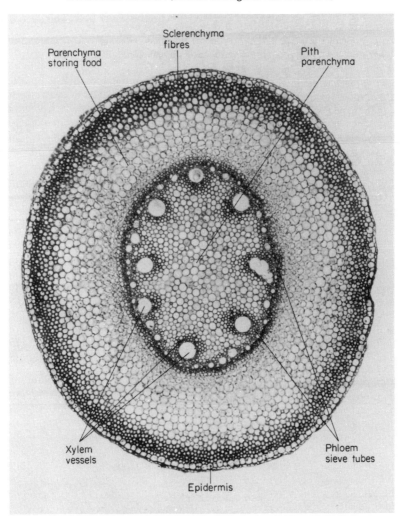

acidic whilst liquids with a pH of more than 7 are **alkaline**. A liquid with a pH of 7 is **neutral**. **Litmus papers** provide a less precise test for pH. If blue litmus paper is dipped in a liquid and the paper turns red the liquid is an acid. If red litmus paper is dipped in a liquid and the paper turns blue the liquid is an alkali.

Temperature

Temperature is a measure of the hotness or coldness of a substance compared with either freezing ice or boiling water. It can be recorded on a scale called the **Celsius scale**. Thermometers show the degree of hotness or coldness of a substance. They do not measure the amount of heat energy. Heat is measured in **Joules** by a calorimeter.

The Celsius (symbol °C) temperature scale is divided into 100 degrees between 0°C (melting ice) and 100°C (boiling water).

Table 1.3 Tests for the identification of chemical substances

Chemical substance	Tests
Water	Turns *white* anhydrous copper sulphate *blue* or cobalt chloride paper from *blue* to *pink*
Carbon dioxide	(i) *Clear* calcium hydroxide solution (lime water) turns *cloudy* (ii) Hydrogen carbonate indicator (thymol blue and cresol red) changes from a *red-purple* colour to *orange-yellow* (iii) Soda lime and sodium or potassium hydroxide absorb CO_2
Oxygen	Relights a glowing splint. Alkaline pyrogallol solution absorbs oxygen
Reducing sugars: glucose, fructose, maltose and ribose (sucrose is a non-reducing sugar)	Benedict's is better than Fehling's solution. Heat the blue solution and sugar together to boiling in a water bath. A *green* to *brown* or *red* precipitate forms. 'Clinistix' is a specific reagent strip test for *glucose*
Starch	*Pale brown* potassium iodide and iodine solution turns a *blue-black* colour
Lipids	(i) A warm sample pressed on paper makes a *translucent* permanent grease mark (ii) Mix oil sample with 3 cm³ ethanol and add 3 cm³ water: a milky *emulsion* forms
Proteins	(i) 'Albustix' reagent strip: colour changes from *yellow* to *green* (ii) 'Biuret' test: add few drops sodium hydroxide and 1% copper II sulphate: a *violet* colour appears
Vitamin C (ascorbic acid)	*Decolourises* a *blue* solution of dichlorophenol indophenol (DCPIP)

Figure 1.13 Temperature scale

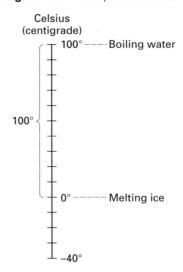

The most common thermometers are:
- **laboratory thermometer**: ranges –10 to +50°C in 0.5°C graduations, –10 to +110°C in 1°C graduations, and –10°C to + 360°C in 2°C graduations;
- **clinical thermometer**: range +35°C to +42°C in 0.25°C graduations;
- **maximum and minimum thermometer**: records the highest and lowest temperatures by metal markers. They are reset with a magnet. Range –10 to +110°C in 1°C graduations.

(See the note on thermometer safety on page 11.)

Figure 1.14   Clinical thermometer

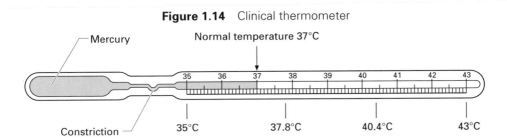

Humidity

Humidity is the relative amount of moisture in the air and is recorded as a relative percentage value by means of either a paper or hair **hygrometer**. Paper and hair are hygroscopic substances. This means they take in water vapour from the air. Normal relative humidity values are between 45 and 75%; dry is 20–30%, wet is 80–100%.

Quantitative observations

Prefixes

Prefixes are used as the first part of the term for units of measurement. For example:
- Mega- (M) = one million times or $\times 10^6$
- Kilo- (k) = one thousand times or $\times 10^3$
- Milli- (m) = one thousandth or $\times 10^{-3}$
- Micro (μ) = one millionth or $\times 10^{-6}$

Number

Counting the numbers of objects or organisms is often necessary when carrying out biological investigations. Small objects, such as daisy flower-parts or different coloured pea seeds, should be placed on a dark-coloured background such as black paper or black cloth. The dark background helps to show up the light coloured objects.

Count in fives, making 'five barred gates' as follows: ⊞, ⊞, ⊞. Thus three 'gates' equal 15 in total, and so on. Each stroke represents one item and each 'gate' represents five items.

Tally counters with push buttons can be used to count large numbers of objects or animals.

Time

Time is measured in seconds (s), minutes (min) and hours (h). There are 60 seconds in one minute and 60 minutes in one hour. There are 60 × 60 = 3600 seconds in one hour. A stopwatch or digital clock can be used to measure time.

Length

The basic unit of length is the millimetre (mm).

10 mm = 1 centimetre (cm)
1000 mm = 1 metre (m)
1000 m = 1 kilometre (km)

A metre rule (100 cm in length) has every centimetre divided into 10 millimetres (mm). When taking a reading it is essential that the eye is vertically over the mark to be read and not to one side which would cause a reading error (see Figure 1.15). Every student should be able to measure the length of objects correctly to within a millimetre (1 mm).

Figure 1.15 Correct scale reading method

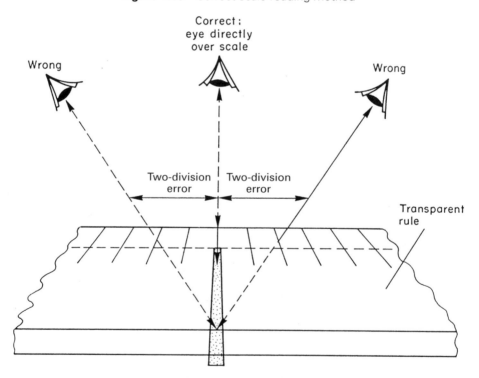

Callipers (see Figure 1.16) can be used to measure thickness or the inside and outside diameters of objects. The open calliper gap is then measured against a metre rule.

Surface area

Surface area is measured in square metres (m²), square centimetres (cm²) or square millimetres (mm²). Most biological specimens have an irregular shape and their surface

Figure 1.16 Combined callipers

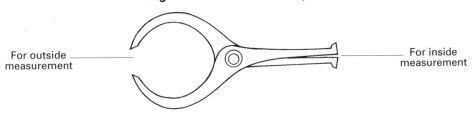

For outside ——————— measurement

For inside ——————— measurement

area is difficult to measure exactly. Two ways of measuring the surface area of a leaf are as follows:

Area by weighing method

1 The surface area of the leaf can be determined by tracing around the outline on to thick card, then cutting out the outline which is then carefully weighed (a grams).
2 A square of the same card is measured and its area is calculated by multiplying the square's length by its width. The square of card is weighed carefully (b grams).
3 The leaf surface area will be found as follows:
   (i) square card weighs b grams and has an area x cm^2
   (ii) leaf outline card weighs a grams and its surface area will be a grams x cm^2 divided by b grams = y cm^2 or area of one leaf surface. As the leaf has an upper and a lower surface the the total leaf surface area = y cm^2 x 2.

Squared graph paper method

Graph paper is made up of 1 millimetre squares (mm^2) and larger 1 centimetre squares (cm^2).
1 The leaf is placed on the graph paper and its outline is drawn on the paper.
2 The number of large (1 cm^2) squares within the leaf tracing is counted, then every small (1 mm^2) square that does not form part of a large cm square is counted by the 'gate' method.
3 The leaf area can then be calculated as follows:
   (i) Total number of large (cm^2) squares = X
   (ii) Total number of small (mm^2) squares outside of the large squares = Y
   (iii) Total area of one leaf surface = $X + \dfrac{Y}{100} cm^2$

Since the leaf has two surfaces the total leaf surface area = $2 \times X + \dfrac{Y}{100} cm^2$.

Quadrats are wood or metal frames measuring one square metre and are used to mark out areas of ground in ecological studies.

Volume

Liquid volume is mainly measured by means of a graduated **measuring cylinder** (see Figure 1.17a). Graduated **syringes** (Figure 1.17b) can also be used to measure volumes of liquids or gases. They are available as 10 cm^3 capacity in 0.2 cm^3 graduations, 50 cm^3 capacity in 1 cm^3 graduations and up to 1000 cm^3 in 10 cm^3 graduations. One litre = 1000 cm^3.

The volume of irregular shaped, non-living bodies can be determined by the **displacement method**. The object is lowered into a measuring cylinder half filled with water (a cm^3) until completely submerged. The new water level is noted as b cm^3. The volume of the object will then be obtained by subtracting the reading for the half-filled cylinder a cm^3 from b cm^3. Alternatively a displacement can may be used (Figure 1.17c).

Figure 1.17 Volume measurement

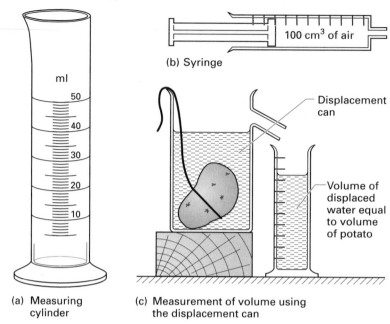

(b) Syringe

100 cm³ of air

Displacement can

Volume of displaced water equal to volume of potato

(a) Measuring cylinder

(c) Measurement of volume using the displacement can

Reading the liquid level in a measuring cylinder is an important technique in which the eye must be level with the lower liquid level of the meniscus as shown in Figure 1.18, otherwise reading errors will occur.

Figure 1.18 Volume scale reading

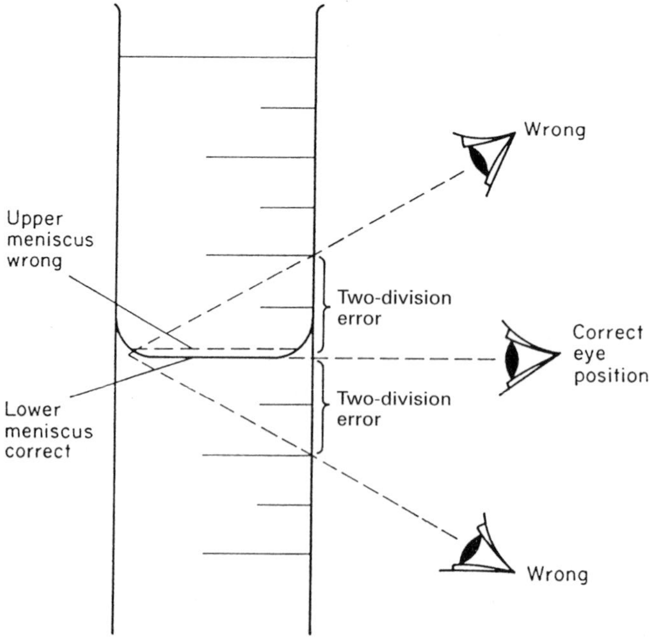

Wrong

Upper meniscus wrong

Two-division error

Correct eye position

Lower meniscus correct

Two-division error

Wrong

Mass or weight

The mass of an object can be determined by a spring balance. A small spring balance has a capacity of 100 g graduated in 1 g divisions, the specimen being supported on a small pan attached to a balance hook. Larger capacity spring balances are available which weigh up to 1000 g (1 kg). A simple lever arm balance with a range up to 250 g or 1000 g in 1 g divisions (see Figure 1.19) will weigh objects directly with reasonable accuracy.

Figure 1.19 Dual range lever arm balance

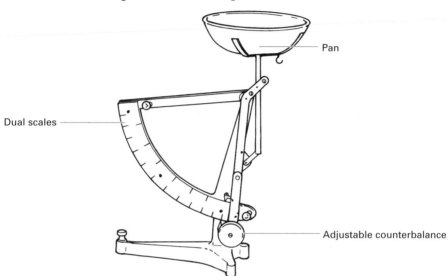

Double pan balances require a box of weights. They weigh to an accuracy of 10 mg or 0.01 g. Every student should be able to weigh an object to an accuracy of 1 g, i.e. if the accurate weight of a potato is 52 g, then a value of either 51 or 53 g is acceptable.

Solution concentration

This figure is generally indicated as a percentage weight (mass) to volume or % w/v. A known weight (mass) of substance is dissolved in 100 cm³ of water. For example, 5 g of substance or solute dissolved in 100 cm³ of water (the solvent) will make a 5% w/v solution.

A 0.5% solution could be made by dissolving 5 g of solute in 1000 cm³ of water; this is more convenient than trying to weigh out 0.5 g of substance.

Different solution strengths of common salt from 0.5, 1, 2 and 10% would be prepared using a simple balance and measuring cylinder. The solutions would then be placed in labelled bottles.

Surface area to weight or volume ratio

The relationship between surface area and volume is illustrated in Figure 1.20.
As plants and animals become larger, their **surface area** becomes *smaller* relative to their **volume**. Similarly, the distance from the centre of the body increases with increasing body size as the cube models show in Figure 1.20.

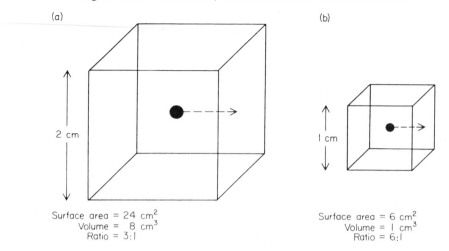

Figure 1.20 Relationship between surface area and volume

(a)

2 cm

Surface area = 24 cm^2
Volume = 8 cm^3
Ratio = 3:1

(b)

1 cm

Surface area = 6 cm^2
Volume = 1 cm^3
Ratio = 6:1

Heat energy

The **heat energy** contained within a substance, e.g. food or a mammal, is measured in **Joules** (J), kilojoules (kJ) and megajoules (MJ). It depends on the body's mass (weight) as well as its temperature.
- one kilojoule (kJ) = 1000 joules
- one megajoule (MJ) = 1000 kJ = 1 000 000 J
Various other methods can be used to measure heat energy using a calorimeter.

Pressure

Pressure is a force exerted by solids, liquids and gases, in or on the bodies of living organisms. The unit of pressure is the **Pascal** (Pa). One kilopascal, kPa = 1000 Pa. Weight per unit area, or grams per square millimetre is an alternative way of expressing pressure measurement.

Pressure changes are indicated by a **manometer** gauge. This is a U-shaped glass tube containing coloured water (see Figure 1.21).
- Equal levels in the manometer gauge show pressure inside the apparatus is equal to the outside air pressure.
- If the level towards the apparatus rises this shows the pressure outside the apparatus is greater than that inside.
- If the level away from the apparatus rises this shows that the pressure inside the apparatus is greater than the surrounding air pressure.
Differences in levels in the manometer are measured carefully with a ruler in centimetres (cm).

Building up your practical skills

When you begin practical work you must aim to use the apparatus, instruments and equipment *safely* and develop your observational skills such as those shown in Figures 1.15 and 1.17. This will give you a good foundation for developing your practical work further.

You must make and record enough observations to fulfil the aim of your investigation. As your skills and confidence in practical work grow you must aim to work out a routine for taking and recording your measurements, making them as accurate as

Figure 1.21 The manometer gauge

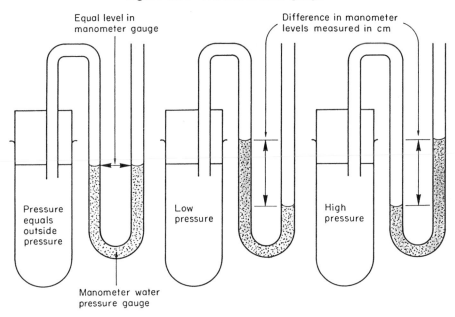

Equal level in manometer gauge

Difference in manometer levels measured in cm

Pressure equals outside pressure

Low pressure

High pressure

Manometer water pressure gauge

possible. If you feel the measurements need checking, repeat them. Make sure that your measurements cover a wide enough range for your purposes and again make sure you have enough data for analysis and drawing conclusions in the later stages of your investigation.

1.3 Data recording skills

All scientific data and information which have been obtained by qualitative and quantitative observations must be carefully recorded, displayed or presented in a special ring file, folio or notebook; this is a means of **information storage**. Alternatively data may be stored on specialised computer software.

Diagrams

Diagrams are essential records for describing specimens or laboratory apparatus. They are, in themselves, descriptions and do not need written descriptions, apart from **annotations** (labels) which are an essential part of any diagram. Similarities or differences between specimens can be illustrated in diagrams by drawing attention to these points by means of prominent arrows.

Line diagrams

These are drawn with a **single line** with a sharp pencil and show a clear outline shape of a specimen. They are drawn in two dimensions with *no shading*. They are not supposed to be artistic sketches showing perspective, three dimensions or background scenery (see Figure 1.22). Diagrams of laboratory apparatus should be drawn with a ruler, and stencils may be used for apparatus outlines.

Figure 1.22   Line diagram drawing

No! Too artistic – this is a 3D drawing

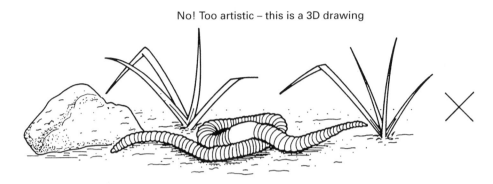

OK! It's diagrammatic

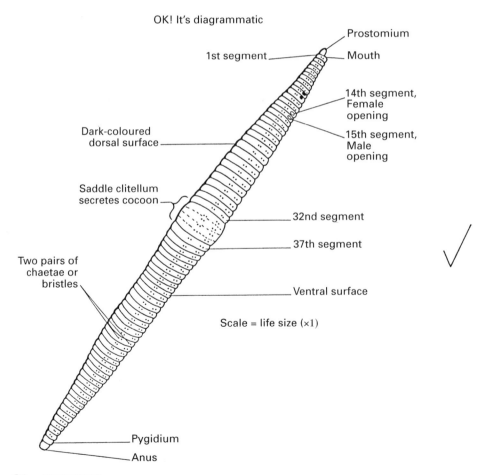

Plan diagrams

These can be compared to 'maps', showing areas where different tissues are found. This type of diagram is used for plant stem, root and leaf structure (see Figure 1.23).

Figure 1.23   A plan diagram

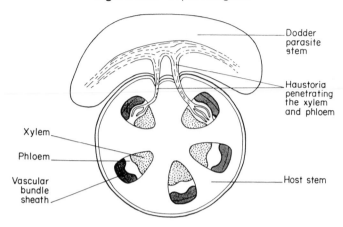

Labelling of *all* diagrams must be in ink with clear pointers or lines connecting the label with the named part of the apparatus or organism.

Scale size

- Life size has a scale equal to ×1.
- Magnified size has scales shown in whole numbers greater than one, e.g. ×3, ×50 or ×2000, etc. For example a photomicrograph of a human flea at scale ×40 will be 40 times its life size.
- Reduced size has scales shown in decimal values, for example ×0.5 is equal to half life size, ×0.1 is one tenth of life size.

Scale calculations

These are calculated from actual measurements made from photographs or diagrams provided in the examination.

- **Magnification** or **reduction** calculation is done by making actual measurements of length from a photograph or diagram, then *dividing* by the life size provided in the question. For example:

$$\frac{\text{diagram measurement}}{\text{life size}} = \text{magnification or reduction scale}$$

$$\frac{\text{diagram measurement}}{\text{life size provided}} = \frac{4.0 \text{ mm}}{0.2} = \times20$$

- **Life size** calculation is done by making actual measurements from a photograph or diagram and dividing by the magnification or reduction scale provided in the question. For example:

$$\frac{\text{diagram measurement}}{\text{magnification or reduction}} = \text{life size}$$

$$\frac{\text{measured length}}{\text{magnification}} = \frac{4.0 \text{ cm}}{3\times} = 1.33 \text{ cm life size}$$

Microscope eyepieces can be fitted with graticules (grids) to estimate the size of objects viewed under the microscope.

Tables

Numerical tables must be recorded neatly and accurately. The numbers must be written correctly showing the decimal point, for example 7.8, not 7 8.

Symbols for the different units of measurement must be shown correctly in the table heading, for example °C, g, cm, cm^2, cm^3. Data values are incomplete without the measurement unit symbol.

The data in Table 1.4 refer to the numbers of certain plants in a crop having a certain height.
- In this experimental investigation, practical measurement of the numbers of plants was by the 'gate' or tally count method, and height was recorded in centimetres with a 2 metre rule.
- The data are arranged in a neat tabular way showing the unit of height measurement (cm).

Table 1.4   Tabular record of number of plants and their heights

Number of plants	Height/cm
2	55
3	60
5	65
10	70
13	75
9	80
25	85
30	90
34	95
41	100
56	105
32	110
22	115
8	120
3	125

Calculations

Calculations are part of data processing and are used to find the meaning of data listed in a table. The following calculations are based on the data in Table 1.4.

Average values

Average values are calculated by adding together all the plant number or height values and dividing by the number of groups or sets. You will see in Table 1.4 that there are 15 sets or groups of plant heights and plant numbers. To calculate the average number of plants in a group:
1 **Add** all the plant numbers together in the results table from 2 +3 + 5 ... to 22 + 8 + 3 using a calculator. Total = 293.
2 **Divide** this by the total number of sets, i.e. 15. Therefore the average number of plants in each group = 293 ÷ 15 = 19.53.

To determine the average height of all the plants in the experimental investigation:
1 **Add** together all the plant height values in Table 1.4 from 55 cm + 60 cm ... to 120 cm + 125 cm. Using a calculator this totals 1350 cm.
2 **Divide** this total value by the number of sets, i.e. 15. Therefore average plant height = 1350 ÷ 15 = 90.0 cm.

Ratios

A **ratio** is the numerical relation that one quantity bears to another. The investigators whose results are shown in Table 1.4 grouped all plants of 85 cm and under as **dwarf plants** and the remainder (over 85 cm in height) were called **tall**. By this calculation 67 are dwarf (85 cm or under in height) and 293 – 67 = 226 are tall (over 85 cm in height).

Ratios are calculated by dividing the smallest number into the largest number:

Ratio of tall to dwarf plants = 226 tall ÷ 67 dwarf = 3.4 to 1

This is also written with the symbol ':' as shown:

Ratio of tall to dwarf plants = 3.4:1

Percentages

Percentage value is a part of one hundred expressed in hundredths. Percentages are calculated from the following relationship:

$$\frac{\text{number of certain quantity}}{\text{total number of all quantities}} \times 100 = \text{per cent } (\%)$$

For example, the percentage of dwarf plants in the investigation will be

$$\frac{67 \text{ dwarf}}{293 \text{ (total of tall and dwarf plants)}} \times 100 = 22.87\%$$

or out of all the plants, 22.87% are dwarf and 100% – 22.87%, or 77.13% are tall plants.

Pictorial data display

Data can be displayed by the following visual methods:

Pie charts

Pie charts are circular diagrams cut by radii into segments or sectors (pie slices) illustrating relative magnitudes or frequencies. A circle is composed of 360° and the angle of each sector can be measured with a protractor.

The calculations for pie diagrams are based on the following relationship:

$$\frac{\text{individual quantity}}{\text{total quantity}} = \frac{\text{degrees in sector}}{360° \text{ in circle}}$$

Thus the total quantity fills the whole circle, and individual component quantities occupy sectors. From Table 1.4, the number of dwarf plants, i.e. 67, that are 85 cm or under in height, can be shown as a pie-diagram sector:

$$\frac{67}{293 \text{ (total quantity)}} = \frac{\text{degrees in sector}}{360°}$$

$$\text{Degrees in sector} = \frac{360° \times 67}{293}$$

67 dwarf plants = 82.3°

A large circle is drawn and the 82° sector is marked off to produce the complete pie chart as shown in Figure 1.24.

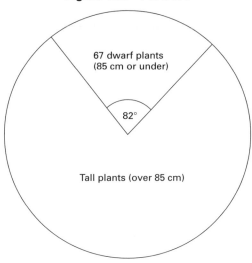

Figure 1.24 Pie chart

67 dwarf plants
(85 cm or under)

82°

Tall plants (over 85 cm)

Block diagrams

Block diagrams display data by means of narrow blocks of equal width, on squared or graph paper.

Histograms

Histograms are used for displaying data concerning one observed variable for one kind of organism, for example plant height, given in Table 1.4. The steadily increasing quantities, namely plant height figures, are arranged on the horizontal line or **X axis**, and the changing or fluctuating values for plant numbers are arranged on the vertical line or **Y axis**, as shown in Figure 1.25. In a histogram the edges of the boxes touch each other.

Bar charts

Bar charts are used for displaying data concerning more than one variable; for example the number of *different* plant species killed by the *same* weedkiller (see Figure 1.26). In a bar chart the blocks *do not* touch each other.

Graphs

Graphs are diagrams that show relationships between two changing or variable quantities. The graph must be drawn on squared paper; the horizontal line is used to record the steadily increasing quantities, such as time, whereas the vertical line is for variable fluctuating quantities, such as weight or temperature.

The following data concern changes in body weight of an insect over a time period of 24 days. Two variables are present, namely (1) body mass (weight) and (2) time. The changes in body mass with time are shown in Table 1.5.

Table 1.5 The change in body mass of an insect over 24 days.

Time/days	1	3	6	9	12	15	18	21	24
Body mass/mg	100	100	200	230	350	530	600	760	1200

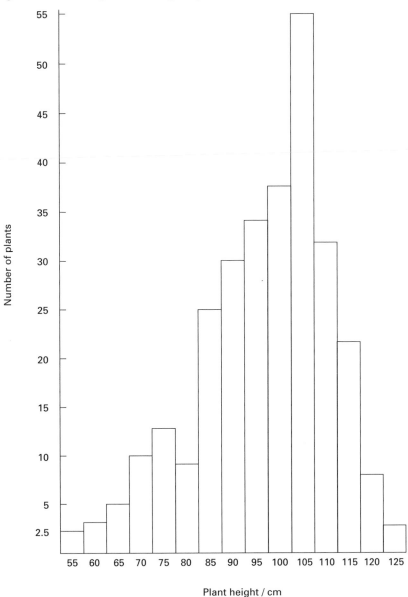

Figure 1.25 Histogram showing height distribution among one type of plant

Since time changes steadily (as an independent variable), we plot it on the X axis or horizontal line. Mass (weight) is plotted vertically on the Y axis since it shows fluctuation in the early data (as a dependent variable). Each mass and time value is marked neatly on the graph with a cross (a dot inside a circle may be used too) and each plotted position is joined by connecting lines to form a curve as shown in Figure 1.27.

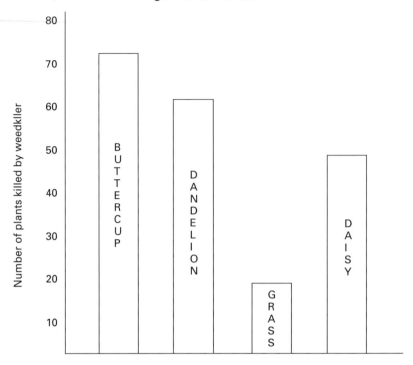

Figure 1.26 Bar chart

Weed species

1.4 Extracting information

Experimental data must be analysed, interpreted and conclusions drawn from them in order to give meaning to the results.

Information from graphs

Form or pattern

The graph in Figure 1.27 has a curve showing a distinct form or pattern. Parts labelled X and Y are very steep; parts labelled A, B and C are less steep; part C is almost horizontal. These curve forms or gradients have specific meanings:

- steep curves X and Y mean a *large* increase in mass occurs over a *short* period of time;
- less steep curves A and B mean a *small* increase in mass occurs over a *longer* period of time.
- horizontal curve C means *no change* in mass over a period of time.

Calculations

Calculations of mass changes can be performed for the different parts of the graph as follows:

Mass (weight) increase for A
= 230 − 200 = 30 mg

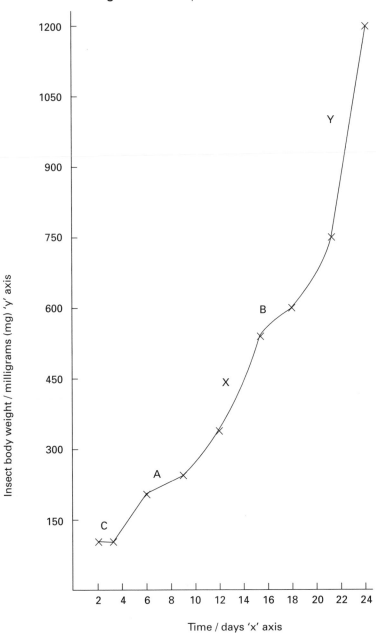

Figure 1.27 Graph construction

Percentage mass (weight) increase

$= \dfrac{30}{200} \times 100 = 15\%$

Mass (weight) increase for B
$= 600 - 530 = 70$ mg

Percentage mass (weight) increase

$$= \frac{70}{530} \times 100 = 13\%$$

Other graph portions show large mass increases. Between day 12 and 15 (X) the increase is

$530 - 350 = 180$ mg

The percentage mass increase for this time

$$= \frac{180}{350} \times 100 = 51\%$$

Between days 21 and 24 (Y) the increase

$= 1200 - 760 = 440$

The percentage mass increase for this time

$$= \frac{440}{760} \times 100 = 58\%$$

Interpolation and extrapolation

This is a process of filling in missing values. If the values are filled in between known values it is called **interpolation**. If the values are filled in to extend the range of known values it is called **extrapolation**.

The information in Table 1.6 is displayed in the graph in Figure 1.28. The data between 5°C and 15°C have been filled in or *interpolated* along the expected gradient, and the curve extended from 45°C or *extrapolated* in order to forecast changes.

Table 1.6   Information for the graph in Figure 1.29.

Temperature / °C	Time for colour change to occur / s
0	93
5	60
15	23
23	10
35	6
45	18

Optimum values

Optimum values for different conditions, e.g. temperature, are those most favourable for a certain biological process, and show as a peak or trough on the curve.

Information from tables

Maximum and minimum values

Tables of data provide such information, for example maximum (greatest) and minimum (least) nutrient content can be extracted from a food composition table (see Table 1.7). For example:

- **Water**   maximum value = cabbage   92 g
           minimum value = milk chocolate   1 g

Figure 1.28 Graph interpolation and extrapolation

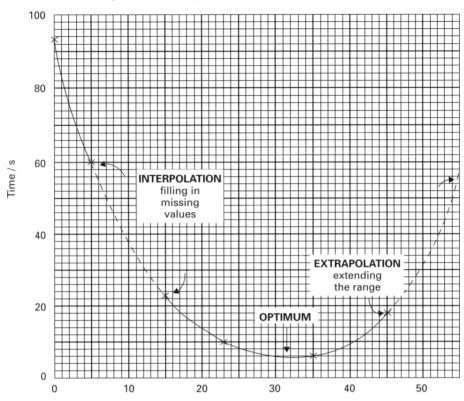

- **Protein** maximum value = roasted peanuts 30.0 g
  minimum value = apple 0.3 g

These tables can provide information for low lipid (fat) diets, or high protein diets.

Data variation

The values recorded in the data may vary due to two factors:

Personal error

This is caused by inability to measure accurately. For example, poor skills in reading scales (see Figures 1.15 and 1.17). Personal error due to poor measuring skills would show as inconsistencies in a graph curve such as that shown in Figure 1.29.

Biological variation

Biological variation in composition and function is found in all living organisms. For example, food composition varies as follows: between varieties of apples – some are sweet, some are less sweet; in salmon – more lipid is found near the head compared to the tail; people vary in body weight and height.

Table 1.7 Composition of 100 g edible portion of some foods. (Reproduced by permission of CIBA–Geigy from Documenta Giegy Scientific Tables)

Food	Water (g)	Protein (g)	Lipid (g)	Carbo-hydrate (g)	Fibre (g)	Available energy value (kJ)	Calcium (mg)	Iron (mg)	Vitamins					
									A (μg)	B_1 (mg)	B_2 (mg)	Nicotinic acid (mg)	C (mg)	D (μg)
Apple, sweet	84	0.3	0.4	15	0.9	244	6	0.3	–	0.04	0.02	0.2	30	–
Orange	87	0.9	0.2	11	0.8	190	33	0.4	8.0	0.08	0.03	0.2	49	–
Cabbage	92	1.6	0.1	6	1.0	105	43	0.6	50.0	0.06	0.05	0.2	60	–
Peas (canned)	82	3.4	0.4	13	1.3	290	25	1.8	50.0	0.11	0.06	0.9	10	–
Potato chips	3	6.7	37.0	49	1.1	2285	25	1.8	–	0.18	0.11	3.2	10	–
Peanuts (roasted)	5	30.0	46.0	18	3.3	2350	74	1.9	–	0.30	0.15	21.0	–	–
Bread, wholewheat	37	9.3	2.6	49	1.5	1010	96	2.2	–	0.30	0.13	3.0	–	–
Cornflakes	4	7.9	0.7	80	0.6	1500	10	1.0	–	0.16	0.08	1.6	–	–
Rice (cooked)	74	2.2	0.1	23	0.1	420	8	0.2	–	0.01	0.01	0.4	–	–
Chocolate, milk	1	6.0	33.5	54	0.5	2275	216	4.0	6.6	0.10	0.04	0.8	–	–
Jam	28	0.5	0.3	71	0.6	1165	12	0.3	2.0	0.02	0.02	0.2	6	–
Butter	15	0.6	81.0	0.4	–	3007	16	0.2	995.0	–	–	0.1	–	1.26
Cheddar cheese	37	25.0	32.0	2.0	–	1670	680	0.9	420.0	–	0.83	1.6	–	0.35
Milk (pasteurised)	87	3.3	4.0	4.9	–	270	125	0.1	44.0	0.04	0.15	0.9	1.8	0.05
Egg, whole (raw)	74	12.8	11.5	0.7	–	660	54	2.7	140.0	0.12	0.34	0.1	–	3.4
Bacon, medium fat	13	25.0	55.0	1.0	–	2550	25	3.3	–	0.48	0.31	4.8	–	–
Liver, beef (raw)	70	20.0	4.0	3.6	–	570	8	12.6	6000	0.27	2.80	16.0	31	0.75
Herring (raw)	73	19.0	8.0	–	–	570	20	1.1	45	0.01	0.33	4.0	–	20–100
Cod (raw)	82	16.5	0.4	–	–	295	18	0.9	–	0.10	0.07	2.17	2	–

Figure 1.29 Experimental error

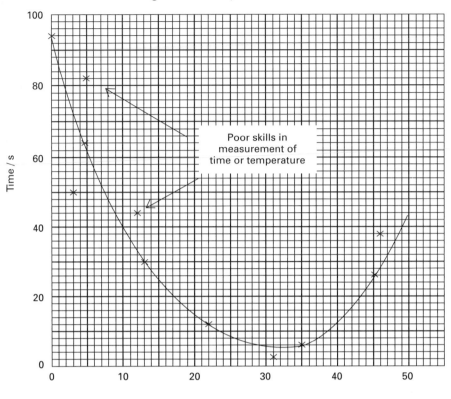

1.5 Looking at the evidence

What do the results show? In sections 1.3 and 1.4 you can see how the data or information can be presented and how further information can be extracted from it. After this has been done you must see if the information can be simplified by looking for a **trend** or a **pattern**. Whether a trend or pattern can be seen or not you must try to draw a **conclusion** from the evidence you have collected. It will help to have revised your other work on this topic before you draw your conclusion so you can see how your evidence fits in with what you have previously done. Think carefully before you write out your conclusion. Think about what you have found out in simple terms then bring in information from your other topic work to help you write the conclusion. For example, you may have found that plants in bright light grow better than those in dim light. If this is the case, think about other work you have done on this topic. Bring in your knowledge of photosynthesis to explain the result and draw the conclusion. Be very clear about the scientific words you use and look at the glossary at the back of the book to help you check their meaning.

When you have completed your conclusion, compare the results with your prediction. Look to see if the results support your prediction or undermine it. In either case work out an explanation and write it down. Again, use your science knowledge and understanding to express your views.

Evaluating the evidence

Before the investigation is complete you must examine the way you carried out your practical work. You must look for results which do not fit the trends or patterns that you have observed. These results are called **anomalous results**. If you find any you must review the way you carried out your investigation to see if they may have been caused by the way you handled equipment or took measurements. During the course of the investigation you may have thought of ways to improve the procedure. At the end you can now make a note of how you would improve the way you carry out the investigation in future. As you look through the investigation at the end you may have further ideas on how you would improve it and these too should be noted down. You must also ask yourself about the reliability of the results and think about ways in which errors may have crept into your investigation.

When you are satisfied that you have evaluated your results against the way you carried out your investigation, rate the firmness of your conclusion. If there were errors in your work, think about how valid the conclusions are and consider repeating your investigation with modifications.

If you are satisfied that your conclusions are firm, think of ways of providing additional evidence for them or think of ways to develop the enquiry further to discover more.

Summary

- Discoveries are made through the scientific investigation process which has many components. (▶ 3)
- The components can be placed in the following groups – planning experimental procedures, obtaining evidence, analysing evidence and drawing conclusions, evaluating evidence. (▶ 5–38)
- There is a range of instruments used in biological investigations. (▶ 12–25)
- There are safety issues which must be considered in all practical work. (▶ 6–11)

Body Form and Structure

The body form and structure of living things

2

Objectives

When you have completed this chapter you should be able to:
- identify the parts of a cell
- compare a plant cell and an animal cell
- distinguish a tissue from an organ
- recognise the five kingdoms of living things and explain how they are sub-divided
- explain the concept of species
- use a key

Genetic material provides instruction to make an organism's **body form**; different genetic material produces the vast **diversity** of body form seen amongst all living things.

2.1 The features and diversity of cells

All living organisms with the exception of viruses are made from genetic material contained within cells. Viruses are non-cellular. With the exception of bacteria and viruses, all living organisms share a similar cell structure. The features of this structure are:
- **genetic material** – enclosed within the **nucleus**;
- **cytoplasm** – a thick, transparent liquid;
- **organelles** – structures which are present in the cytoplasm.

The parts of a cell

The major parts of a cell are illustrated in Figures 2.1, 2.2 and 2.3.
- **The nucleus:** The nucleus contains the genetic material consisting of a nucleolus and chromosomes enclosed within a porous envelope. Nucleus means 'little nut'.

Figure 2.1 The parts of a cell

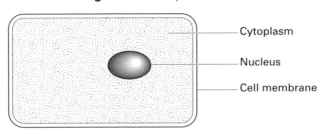

Cytoplasm

Nucleus

Cell membrane

- **The cytoplasm:** This is a thick, transparent liquid in which most of the chemical reactions involved in life processes take place. It fills the cell except for the places where vacuoles are present.
- **The cell surface membrane** (plasma membrane): This is composed of lipids and proteins and encloses the cell. It is a partially permeable membrane and controls the movement of materials into and out of the cell.
- **Organelles:** Small structures called organelles are present in the cytoplasm. A **mitochondrion** is a sausage-shaped organelle in which chemical reactions take place to release energy from food. In a very active cell such as a liver cell there may be up to 1000 mitochondria in the cytoplasm.

Animal cells

Figure 2.2 A typical animal cell (a cell from the human mouth lining)

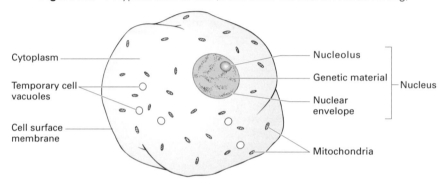

Animal cells are generally smaller than plant cells. They have an irregular shape. **Vacuoles** may not be present but when they do occur they are small, numerous and temporary. Granules of the the carbohydrate **glycogen** are found in the cytoplasm.

Plant cells

Figure 2.3 A typical plant cell (a palisade leaf cell)

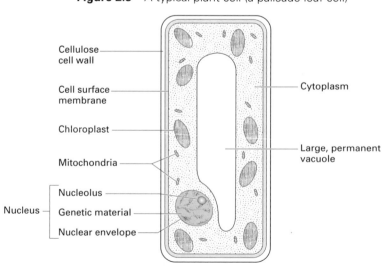

Plant cells are generally larger than animal cells. They have a regular cell shape due to the rigid **cell wall**. This is made of cellulose and encloses the cell surface membrane. A plant cell has a large permanent vacuole. It contains cell sap which is important in the transfer of materials inside the cell. **Chloroplasts** are organelles which are only found in green plant cells. They contain the pigment called **chlorophyll** which gives the plants their green colour and traps energy from sunlight for food production. Chloroplasts are not found in all plants cells, for example the storage cells of the potato tuber and the onion bulb. Granules of the carbohydrate **starch** may be found in the cytoplasm. (See also Figure 3.8 on page 79.)

Questions

1 How is a plant cell , e.g. a leaf cell, similar to an animal cell, e.g. a cell from the human mouth lining?
2 How is a plant cell different from an animal cell?

Cell diversity

Many tiny living things have a body made from only one cell. Some **algae** (see page 49) are single-celled and **Rhizopods** (see page 49) are single-celled or **unicellular**. In these living organisms the cell has to perform *all* the life processes to keep the organism alive.

Most plants and animals have bodies made from many cells. They are called **multicellular**. In these organisms the cells perform particular tasks in the organism's life. Here are a few examples.

Animal cell diversity

Sperm cell

Figure 2.4   A sperm cell

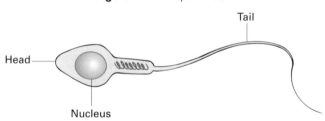

Tail

Head

Nucleus

This cell is specialised to transfer the genetic material from the male to the female. It has a streamlined shape to move easily through the liquid surrounding the egg. The tail thrashes to move the sperm forwards. The sperm's cytoplasm is greatly reduced to keep the cell bulk to a minimum for easy movement. It draws in energy-rich substances from the surrounding liquid to provide the power to move the tail.

See page 264 for the role of the sperm cells in reproduction in animals.

Effector nerve cell

The animal effector (motor) nerve cell receives signals and transports them over a long distance through the body, eventually passing them on to the muscle. The signal arrives as a chemical at the tips of the dendrites, passes along the axon body as an electrical impulse and leaves the cell as a chemical at the motor end plates. See page 207 for more details.

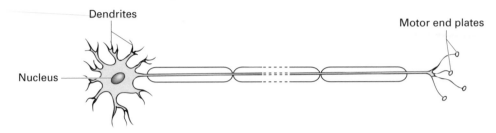

Figure 2.5 An effector nerve cell

Dendrites

Motor end plates

Nucleus

Plant cell diversity

Root hair cell

Figure 2.6 The root hair cell of a flowering plant

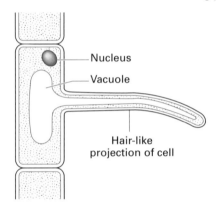

Nucleus

Vacuole

Hair-like
projection of cell

This cell is long and thin. Its surface provides a large area in proportion to it's volume for the absorption of water and its thinness allows the water to enter quickly. The cell does not contain chloroplasts as it grows in the darkness of the soil (see page 116).

Palisade leaf cell

This cell is cylindrical so that many similar cells can be packed together in the upper part of the leaf where most light is received (see Figure 2.3). It contains a large number of chloroplasts which are moved by the cytoplasm into the best positions to receive the maximum amount of light falling on the cell. The cell is specialised to trap as much energy from sunlight as possible to manufacture food (see page 79).

Questions

3 How are sperm cells specialised for movement?
4 Why are nerve cells very long?
5 Why are root hair cells not short and thick?
6 How can you tell a palisade cell is specialised to trap energy from sunlight?

2.2 Diversity of tissues and organs

Tissues

In multicellular organisms cells are arranged into groups called **tissues**. Each tissue has a particular function in the life of the organism (see Figure 9.10 on page 230).

Muscle tissue

Figure 2.7 Smooth muscle tissue in the stomach wall

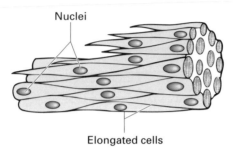

Muscle cells of animals form an **effector tissue** which can contract or get shorter. When the muscle tissue in the **biceps** muscle contracts, it pulls on the bones in the lower arm and raises them. Muscular tissue enables part or the whole of the animal's body to move (see page 232).

Glandular tissue

Figure 2.8 Glandular tissue

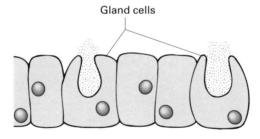

Glands are effector tissues whose cells **secrete liquids** as a response action. In the mouth, tissues of saliva-secreting cells make a juice which moistens food and digests starch. See Section 8.6 for more details.

Xylem tissue

In plants, water is transported through the roots and the shoot by **xylem tissue**. It is made from columns of dead cells which have lost their upper and lower walls, forming long tubes (see Figure 5.5 on page 114).

Figure 2.9 Xylem tissue

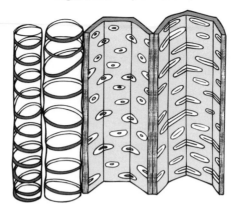

Questions

7 What is the difference between muscular tissue and glandular tissue?
8 How is xylem tissue specialised for transporting water?

Organs

Groups of different tissues grow together and form **organs**. For example, in the wall of the alimentary canal you can find glandular tissue to digest food, muscular tissue to move the food along and connective tissue to hold all the different tissues together (see Figure 4.9 on page 98).

Figure 2.10 Cross-section through the alimentary canal

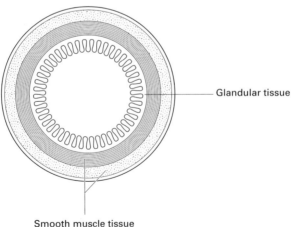

Glandular tissue

Smooth muscle tissue

2.3 The features and diversity of body forms

There is a huge **diversity** of living things on the Earth – over two million different kinds of living organism. The differences in their body structure and form are due to the different **genetic material** in each organism.

The vast diversity of living organisms can be classified on the basis of organisms sharing certain structural features or **characteristics** seen in their body form, for example the appearance of the skin, the presence of limbs or wings and the presence of chlorophyll in the leaves or stems. The mature or adult forms of different organisms are described in this chapter. The younger forms of the organism may differ from the adult, for example there are many differences between a tadpole and a frog or a maggot and a blow-fly.

Genetic material and classification

The external appearance of an organism's **body** is due to its genetic material. This provides the patterns for making the different proteins for body structure such as collagen in bones, or for body functions, e.g. enzymes for digesting food. Each species has its own specific kind of genetic material. In reptiles, the genetic material which controls body covering directs the body to make scales. In birds, the genetic material directs the body to make feathers and in mammals it directs the body to make hairy skin.

The distinctive body covering of an organism is just one characteristic which can be used in classification. Many characteristics are used to classify organisms. Some are **external** and **structural**, giving the body its form. Others are **internal** and **functional** and are involved in keeping the organism alive.

Classifying living things

Living organisms are classified according to their external and internal features. In the largest groups, called **kingdoms**, the organsims in the group share a few *major* features. Each kingdom is split into a number of smaller groups called **phyla** (singular = **phylum**). The members of each phylum have *more* features in common. The organsims in each phylum are divided up into smaller groups called **classes** according to the features they possess. This division into smaller and smaller groups continues with further grades of classification called **order**, **family**, **genus** and **species** (see also page 63).

Questions

9 Arrange all the names of the groups in order starting with Kingdom and ending with species.

In addition to the seven main groups in the classification system, a group may be considered to be larger than one grade yet smaller than another. For example, a group may be considered to be larger than a class but smaller than a phylum. This kind of group has the prefix 'super' added to it. A group larger than a class would be a **super-class** (see arthropods, page 58). In a similar way, if a group is thought to be just a little smaller than one grade it may have the prefix 'sub' added to it. for example, a class may be divided into **sub-classes**.

The classification of living organisms

All living organisms are grouped into five **kingdoms**, which are described on the following pages. One exception to this, the viruses, cannot be placed in any of the five kingdoms and is classified separately.

Viruses

Figure 2.11 The structure of viruses (not to scale)

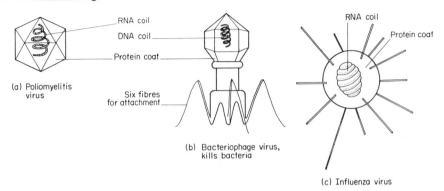

A virus is a particle (virion) that is only visible through an electron microscope. It consists of genetic material (DNA and RNA) surrounded by a protein coat. Viruses can only live as parasites in other living organisms, where they cause disease. Examples of diseases caused by viruses in humans are influenza, rubella, the common cold, poliomyelitus, and AIDS (see page 360).

Kingdom Prokaryotae

Prokaryotes are unicellular organisms with cell walls but no membrane-bound nucleus and no membrane-bound organelles. This kingdom includes bacteria.

Figure 2.12 Electron micrograph of bacteria (Upjohn Ltd)

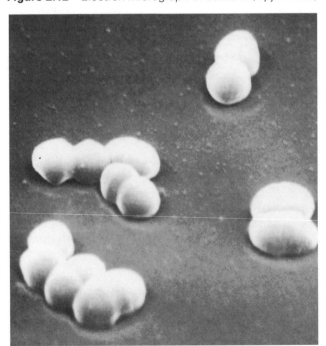

There are over 1000 species of bacteria and most can be recognised by their body shape as revealed by the electron microsope.

Figure 2.13   Bacterium body structure

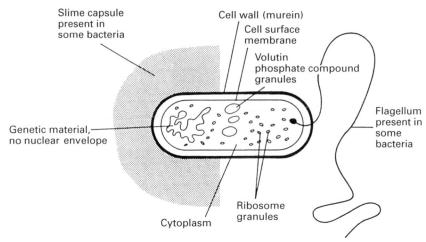

The genetic material does not have a nuclear envelope and is surrounded by cytoplasm and a cell wall. Some species of bacteria cause disease. Examples of diseases caused by bacteria in humans are tetanus, whooping cough, salmonella and gonorrhoea.

___ Question _____

10 How are viruses different from bacteria?

Kingdom Protoctista

There are over 50 000 species of protoctists. They are neither plants, animals nor fungi and all have genetic material surrounded by a nuclear envelope. They are divided into several phyla.

Phylum Rhizopoda

Amoeba is a **rhizopod**. It has **pseudopodia** (false feet) which are a characteristic of the phylum. Its body is made up from just one cell (see Figure 2.14).

Phylum Chlorophyta (green algae)

Chlamydomonas and *Spirogyra* are two examples of organisms in this phylum. *Chlamydomonas* has a body made up from just one cell. *Spirogyra* has a body in the form of a thread or filament. This is made from many similar cells joined together. The cells of both organisms have green pigments in their chloroplasts and a protective cell wall (see Figure 2.15).

___ Question _____

11 In what ways is *Amoeba* different from *Chlamydomonas*?

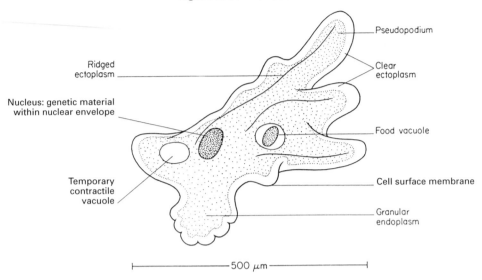

Figure 2.14 *Amoeba*

Pseudopodium

Ridged ectoplasm

Clear ectoplasm

Nucleus: genetic material within nuclear envelope

Food vacuole

Temporary contractile vacuole

Cell surface membrane

Granular endoplasm

|———————————— 500 μm ————————————|

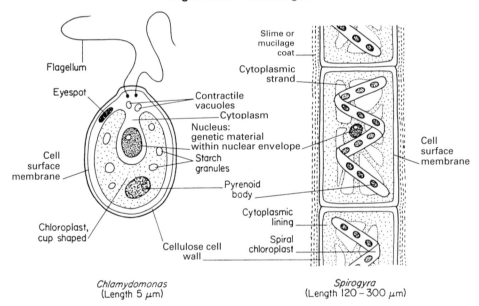

Figure 2.15   Green algae

Flagellum

Eyespot

Slime or mucilage coat

Cytoplasmic strand

Contractile vacuoles

Cytoplasm

Nucleus: genetic material within nuclear envelope

Starch granules

Cell surface membrane

Cell surface membrane

Pyrenoid body

Chloroplast, cup shaped

Cytoplasmic lining

Cellulose cell wall

Spiral chloroplast

Chlamydomonas
(Length 5 μm)

Spirogyra
(Length 120 – 300 μm)

Kingdom Fungi

There are over 50 000 species of fungi which range in size from microscopic **moulds** and **yeasts** to large **mushrooms** and **toadstools**. All are non-green because they do not possess chlorophyll.

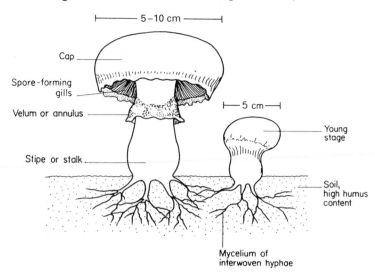

Figure 2.16 Field mushroom (*Agaricus campestris*)

5–10 cm

Cap

Spore-forming
gills

Velum or annulus

5 cm

Young
stage

Stipe or stalk

Soil,
high humus
content

Mycelium of
interwoven hyphae

In this kingdom the genetic material is contained within a nuclear envelope. The protective cell wall is composed of **chitin** or **fungal cellulose** which is different from green plant cellulose. Thread-like **hyphae** form the fungus body called the **mycelium**. Many fungi are **parasites**, obtaining their nutrients from a living host. Others, like the field mushroom and moulds, are **saprophytes**, helping to **decompose** organic material.

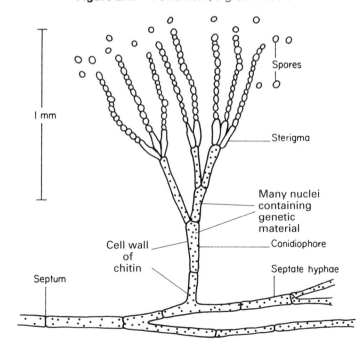

Figure 2.17 *Penicillium*, a green mould

Spores

1 mm

Sterigma

Many nuclei
containing
genetic
material

Cell wall
of
chitin

Conidiophore

Septate hyphae

Septum

Kingdom Plants

The plant kingdom consists of over 400 000 species. They all share the following characteristics:
- the body is composed of many cells – it is **multicellular**;
- cell walls are made of **cellulose**;
- **chlorophyll** is present in all plants;
- genetic material is contained *within* a nuclear envelope.

The main phyla are described here.

Phylum Bryophyta

The members of this phyla have a plant body that is made up of tiny leaves (mosses) or a flat strap or **thallus** (liverworts). Neither mosses nor liverworts have true roots but hold themselves in place by slender **rhizoids**. They produce **spores** from spore capsules.

Figure 2.18 Bryophytes

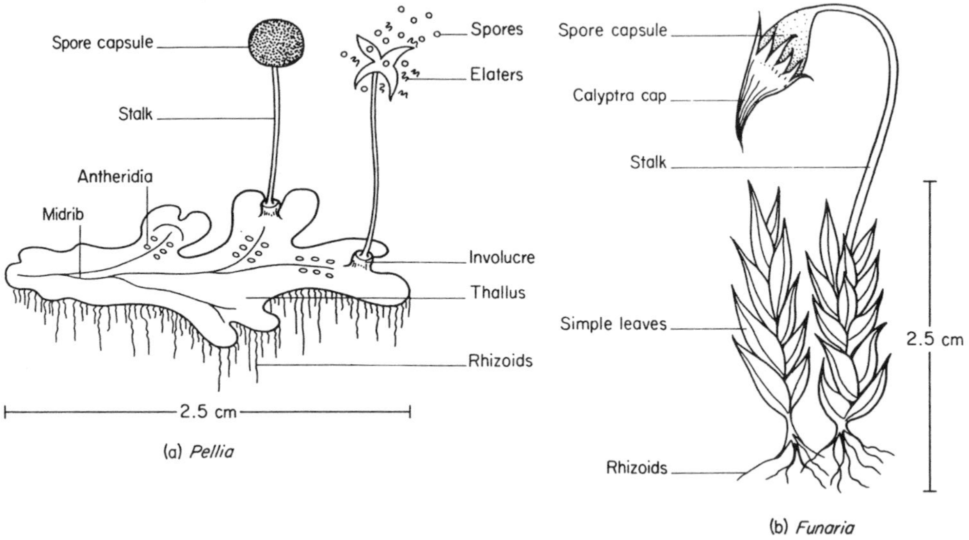

(a) *Pellia*

(b) *Funaria*

Phylum Filicinophyta (ferns)

The members of this phyla have a plant body composed of a large leaf-like **frond** which is coiled up in the bud like a violin head. The plants have an underground stem called a **rhizome**, with small adventitious roots.

Phylum Coniferophyta

The members of this group are known as the **conifers**. Most produce seeds inside cones which open to release them. All are trees or shrubs.

Phylum Angiospermophyta (flowering plants)

There are over 250 000 species of flowering plants. They all share these characteristics:
- The body is composed of a distinct overground stem supporting leaves and flowers, together with an underground root system.
- Flowers produce fruits enclosing seeds.

There are two classes of flowering plants – **monocotyledons** and **dicotyledons** (see Table 2.1).

Figure 2.19   Ferns

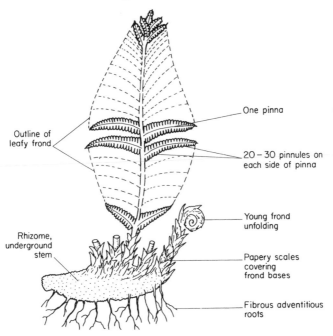

One pinna

Outline of
leafy frond

20 – 30 pinnules on
each side of pinna

Young frond
unfolding

Rhizome,
underground
stem

Papery scales
covering
frond bases

Fibrous adventitious
roots

Figure 2.20   Male and female cones of a conifer, *Pinus sylvestris*

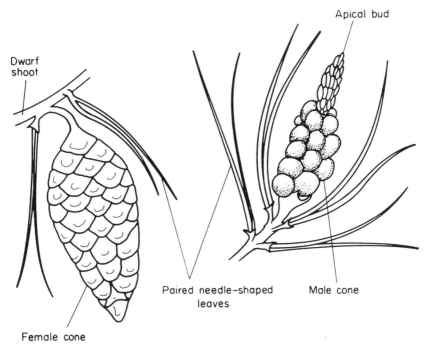

Apical bud

Dwarf
shoot

Paired needle-shaped
leaves

Male cone

Female cone

Figure 2.21 Structure of a typical dicotyledon (wallflower)

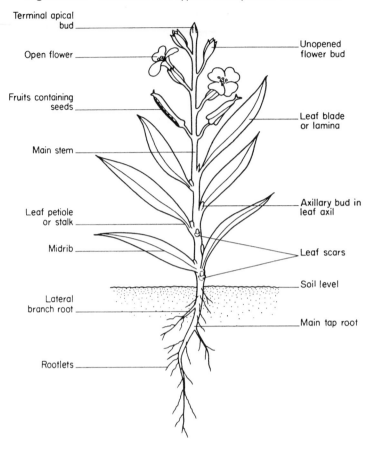

Terminal apical bud

Open flower

Fruits containing seeds

Main stem

Leaf petiole or stalk

Midrib

Lateral branch root

Rootlets

Unopened flower bud

Leaf blade or lamina

Axillary bud in leaf axil

Leaf scars

Soil level

Main tap root

Table 2.1 The two classes of flowering plants

Class	Monocotyledons	Dicotyledons
Species:	over 75 000	over 225 000
Seeds:	one seed leaf	two seed leaves
Leaves:	narrow and parallel-veined	broad and net-veined
Body form:	herbs	herbs, shrubs and trees
Examples:	grasses, lilies, crocuses	buttercups, privet, sycamore

Question

12 Construct a table of the plant kingdom. Set out the characteristics of each group under the headings of the kingdom, phyla and classes. Use this layout to help you:

Kingdom: Plants
Phyla: Bryophytes Ferns Conifers Angiosperms
Classes: Monocotyledons Dicotyledons

Figure 2.22 Structure of a typical monocotyledon (foxtail grass)

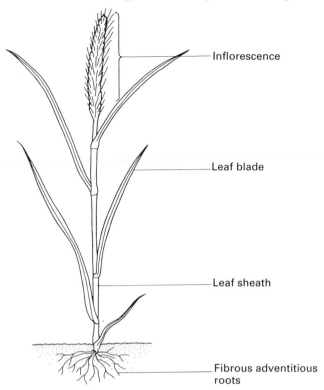

Inflorescence

Leaf blade

Leaf sheath

Fibrous adventitious roots

The Animal Kingdom

The animal kingdom consists of over one million species. They share the following characteristics:
- the body is multicellular;
- no cellulose cell wall is present;
- no chlorophyll is present;
- a mouth is present;
- nitrogenous waste is produced;
- genetic material is contained within a nuclear envelope.

The main animal phyla are described here.

Phylum Cnidaria

One opening leading to the gut, e.g. *Hydra*, jellyfish and sea anemones.

Phylum Nematoda (roundworms)

Thread-like bodies; found in soil, aquatic habitats and as parasites.

Phylum Platyhelminthes (Flatworms)

The animal's body is flat and is *not* divided into segments. A mouth is usually present but an anus is absent. The tapeworm is an example of a flatworm. It has a tiny head with hooks and suckers to hold it in place in the gut of its host and produces flat **proglottids** which contain the reproductive organs. These form a tape-like body several metres long in the host's intestine.

Figure 2.23 Tapeworm

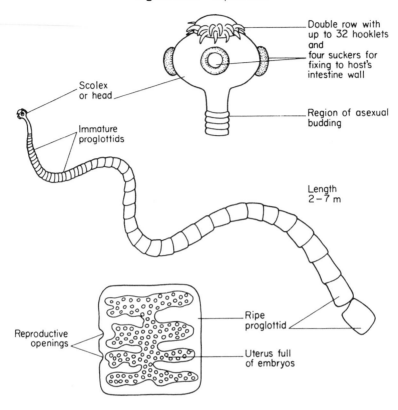

Scolex or head

Double row with up to 32 hooklets and four suckers for fixing to host's intestine wall

Region of asexual budding

Immature proglottids

Length 2–7 m

Reproductive openings

Ripe proglottid

Uterus full of embryos

Phylum Annelida

The body is long and thin and divided into segments. It is covered with moist skin. The earthworm is a an annelid that lives in the soil. It has bristles called **chaetae** on its segments which help it to move. It has no distinct head and a structure called a **saddle** which is connected with reproduction. The leech is a freshwater annelid and the lugworm is a marine annelid which can be found on sandy beaches.

Question

13 How is a flatworm different from an annelid worm?

Phylum Mollusca

Molluscs have soft bodies which are *not* divided into segments. They have a distinct head and a foot and the body is covered in moist skin. Most species have a chalky **shell**. Snails have one coiled shell, cockles and mussels have two flat shells. The garden snail is a mollusc that lives on land. It has *two* pairs of tentacles. The larger pair each have an eye at the tip. Freshwater snails only have *one* pair of tentacles.

Question

14 How is an annelid different from a mollusc?

Figure 2.24 Earthworm

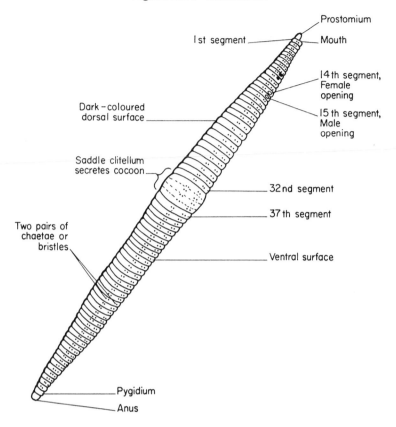

Prostomium

1st segment — Mouth

14th segment, Female opening

15th segment, Male opening

Dark-coloured dorsal surface

Saddle clitellum secretes cocoon

32nd segment

37th segment

Two pairs of chaetae or bristles

Ventral surface

Pygidium

Anus

Figure 2.25 Garden snail

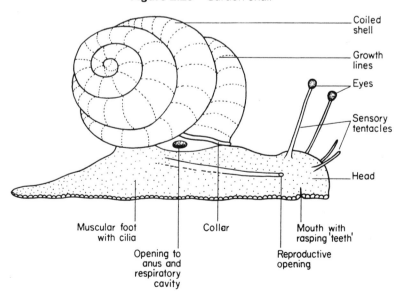

Coiled shell

Growth lines

Eyes

Sensory tentacles

Head

Muscular foot with cilia

Collar

Mouth with rasping 'teeth'

Opening to anus and respiratory cavity

Reproductive opening

THE BODY FORM AND STRUCTURE OF LIVING THINGS   57

Phylum Arthropoda

This large phylum has over 750 000 species which share the following characteristics:

- hard outer skeleton called an **exoskeleton**;
- a segmented body;
- tubular, jointed, paired limbs or appendages.

There are four main super-classes (another division of the classification system).

Super-class Crustacea (crabs, lobsters, shrimps)

Crustaceans have an indistinct head with *two* pairs of antennae. They are found mainly in freshwater and marine environments. The marine shrimp has a large shell-like structure covering the front of its body, called a **carapace**. The shrimp has 10 pairs of appendages divided into five pairs of walking legs and five pairs of swimmerets.

Figure 2.26 Marine shrimp

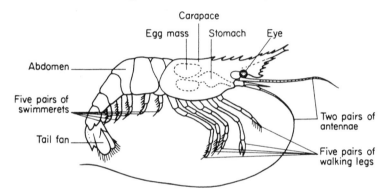

Class Chilopoda (centipedes and millipedes)

The body has a distinct head with a pair of jaws. It has many pairs of legs all along the body. These animals are found on land where they generally live in the soil. The soil millipede has between 130 and 200 pairs of legs.

Figure 2.27 Millipede

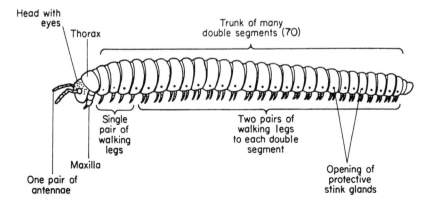

Class Insecta (beetles, ants, butterflies)

The greatest number of arthropod species are insects. The body is divided into three distinct regions: **head, thorax** and **abdomen**. The head has a pair of antennae and a pair of eyes. It also has three pairs of mouthparts. The thorax has *three* pairs of legs and in winged insects it may have up to *two* pairs of wings. Insects are found on land and in fresh water. Many can fly.

Figure 2.28 Grasshopper

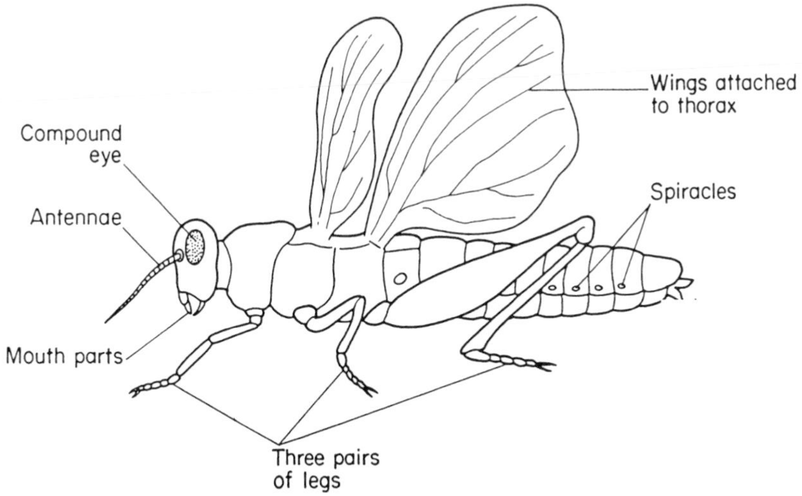

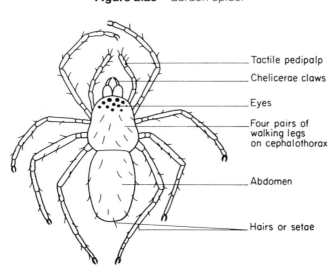

Compound eye

Antennae

Mouth parts

Wings attached to thorax

Spiracles

Three pairs of legs

Class Arachnida (spiders, scorpions, mites)

Figure 2.29 Garden spider

Tactile pedipalp

Chelicerae claws

Eyes

Four pairs of walking legs on cephalothorax

Abdomen

Hairs or setae

The body is divided into **cephalothorax** (fused head and thorax) and **abdomen**. Four pairs of legs are joined to the cephalothorax. Most arachnids live on land but a few live in fresh water.

Phylum Echninodermata

This phylum consists entirely of marine animals which have a tough, thick skin with spines and plates. There is a ventral mouth and the adult body may have five rays. Examples are starfish (Figure 2.30), sea urchins, feather stars, brittle stars and sea cucumbers.

Figure 2.30 *Asterias*, a starfish

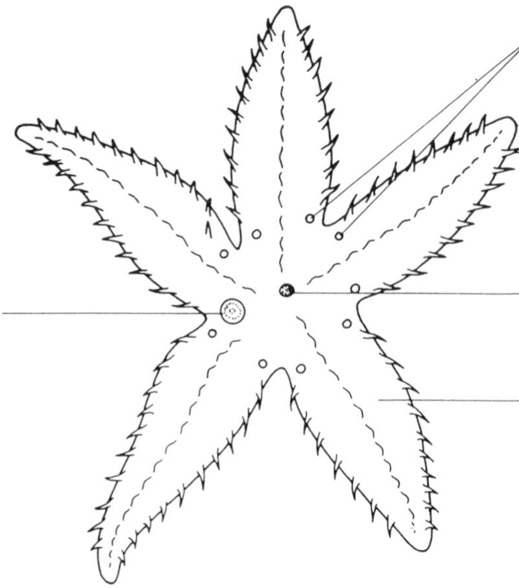

Genital openings

Spines

Anus

Arm

Opening into water vascular system

Questions

15 How is a crustacean different from a chilopod?
16 How is an insect different from an arachnid?
17 What do a crustacean and an insect have in common?

Phylum Chordata

This large phylum has over 40 000 species found in every habitat – aquatic, terrestrial (land) – and some can fly.

Chordates share the following characteristics:
- The body is divided into a **head** and a **trunk**. A **tail** is present during early development and sometimes afterwards.
- a tubular **nerve cord** or **spinal cord** is found on the dorsal side (the back).
- **vertebrae** are present in many chordates – these are bones which protect the spinal cord.

The six major classes are described here.

Class Chondrichthyes (cartilaginous fish)

These fish have an internal skeleton or **endoskeleton** made of flexible **cartilage**. The mouth is found on the lower or ventral surface. The **fins** are fleshy. **Gills** have separate, slit-like openings. Almost all cartilaginous fish are marine. The dogfish, with its rough skin, is a typical example of a cartilaginous fish.

Figure 2.31 Dogfish

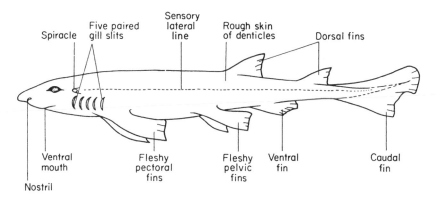

Spiracle
Five paired gill slits
Sensory lateral line
Rough skin of denticles
Dorsal fins
Ventral mouth
Nostril
Fleshy pectoral fins
Fleshy pelvic fins
Ventral fin
Caudal fin

Class Osteichthyes (bony fish)

These fish have an endoskelton composed of rigid **bone**. The mouth is at the tip of the body, called a terminal mouth. The fins are scaly and supported by bony rays. The gills have a flap-like covering called the **operculum**. Bony fish are found in marine and freshwater environments. The stickleback is an example of a freshwater bony fish. It has a slimy, scaly skin.

Figure 2.32 Stickleback

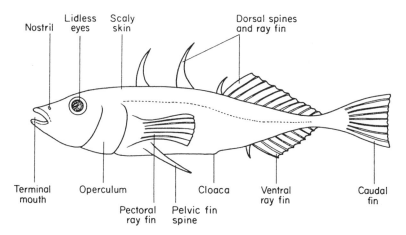

Nostril
Lidless eyes
Scaly skin
Dorsal spines and ray fin
Terminal mouth
Operculum
Pectoral ray fin
Pelvic fin spine
Cloaca
Ventral ray fin
Caudal fin

Class Amphibia

Amphibians have soft, **moist skin**. It is permeable and can be used as a breathing organ. The hind limbs have **webbed** digits. The life cycle involves living in water and on land. Many shell-less eggs are laid in water and the young that hatch from them live in water. They have gills and are known as larvae or **tadpoles**. The larvae change into adults in a process called **metamorphosis**. Adult amphibia have lungs and live on land. The frog is an example of an amphibian.

Figure 2.33 Frog

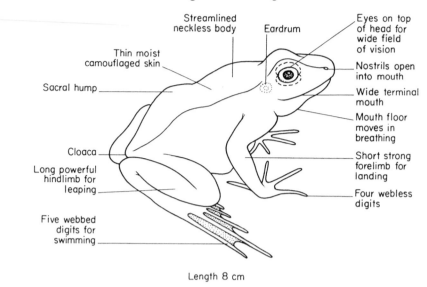

Streamlined
neckless body

Eardrum

Eyes on top
of head for
wide field
of vision

Thin moist
camouflaged skin

Sacral hump

Nostrils open
into mouth

Wide terminal
mouth

Mouth floor
moves in
breathing

Cloaca

Short strong
forelimb for
landing

Long powerful
hindlimb for
leaping

Four webless
digits

Five webbed
digits for
swimming

Length 8 cm

Class Reptilia

Reptiles have a **dry, scaly skin** which is not permeable. They do not have webbed limbs and some reptiles, the snakes, are limbless. Reptiles breathe with lungs. They develop from soft-shelled eggs which are laid on land. Most reptiles live on land but some live in water, returning to the water surface to breathe.

Figure 2.34 Lizard

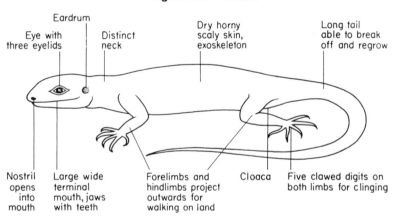

Eardrum

Eye with
three eyelids

Distinct
neck

Dry horny
scaly skin,
exoskeleton

Long tail
able to break
off and regrow

Nostril
opens
into
mouth

Large wide
terminal
mouth, jaws
with teeth

Forelimbs and
hindlimbs project
outwards for
walking on land

Cloaca

Five clawed digits on
both limbs for clinging

Length 18 cm

Class Aves (birds)

The skin of birds is covered in feathers. The forelimbs are modified to form wings and the hind limbs are covered in scales. Birds have lungs and a constant body temperature. This is sometimes described as being **warm-blooded**. Birds lay hard-shelled eggs. All birds nest on land and are land-dwelling but a few spend most of their time in

Figure 2.35 Pigeon

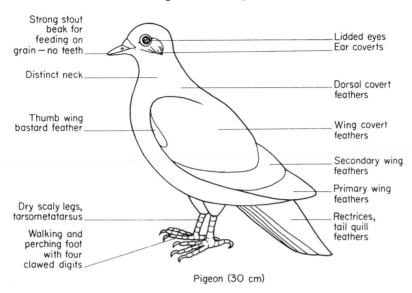

Strong stout beak for feeding on grain — no teeth

Distinct neck

Thumb wing bastard feather

Dry scaly legs, tarsometatarsus

Walking and perching foot with four clawed digits

Lidded eyes
Ear coverts

Dorsal covert feathers

Wing covert feathers

Secondary wing feathers

Primary wing feathers

Rectrices, tail quill feathers

Pigeon (30 cm)

the air or on the surface of water. The pigeon is an example of a bird that lives on land. It has feet which are adapted for walking and perching.

Class Mammalia

The skin of mammals has hair in follicles. Female mammals have **mammary glands** which provide milk to feed the young. Lungs are present and a diaphragm inside the trunk divides it into the thorax and the abdomen. Mammals have a constant body temperature and, like birds, are sometimes described as being warm-blooded. Female mammals produce one or a small number of eggs at one time. The eggs are tiny and almost all are without shells. The eggs develop *inside* the female. Two exceptions are the eggs of the duck-billed platypus and the spiny anteater which have shells and are laid; the young develop inside the eggs until they hatch. In **marsupial** mammals such as the kangaroo, the young are born into a **pouch** at an early stage in their development and continue to develop inside the pouch. In most mammals the young stay inside the mother until they are more fully developed. The organ of attachment between the developing young and the mother is called the **placenta**, so these mammals are called **placental** mammals.

The rabbit is an example of a placental mammal (see Figure 2.36).

The smallest classification group – the species

A **species** is a group of organisms which resemble each other closely. For example, the domestic dog.

- Members of a species have closely similar genetic material and physical characteristics.
- Members of a species can **interbreed** to produce live offspring, seen in the different breeds of dogs.

Domestic cats have a different genetic make-up to dogs and as a separate species are unable to interbreed with dogs.

Figure 2.36 Rabbit

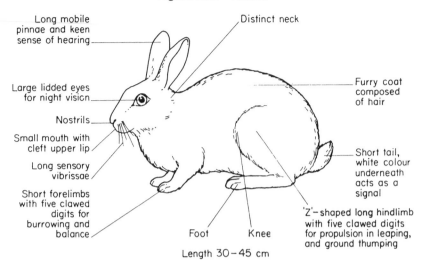

Long mobile pinnae and keen sense of hearing

Distinct neck

Large lidded eyes for night vision

Nostrils

Small mouth with cleft upper lip

Long sensory vibrissae

Short forelimbs with five clawed digits for burrowing and balance

Furry coat composed of hair

Short tail, white colour underneath acts as a signal

Foot   Knee

'Z'-shaped long hindlimb with five clawed digits for propulsion in leaping, and ground thumping

Length 30-45 cm

A **population** is a number of organisms of the same species, for example a pack of wolves, a school of whales or a field of wheat. All the members of the population can interbreed because they have similar genetic material.

The individual and organism variation

The individual organism is *one member* of a certain species, that is one dog or one human being. It is the smallest unit or component of a population.

Each individual organism has a slightly different specific genetic make-up when compared with other members of the species population. This genetic difference can be shown in specific genetic material types or profiles possessed by human beings and can also be seen in different fingerprints. Every human being has their own specific type of genetic material. These types are also called DNA 'profiles' or 'finger prints' (see page 303).

Variation describes the differences between **individuals** of the same species; these variations are seen among the domestic dog species, which exists in about 400 different varieties or **breeds**, each having a distinct body form, ranging from the large borzoi to the small dachshund. It is also in plant species, for example red and yellow varieties of roses.

The domestic dog breeds have similar DNA (genetic material), allowing all varieties or breeds to interbreed; consequently the different breeds of domestic dog are biological variations within the same species. They share the following dog-like characteristics of the species:
- fairly long legs; stand on their toes which have blunt claws;
- face has a long muzzle or projecting mouth and nose;
- body is strongly muscular with a tail; the hair covering long or short – often a feature of the different breed or variation;
- all are excellent runners; some are used in hunting
- have an alert look.

Variation will be seen throughout a population, for example the human species population shows differences in colour of skin, eyes and hair, also differences in height from dwarf to very tall. (See also chapter 12.)

64   *BIOLOGY*

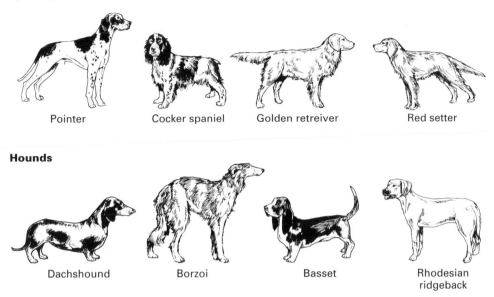

Figure 2.37 Varieties or breeds of dog

Sporting breeds

Pointer Cocker spaniel Golden retreiver Red setter

Hounds

Dachshound Borzoi Basset Rhodesian ridgeback

Using a key

A key is a series of paired questions or descriptions which can be used to identify plants and animals. It can be used to put living things in groups as large as phyla and classes or to identify them as species.

Key to the classes of the phylum Chordata

1	**a** Hair present	**Mammal**
	b Hair absent	Go to 2
2	**a** Feathers present	**Bird**
	b Feathers absent	Go to 3
3	**a** Dry, scaly skin	**Reptile**
	b Other type of skin	Go to 4
4	**a** Moist, scale-less skin, four limbs	**Amphibian**
	b Limb-less	Go to 5
5	**a** Fins, cycloid scales, bony skeleton	**Bony fish**
	b Fins, plate-like scales, cartilaginous skeleton	**Cartilaginous fish**

Question

18 (a) Use the key to identify the animals in Figure 2.31 to 2.36.
   (b) List the steps followed when using the key for each animal.

Summary

- The basic components of a cell are nucleus, cytoplasm and cell membrane. (▶ 41)
- A typical plant cell differs from an animal cell by having a cellulose cell wall, vacuole and chloroplasts (▶ 42)
- Cells are specialised for different tasks. (▶ 43)
- Organism diversity is seen in the *great* differences between different species. (▶ 47–63)
- There are five kingdoms of living things. They are Prokaryotes, Protoctists, Fungi, Plants and Animals. Each kingdom is divided into sub-groups. (▶ 47–63)
- The species is the smallest group in this classification. (▶ 63)
- Organism variation is seen in the *small* differences between individuals of the same species. (▶ 64)
- Living things can be identified by the use of keys. (▶ 64)

Life Support Processes

Green plants: the food producers

Objectives

When you have completed this chapter you should be able to:
- describe the process of photosynthesis
- set up experiments to investigate photosynthesis
- explain how external factors affect photosynthesis
- describe the structure of a leaf
- explain how the leaf is adapted for photosynthesis
- explain the uses of the products of photosynthesis
- understand the importance of minerals in the development of the plant

3

Green organisms trap the sun's energy to power the Earth's open system (see Section 14.2).

Twenty five per cent of all known living species have green chloroplasts and make their own food from **light energy, water, carbon dioxide** and certain **mineral ions**. The food made by green plants, also called **producers**, is used by the remaining 75 per cent of species – animals, fungi and some bacteria, also called **consumers**.

The manufactured food is used by green plants to make
- new **genetic** material
- new cell **structural** material
- new **functional** materials, e.g. enzymes.

It also provides a source of energy for all living organisms.

3.1   Photosynthesis

Photosynthesis is the process by which green plants make food using **radiant** or **light energy**. The light used may be natural, i.e. from the sun, or artificial, e.g. from an electric lamp. The chemical reaction which takes place in photosynthesis can be represented by this word equation:

carbon dioxide + water → glucose + oxygen

This takes place in the presence of chlorophyll and light energy. Carbon dioxide and water are the *reactants* in this process and glucose and oxygen are the *products*.

This chemical symbol equation provides a summary of the **reactants** and **products** of photosynthesis:

$$6CO_2 + 6H_2O \xrightarrow[\text{chlorophyll}]{\text{light energy}} C_6H_{12}O_6 + 6O_2$$

This reaction is a summary of many reactions which take place inside the green plant cells. These reactions are controlled by **intracellular enzymes**.

The reactants of photosynthesis

The following are the essential reactants necessary for photosynthesis to occur:

- **Carbon dioxide** from the air or dissolved in fresh water or sea water, to provide the chemical **carbon**.
- **Water** from the soil or aquatic habitats to provide the chemical elements **hydrogen** and **oxygen**.
- **Radiant light** from the sun or from artificial light from oil or electric lamps; this provides the **energy** for the chemical processes to occur. Note: only about 1% of solar radiation is absorbed; the remaining 99% is reflected into space (see Figure 14.7).
- Green **chlorophyll** pigments found in plant cell **chloroplasts** trap and absorb light energy.
- Intracellular **enzymes** catalyse the processes which build up the food molecules.
- **Heat energy** is needed for the chemical reactions to occur and to provide the optimum temperature for efficient enzyme function.

The products of photosynthesis

- **Oxygen gas** is the product essential for release of energy in other living organisms. All the oxygen found in sea water (0.85%), fresh water (1.0%) and air (20%) has come from the process of photosynthesis in green plants, certain green protoctists and blue-green bacteria (see Table 14.2).
- The carbohydrate sugar, **glucose**, is the high energy organic food needed by all living things. Glucose is rapidly changed into starch in most green plants by the intracellular enzyme phosphorylase.

A controlled experiment to show how glucose can be changed into starch

1 Prepare a 5% solution of glucose phosphate by dissolving 1 g in 20 cm^3 of distilled water. Place the solution in a bottle fitted with a dropper. Note: Use with care as it is expensive to buy.

2 Extract the phosphorylase enzyme from fresh, new potato by crushing and grinding the potato and filtering the juice. Collect the filtrate and filter again through a new, clean filter paper. Test for the absence of starch by adding iodine solution to a small sample, which should *not* turn blue-black if starch is absent. Repeat the filtration if there is any sign of starch present. Boil and cool some of the extract for the control experiment.

3 Take a large, clean, white, glazed wall tile and add four separate drops of the glucose phosphate solution in a row along the top of the tile. Then add two drops of fresh potato enzyme extract to each drop and leave for 2 minutes.

4 Place four drops of glucose phosphate solution at the bottom of the tile for the control experiment then add drops of boiled and cooled potato extract.

5 Test each drop on the tile with iodine solution after the 2 minutes and notice which row of drops turns blue-black in colour, showing that starch has been formed. Starch should only form where the fresh potato enzyme was used, and there should be none where the boiled or destroyed enzyme was used.

The reaction can be represented by this word equation

$$\text{Glucose (phosphate)} \xrightarrow[\text{enzyme}]{\text{phosphorylase}} \text{starch} + \text{water}$$

Potato cells contain the intracellular enzyme called phosphorylase. It converts glucose into starch and water by a type of chemical reaction called a **condensation reaction** (removal of water).

Starch forms granules in storage cells. It is **insoluble** and does not affect the cell sap concentration. This stops changes in the cell sap due to osmosis (see page 111). If the cell stored glucose, which is **soluble**, this would *increase* the concentration of the cell sap and water would move in from surrounding cells by osmosis and prevent them from functioning properly. When glucose is required by the plant, the enzyme **amylase** acts on the starch. **Hydrolysis** takes place (addition of water) and starch molecules are changed into glucose molecules.

Questions

1 What is the word equation for photosynthesis?
2 What is the purpose of chlorophyll?
3 From what is starch produced?

The two stages of photosynthesis

There are two main stages of photosynthesis – the **light stage** and the **dark stage**.

- **Light stage:** Light energy is absorbed by chlorophyll and water is split into hydrogen atoms and oxygen gas. This reaction does not involve enzymes and is not affected by temperature.
- **Dark stage:** Does not require light; molecules of sugar are built up and water is also produced. This reaction involves enzymes and is affected by temperature.

3.2 Experiments to investigate photosynthesis

Evidence that photosynthesis has occurred can be seen in the formation of starch or in the production oxygen gas.

The starch test

1 Immerse either a whole leaf, or discs punched out of a leaf with a large-diameter cork borer, in boiling water for half a minute.
2 Make sure all Bunsen flames are turned off, then immerse the leaf or leaf discs in hot ethanol (methylated spirits) in a heated water bath for several minutes or until the green colour has been removed from the leaf.
3 Dip the leaf or leaf discs into hot water to soften them.
4 Treat the leaf or leaf discs with iodine solution and rinse away the excess in cold water.

Boiling water destroys the leaf cell permeability allowing ethanol to penetrate and dissolve the green chlorophyll pigments from the chloroplasts. Hot water softens the tissues. The colourless leaf then stains blue-black showing the presence of starch granules.

Destarching plants

Experiments are performed using potted flowering plants such as *Pelargonium* (geranium) and *Tradescantia*. Any starch present in the leaves is removed by keeping the plants in total darkness for 48 hours before any experiment. Any starch found in a leaf after an experiment must therefore have formed under the experimental conditions.

Etiolation is a different process involving growing green plants in the dark to develop yellow leaves lacking chlorophyll.

Questions

4 Why must the leaf be dipped in boiling water before immersing it in ethanol?
5 Why must all flames be extinguished before ethanol (methylated spirit) is used?
6 A leaf in hot ethanol becomes crisp. How does dipping it in hot water help to spread out the leaf for testing with iodine solution?
7 What is the difference between a destarched plant and an etiolated plant?

Controlled experiments

The following experiments show the important role in photosynthesis of different factors: carbon dioxide, light and chlorophyll.

The need for carbon dioxide

1 Destarch a potted plant.
2 Enclose one leaf in carbon dioxide-free air over soda lime or potassium hydroxide (see Figure 3.1).
3 Expose to light for 6 hours then perform the starch test.

Figure 3.1 To show that carbon dioxide is needed for photosynthesis

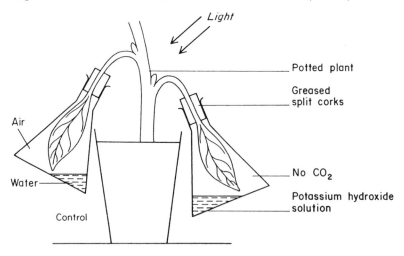

Control

1 Use the same plant.
2 Enclose one leaf in air over water (see Figure 3.1).
3 Expose to light for 6 hours and then perform the starch test.

The need for light

Figure 3.2 To show that light is needed in photosynthesis

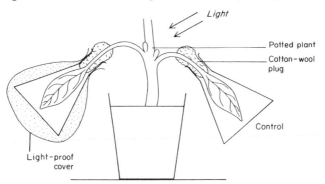

1 Destarch a potted plant
2 Keep one leaf in a light-proof flask for 6 hours then perform the starch test.

Control

1 Use the same plant.
2 Keep one leaf in a clear glass flask for 6 hours then perform the starch test.
Only about 1–5% of the light falling on a green leaf enters the leaf by refraction for use in photosynthesis; the remaining 99–95% is reflected from the leaf surface.

The need for chlorophyll

Figure 3.3 Variegated leaves of *Hosta* (f8 Imaging)

1 Destarch a potted plant with variegated leaves.
2 Trace the pattern on a variegated leaf.
3 Expose the leaf to light for six hours then perform the starch test.

Control

The leaf-pattern tracing is the control. It can be used to compare the position of the green regions with the regions stained with iodine.

Questions

8 What reagent is used to test for starch?
9 What colour change is seen if starch is present?
10 How is a variegated leaf different from a normal leaf?
11 What effect does each factor (carbon dioxide, light and chlorophyll) have on photosynthesis?

The production of oxygen

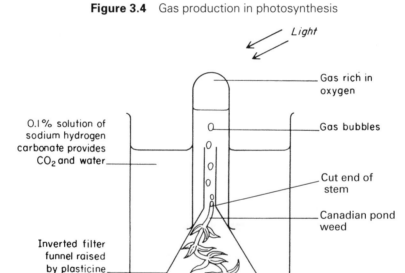

Figure 3.4 Gas production in photosynthesis

1 Place a cut stem of Canadian pond weed (*Elodea canadensis*) under an inverted glass filter funnel in 0.1% sodium hydrogen carbonate solution. Leave the apparatus in a well-lit position in a water bath maintained at 30°C for 3 days.
2 Test the gas collected in the inverted, water-filled test tube with a glowing splint.

Control

1 Assemble a second set of apparatus as before but place in total darkness for 3 days.

2 Note the amount of gas that collects and test for oxygen.
Allow a sufficiently large sample of gas to collect in the inverted tube. The chemical composition of the gas is found by inserting a **glowing** wood splint. If it **relights**, it shows the presence of oxygen. If the gas is analysed using a syringe and alkaline pyrogallol solution it will be shown to contain more oxygen than a similar volume of room air.

The gas produced in photosynthesis is not pure oxygen but a mixture of gases, mainly oxygen, nitrogen and carbon dioxide in lesser proportion than in room air. The rate of photosynthesis is indicated as the number of gas bubbles produced per minute from the cut end of the stem.

3.3 Using the products of photosynthesis

Sugars and phosphates are used to make new **genetic material**. Genetic material is composed of **phosphates**, certain compounds of nitrogen (called **organic bases**) and a sugar called **ribose**. Consequently photosynthesis is a vital **life support** process.

- The glucose may be used as an **energy source** in respiration. The energy released is used in the life processes in the cell. Carbon dioxide and water are produced.
- The glucose may be converted into **starch** molecules and stored in the roots, stems or leaves.
- The glucose may be converted into molecules of **cellulose** and used to form the cell walls of the plant.
- The glucose may be converted into **sucrose** molecules and stored in fruits where it attracts animals to feed on them and disperse the seeds.
- The glucose may be converted into **lipids** and stored in seeds.
- The glucose may be joined with certain **minerals**, especially nitrate, to make amino acids and proteins.
- The oxygen is used in **respiration** by the plant cells and forms an important component of the air (see Table 14.2).

3.4 External factors and their affects on photosynthesis

The experimental set up used to show that photosynthesis produces oxygen (see Figure 3.4) can be used to determine the effect of changing one external factor, whilst others remain constant. The environmental components or factors are light, carbon dioxide and temperature.

Light intensity

An electric lamp (25 W) can be used to illuminate the apparatus at different distances. This provides a means of presenting light of different intensities to the plant. Alternatively a range of bulbs, for example 40 W and 60 W, may be used from the same distance. The temperature is maintained at 30°C and a 1% solution of sodium hydrogen carbonate is used. A glass tank or a beaker full of water is placed between the apparatus and the lamp to act as a heat shield.

Carbon dioxide

The concentration of carbon dioxide can be varied by using a range of concentrate solutions of sodium hydrogen carbonate which provide from 0.01 to 1.0% of carbon dioxide. Distilled water is used to provide zero concentration. Both the temperature and the light intensity are maintained at constant levels throughout the experiments. The rate of photosynthesis increases up to a certain concentration of carbon dioxide, above which there is no further increase.

Temperature

The effect of varying the temperature can be investigated if the light intensity and carbon dioxide concentration are kept constant. The temperature can be varied by using a thermostatically controlled water bath. This apparatus can show the effect of temperature change on the rate of photosynthesis.

The rate of photosynthesis rises as the temperature rises and reaches an optimum at 35°C, after which the rate falls and stops above 40°C when the enzymes are destroyed. A rise in temperature of 10°C below the optimum temperature *doubles* the rate of photosynthesis.

Question

12 Sketch a graph to show how the light intensity in a field changes over a 24 hour period. Start the graph at midnight and finish at midnight. Draw on the graph how you think the concentration of carbon dioxide changes over the 24 hour period. (Remember plants produce carbon dioxide in respiration as well as take it in during photosynthesis.)

Limiting factors

A **factor** is a component or part of the environment. Temperature, light intensity, carbon dioxide concentration and the concentrations of certain mineral ions are factors that affect the *rate* or speed of photosynthesis.

A **limiting factor** is any factor which by its presence, absence, increase or decrease, affects a life process (or metabolic process) in an organism. The rate of a metabolic process is controlled by that factor which is in least supply. If all the factors are present in excess, the rate of photosynthesis will be the maximum possible. If one factor is not present in excess the rate of photosynthesis will be affected. For example, too low a light intensity in the presence of an excess of the other factors slows the rate of photosynthesis. This rate will remain until the light intensity is increased to its optimum level. The optimum temperature for photosynthesis is between 20°–25°C. Below 0°C the enzymes cannot work and above 40°C the enzymes are destroyed.

Increasing the temperature has no effect on the rate of photosynthesis when light intensity is low, therefore light intensity is the limiting factor.

The amount of carbon dioxide in the air is 0.03%. This amount limits the rate of photosynthesis if all the other factors are in excess. This amount is raised to 0.1% in glasshouses by installing gas burners which release carbon dioxide as the gas burns. The burners also raise the temperature of the glasshouse so the increase in the two factors raises the rate of photosynthesis. This results in increased crop production. For example, lettuces grown in glasshouses with air enriched with carbon dioxide are larger and have a higher dry weight than lettuces produced outside.

Figure 3.5 Graph showing the effect of temperature and light intensity on the rate of photosynthesis in an alga

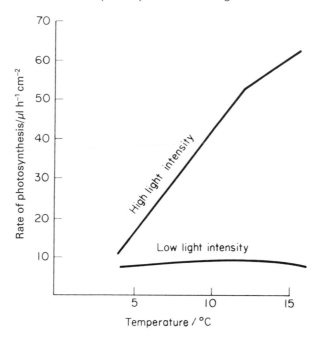

Light quality

The **quality** of light (its wavelength measured in nanometres, nm) plays a vital role in photosynthesis. Green plants can perform photosynthesis only when the available light is within the visible spectrum. They cannot utilise ultraviolet or infra-red light. Red light, and to a lesser extent blue light, are best for green plants. The energy content of the light is measured in joules (J). Its brightness or intensity is measured in kilolux (klx). The duration of light received by a plant (the time of exposure or the day length) is also important in photosynthesis. Green plants show adaptations to variations in light quality, energy content and duration.

3.5 The leaf and photosynthesis

The foliage leaves are the main organs of photosynthesis in a green plant.

Dicotyledon plant foliage leaves are **broad** with **net veins** (see Figure 3.6). They have *differing* upper and lower surfaces, and the upper surface is held at **right angles** to the direction of light. **Monocotyledon** plant foliage leaves are **narrow** with **parallel veins**; upper and lower surfaces are *similar* and the leaves grow **upwards** towards the light.

The surface area to volume ratio of a flowering plant is very large so the leaf can absorb the maximum amount of light.

Leaf internal structure

A typical flowering plant leaf has *three* main regions. They are the **upper epidermis**, a middle region called the **mesophyll** and the **lower epidermis**.

Figure 3.6 The structure of a dicotyledon plant leaf

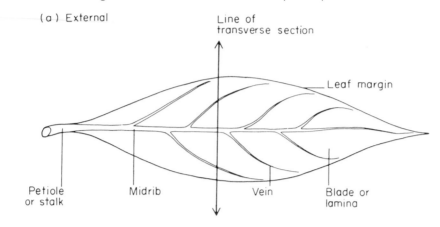

(a) External

Line of transverse section

Leaf margin

Petiole or stalk

Midrib

Vein

Blade or lamina

(b) Internal

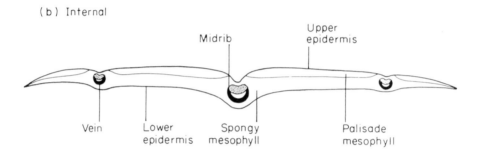

Midrib

Upper epidermis

Vein

Lower epidermis

Spongy mesophyll

Palisade mesophyll

Figure 3.7 Photomicrograph of transverse section through a holly leaf (*Ilex* sp.) (Griffin Biological Laboratories)

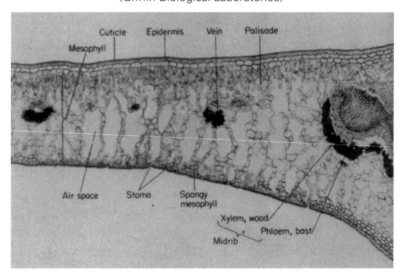

Mesophyll

Cuticle

Epidermis

Vein

Palisade

Air space

Stoma

Spongy mesophyll

Xylem, wood

Phloem, bast

Midrib

Figure 3.8 Internal structure of a dicotyledon leaf

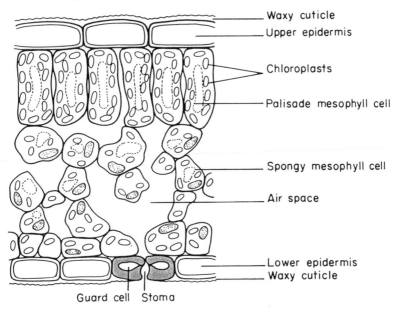

Waxy cuticle
Upper epidermis

Chloroplasts

Palisade mesophyll cell

Spongy mesophyll cell

Air space

Lower epidermis
Waxy cuticle

Guard cell   Stoma

The upper epidermis

This is a single transparent layer of cells which completely covers and protects the upper surface of the leaf blade (lamina). It is covered with a **waxy cuticle** which prevents water loss by evaporation. Each cell has a nucleus and clear cytoplasm which lets the light pass through it to the mesophyll where photosynthesis takes place.

The mesophyll

The cells of the mesophyll region have films of water on their outer surfaces allowing gases to diffuse to and from the leaf air spaces. In dicotyledon plants this part is divided into two layers.

The upper layer of cells is called the **palisade mesophyll**. It lies beneath the upper epidermis and receives the maximum amount of light from the sun's rays. The cells are cylindrically shaped, thin walled with small intercellular air spaces and contain a large number of **chloroplasts**. These are able to move inside the cell by cytoplasmic streaming. In poor light the disc-shaped chloroplasts arrange themselves sideways on to the light near the top of the palisade cell. In bright light the chloroplasts are arranged edge- or end-on near the sides and base of the palisade cells. Most of the photosynthesis taking place in the leaf occurs in this layer.

The layer of cells below the palisade cells is called the **spongy mesophyll**. This is a layer of irregularly shaped, thin-walled cells. They have few chloroplasts and large intercellular air spaces. A large air space is located behind each **stoma** or pore. Excessive water loss is partly reduced by the stomata being positioned on the lower leaf surface away from the direct rays of the sun.

Figure 3.9 Stomata (surface view)

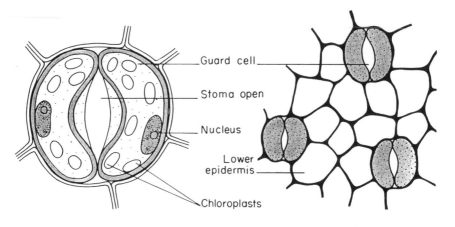

Guard cell

Stoma open

Nucleus

Lower epidermis

Chloroplasts

The lower epidermis

This is similar to the upper epidermis but differs from it by having a great number of pores or **stomata**. Each stoma is composed of two **guard cells** containing a nucleus and a few chloroplasts. The cell wall of the guard cell next to the stoma or pore is thickened.

Leaf adaptations for photosynthesis

- Large surface area for maximum light absorption.
- Leaf arrangement at right angles to the light source with a minimum of overlap in a leaf mosaic pattern.
- Rapid transport to and from mesophyll cells by extensive vein system.
- Numerous stomata allow rapid exchange of the gases, oxygen and carbon dioxide.
- The thinness of the leaf reduces the distances between cells so materials can be transported quickly between them. The thin cell walls of the mesophyll cells allow for rapid diffusion.
- Chloroplasts are usually concentrated in the upper layer, close to the light entering the leaf.

3.6 The chemical composition of a plant

When all the water is removed from a plant, 95.6% of the dry matter is composed of the elements carbon, hydrogen, oxygen and nitrogen. They are joined together to form **organic compounds**. Carbon, hydrogen and oxygen are obtained from carbon dioxide and water – the raw materials of photosynthesis.

The remaining 4.4% of the green plant dry mass consists of the following chemical elements in the form of different **inorganic compounds**: sulphur, phosphorus, calcium, iron, magnesium, potassium, chlorine, aluminium and silicon. The sources of these chemical elements are the soluble salts in the soil water. They are taken into the plant root system as **ions**.

Figure 3.10 Water culture apparatus

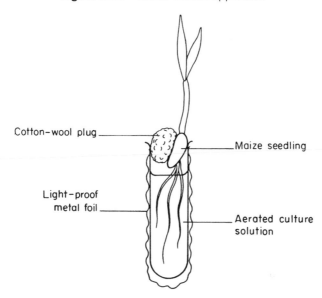

Cotton-wool plug

Maize seedling

Light-proof metal foil

Aerated culture solution

The importance of minerals

The effect of a mineral on the growth and development of a plant can be investigated by growing the plant in an environment that lacks the particular mineral. This is done by growing plants in water or sand **culture solutions**.

A complete culture solution contains all the minerals required for healthy growth. This is used as the 'control' solution. An incomplete solution is one that has one mineral lacking. Several incomplete solutions can be prepared, each with a different mineral lacking. A mineral lacking from an incomplete solution could be nitrogen, calcium, sodium, magnesium, iron, phosphorus or potassium.

The experiment is set up in the following way:

1 A 'control' solution or complete culture is prepared together with seven other incomplete cultures. Each solution and the containing tube is heat-sterilised in a pressure cooker.
2 Fourteen-day-old seedlings of maize, wheat, barley or oats are grown, allowing them to consume their seed food reserves. The seedlings are washed in cold water and supported with their roots dipping into the culture solution. Periodically, the solution is aerated by blowing in a current of sterile air.
3 The outer surface of the cultured tubes is made light-proof, and the tubes with their seedlings maintained together for 3 weeks at a constant temperature of 30°C. The solutions are periodically topped up with fresh, sterile solution.

Aeration of the culture solution is needed for root hairs to respire. Sterilisation prevents the growth of microbes, and light-proofing prevents algal growth. The differences to look for between the seedlings grown in the 'control' and in the incomplete cultures are greenness or the intensity of the chlorophyll colour, size of leaves and plants, root growth and general 'healthy' appearance. The lack of a mineral leads to the development of a deficiency disorder in the plant (see Table 3.1).

Table 3.1 Minerals and plant health

Mineral ion	Purpose	Effect of deficiency
Nitrogen as nitrates or ammonia	Amino acid and protein formation (only producers can make amino acids – animals can not)	Poor growth; chlorosis, a yellow leaf colour due to chlorophyll deficiency
Phosphorus as phosphate	Genetic material, formation; also for reactions in respiration and photosynthesis	Poor growth
Calcium	For cementing plant cells together	Poor stem and root growth
Magnesium	Forms the centre of the chlorophyll molecule	Leaf yellowing – chlorosis
Iron	Is a catalyst for chlorophyll formation	Leaf yellowing – chlorosis; weak stem
Potassium	Chlorophyll formation; also flower and fruit formation	Leaf yellowing – chlorosis

Questions

13 Which mineral is needed for making proteins?
14 Which minerals are needed to prevent chlorosis?
15 Which minerals may be lacking if a plant grows poorly?
16 Which mineral is part of the chlorophyll molecule?

Summary

- Plants make food by photosynthesis. The word equation for this process is carbon dioxide + water → glucose + oxygen (▶ 69)
- Photosynthesis uses energy in sunlight that is trapped in chlorophyll within the chloroplasts. (▶ 70)
- Experiments can be set up to demonstrate aspects of the process of photosynthesis. (▶ 71–75)
- External factors – light, carbon dioxide, temperature – affect the rate of photosynthesis. (▶ 75–77)
- The structure of the leaf is adapted for photosynthesis. (▶ 77–80)
- Minerals are essential for the health of the plant. (▶ 80–82)

4 Nutrition and the food consumers

___ Objectives ___

When you have completed this chapter you should be able to:
- describe the purposes of nutrition
- describe the different kinds of nutrients and identify their sources
- perform simple food tests for carbohydrate, lipids and proteins
- describe dietary balance
- describe the five stages in the processing of food in animals
- describe ingestion in fluid feeders, filter feeders and large particle feeders
- understand how to care for teeth
- understand digestion in the human alimentary canal
- describe the sites of absorption in the human alimentary canal
- describe digestion in herbivorous and carnivorous animals
- describe the action and effects of decomposers
- understand the methods of food preservation
- understand the purpose of food additives

4.1 Introducing nutrients

Foods are **mixtures** of different chemical compounds. They are called **food nutrients** and they have certain functions in the body. Nutrients are used for:
- formation of **genetic material**
- **growth** of organisms
- formation of **body structure**
- **body functions** or metabolism
- **source of energy** in high-energy substances, mainly sugars and lipids.

The energy value of food is measured in kilojoules (kJ).

One kilojoule, kJ = 1000 J
One megajoule, MJ = 1000 kJ = 1 000 000 J

Table 4.1 Average Composition of Natural Whole Foods per 100 g

						VITAMINS			MINERALS			
FOOD GROUPS	WATER g	ENERGY kJ	PROTEIN g	CARBOHYDRATE g	LIPID g	A µg	D µg	C mg	CALCIUM mg	IRON mg	CHOLESTEROL	DIETARY FIBRE
CEREALS	10	1500	10	70	2.5	Trace	NIL	NIL	32	35	NIL	High
VEGETABLES	90	85	1.5	3.6	Trace	Carotenes 900	NIL	33	55	0.8	NIL	Moderate
STARCHY VEGS	70	300	1.5	15	Trace	15	NIL	18	20	0.6	NIL	Mod.
FRUIT	85	200	1.0	10	0.5	250	NIL	30	20	0.5	NIL	Mod.
PLANT ORIGIN FOOD AVERAGE	65	550*	3.5	28*	1.0	290*	NIL	20*	31	1.35	NIL	High to Mod.*
Footnote: FUNGI Yeasts and Mushrooms	80	200	7.0*	5.0	0.5	Trace	NIL	Trace	8	3	NIL	Low
Meat	75	525	18	Trace	6	High to Low	Trace	14 to trace	9	High to 1.0	Medium	NIL
FISH	75	510	17	Trace	6	High to 45	High to 23	Trace to 25		0.5	Medium	NIL
MILK	87	300	3	5	4.5	40	0.1	4	115	0.07	Medium	NIL
EGGS	75	620	12.5	Trace	11.0	140	2.0	Trace	54	1.0	Very high	NIL
ANIMAL ORIGIN FOOD AVERAGE	78	490	12.5*	Trace	7*	4 to High*	20*	4	50*	0.75 to High	Low to High*	NIL

*Indicates HIGH values.

Carbohydrates

Carbohydrates provide 17 kJ/g of energy. The main carbohydrates in food are **starch** and **sugars** (sucrose and glucose from plants and glycogen and glucose from animals). Foods rich in carbohydrates are cereals, root vegetables and fruits. Carbohydrates, especially sugars, provide energy for the body which the body can use straight away.

Lipids

Lipids (oils and fats) provide 38 kJ/g of energy. Foods rich in fats are butter, cheese, full-cream milk, margarine, meat, fish and nuts. Lipids are required to make cell membranes, provide a heat insulation layer in the skin and as an energy reserve.

NB: Mineral oils – liquid paraffin, vaseline, petrol – are not lipids and are unable to provide the human body with energy.

Proteins

Proteins provide 17 kJ/g energy but are not primarily a source of energy as are carbohydrates. Foods rich in proteins are meat, fish, eggs, milk, cheese, bread, cereals, potatoes, peas and beans. Proteins are used for making muscles, bones, skin, enzymes, hormones and antibodies.

Minerals

Calcium

Calcium is a mineral ion. Foods rich in calcium are bread, flour, cheese, milk, vegetables and sardines. Calcium is used by the body to make healthy bones and teeth, to assist blood clotting and for the contraction of muscles.

Iron

Iron is a mineral ion. Foods rich in iron are bread, flour, meat, liver, eggs and green vegetables. Iron is used by the body to form haemoglobin in red blood cells, myoglobin in the muscles and in enzyme actions.

Vitamins

Vitamin A

Vitamin A (retinol) is found in butter, margarines, fish liver oils, liver, milk, carrots, tomatoes and apricots. It is used by the body for growth, night vision and the making of healthy, moist membranes such as those in the nose. It is a lipid-soluble vitamin.

Vitamin C

Vitamin C (ascorbic acid) is found in blackcurrants, potatoes, sprouts, cabbage, tomatoes and citrus fruits. It binds cells together, helps to form blood, blood vessels, bones, teeth and helps to maintain the health of moist membranes. It is a water-soluble vitamin.

Vitmain D

Vitamin D (cholecalciferol) is found in butter, margarines, fish liver oils, eggs, oily fish, cheese and milk. It is also made by the human skin when sunlight shines on it. Vitamin D helps the body to take up calcium and phosphorus. It is important in the formation of teeth and bones. It is a lipid soluble vitamin.

Fibre

Dietary fibre is indigestible material from plant foods. It is made from the cellulose and lignin in the plant cell walls. Foods rich in dietary fibre are cabbage, beans, peas, cereals and wheat bran. Dietary fibre absorbs water in the gut and makes the faeces (see page 103) moist and bulky. It pushes against the gut wall and stimulates the gut wall muscles to push the food along in waves of **peristalsis** (see Figure 4.10). Dietary fibre also makes the gut wall produce mucus which helps the food slide through the gut.

Water

Water is taken into the body in drinks and in many foods. For example, 100 g of apple contains 84 g of water and 100 g of wholemeal bread contains 37 g of water. Water is also produced as a by-product of respiration (see page 143 and Section 6.4). The chemical reactions of life processes take place in water, it transports material around the body and gives support to the inside of cells so they can maintain their shape. Water stores heat and helps the body maintain its constant temperature.

Question

1 Which nutrient is required:
   (a) to make cell membranes?
   (b) to build muscles and make enzymes?
   (c) to provide readily available energy?
   (d) to make haemoglobin?
   (e) for the building of muscles and contraction of muscles?
   (f) to help the body take up calcium and phosphorus?
   (g) to transport materials around the body?
   (h) to help move food along the gut?
   (i) to bind cells together?

Alcohol

Alcoholic drinks contain **ethanol**. The amount of ethanol in an alcoholic drink varies from 3–6% in beers to 30–40% in spirits such as whiskey. Ethanol is not an essential nutrient and we can live without it, although it can provide energy (29 kJ/g), is absorbed directly through the lining of the stomach and stimulates the production of digestive juices. Its adverse effect on health when too much alcohol is consumed places it in a class apart from normal nutrients.

Drinking alcohol reduces the speed of coordination of the body which can lead to accidents. Drinking a large amount of alcohol at one time can lead to coma and death while in some people regular drinking of alcohol can lead to addiction called **alcoholism**. An alcoholic suffers from protein, mineral and vitamin deficiency. Disorders of the nervous system and the muscles develop. Alcohol is a poison and the liver destroys it, but in alcoholics, the liver cells are eventually replaced by fibrous tissue causing **cirrhosis** of the liver; death from liver failure follows.

Questions

2 In what way could alcohol be considered a nutrient?
3 What are the harmful effects of alcohol?

4.2 Food tests

(See also Table 1.3 on page 16).

Test for starch

Starch is detected by putting drops of **iodine solution** on the food.
If starch is present the food turns blue-black.

Carbohydrates

┌─ **Test for sugar (reducing sugar, e.g. glucose)** ─────────────────

The food is mashed up with water and heated with an equal volume of
Benedict's solution in a test tube.
HAZARD: Wear eye protection.
After about two minutes the mixture turns green, yellow, orange, red or brown,
depending on the quantity of reducing sugar present.

Lipids

There are two tests for lipid and oil:

┌─ **Greasy stain test** ─────────────────────────────────

The food is rubbed on thin paper and the paper is held up to the light.
If a lipid or oil is present the paper will have a greasy stain through which the light
shines (translucent).

┌─ **White precipitate test** ─────────────────────────────

The food is shaken with a small amount of ethanol in a test tube. Care is taken to
make sure there are no flames from Bunsen burners near the experiment. The
food and ethanol mixture is poured into a test tube of cold water. Any lipid or oil
present makes a white cloud in the water.

Protein

┌─ **Biuret test** ──────────────────────────────────────

HAZARD: Wear eye protection
A few drops of dilute sodium hydroxide are added to a sample of food mashed
with water.
A few drops of dilute copper sulphate solution are added and a purple colour
appears if protein is present.

┌─ **Questions** ──

4 What food substances are present in a sample which turns blue-black with
iodine solution and leaves a greasy mark on paper?
5 Why must eye protection be worn in (a) the test for sugar, (b) the test for
protein?
6 Two samples of food were tested for sugar. One sample turned green after
heating for two minutes and another sample turned brown after heating for the
same amount of time. What does this tell you about the food samples?

4.3 A balanced diet

A **balanced diet**, or dietary balance, is one in which *all* the nutrients are present in the *correct amounts* to keep the body healthy. It is related to the age, size and activity of the body.

   Malnutrition, or dietary imbalance, arises from either a **deficiency** (undernutrition) or an **excess** intake of nutrients (overnutrition).

Undernutrition – dietary deficiency

Starvation occurs if *all* nutrients are deficient. In young infants, the wasting and emaciation of the body resulting from general starvation is called **marasmus**.

Figure 4.1   Four children of the same age, three of whom suffer from protein deficiency – Kwashiorkor (WHO)

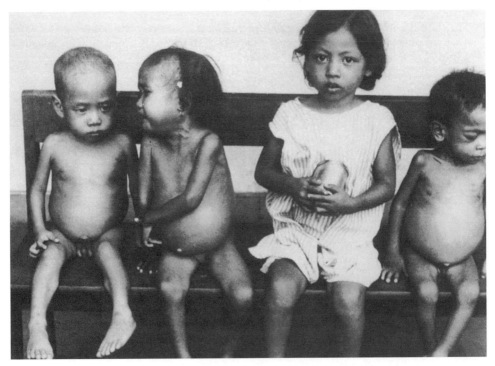

- A lack of protein in the diet of young children leads to **kwashiorkor**. It is seen in weaned children aged 2 years who were previously breast-fed and received their protein from human milk. After weaning, the children were fed on mainly starchy foods without protein. They developed a swollen stomach, muscle wasting and anaemia (see Figure 4.1).
- A lack of calcium in the diet can lead to the bones becoming soft or brittle.
- A lack of iron in the diet produces **anaemia**. The symptoms of anaemia are tiredness, pale skin and breathlessness because the number of red blood cells carrying oxygen is reduced.
- A lack of vitamin A in the diet may lead to the development of **night blindness** (inability to see shapes in dim light) low resistance to infection, poor growth and dry corneas.

- A lack of vitamin C in the diet can lead to **scurvy**. This deficiency disease has the symptoms of swollen, bleeding gums, loose teeth, joint pains, weakness and wounds that fail to heal properly.
- A lack of vitamin D in children's diets leads to softening of the bones and bending of the legs due to the weight of the child's body (**rickets**). The child's teeth also form badly. In adults, a lack of vitamin D leads to softening of the bones called **osteoporosis**.
- A lack of dietary fibre in the diet leads to the formation of hard, dry faeces which pack together in the large intestine and cause **constipation**. There is evidence to suggest that a lack of dietary fibre increases the risk of **bowel cancer**.
- Between 65–70% of the human body weight is due to water. An adult is unable to survive for more than 2 or 3 days without water but can survive for many weeks without food, provided that drinking water is available. A lack of water in the diet leads to **dehydration**.

Overnutrition – dietary excess

A *high* daily intake of **sucrose** (cane sugar) can cause tooth decay.

People who have a diet which contains a large quantity of **animal fat** may suffer from heart disease because the lipids form deposits on the insides of the blood vessels, making it more difficult for the heart to pump the blood around the body. Hard or **saturated** lipids, e.g. lard and meat fats, are considered harmful. Soft, **unsaturated** lipids, such as peanut, fish, sunflower and other vegetable oils, are considered beneficial.

If the intake of energy from energy-providing foods – proteins, carbohydrates and lipids – is in excess of the body's needs, the excess nutrients will not be excreted but will be converted into a form of food storage, as glycogen in the liver and muscles or as fat underneath the skin and around the kidneys. This leads to overweight, fatness or **obesity**. The extra weight increases demands on the circulatory system and can lead to heart disease. Obesity is also believed to be related to the development of sugar diabetes.

Questions

7 What causes kwashiorkor and what are its symptoms?
8 (a) Which members of the population are at risk of kwashiorkor?
  (b) Why are these people at risk?
9 (a) What causes anaemia?
  (b) How can anaemia be cured?
10 (a) Why are diseases caused by a lack of certain vitamins and minerals also known deficiency diseases?
   (b) What deficiency disease has the symptoms of bleeding gums and loose teeth?
   (c) How may this disease be cured?
11 How are rickets and osteoporosis (a) similar, (b) different?
12 (a) How may too much sugar in the diet affect the body?
   (b) Give two ways in which the body is affected.

Balancing the diet

Nutrients

The amounts of nutrients in different foods have been measured so that foods can be selected for the nutrients they contain. Table 4.1 on page 84 shows the average composition of natural whole foods. A balanced diet is achieved by eating a mixture of foods chosen from each of the four main food groups – cereals and bread, milk and milk products, fruits and vegetables, meat and fish.

A **vegetarian** diet includes food from plants and animal products such as milk and egg. It does not contain meat. The extra protein required can be obtained by eating pulses (peas and beans). A **vegan** eats a very strict vegetarian diet which does not even include animal products. This diet is deficient in certain amino acids for growth and repair of the body, vitamin B12 for development of red blood cells, and calcium.

Energy

The daily energy needs of a person depend on his/her age, body size, sex and occupation, as Table 4.2 shows.
- **Age:** The energy needs of growing children are much greater than those for adults on a weight for weight basis. Elderly people require less energy from their food.
- **Body size:** a tall, heavily built body requires more energy than a slightly built, short body.
- **Sex:** Males need more daily energy than females but the needs of females increase in times of menstruation, pregnancy and lactation or breast-feeding.
- **Work:** Physical activity in work varies with the job. Sedentary work, where the person sits down to work (e.g. office work and bus driving) is less energy demanding than labouring work on building sites or in warehouses.

Protein intake

The daily protein needs of different people are shown in Table 4.3.

Minerals and vitamins

The small amounts, **micrograms** (μg) of vitamins A, D, B12 and folic acid, and **milligrams** (**mg**) of all other vitamins and minerals, will naturally be available provided a mixed diet of the correct energy composition is taken.

Questions

13 Use Table 4.1 on page 84 to answer questions (a)–(d).
  (a) Which foods provide (i) highest energy content, (ii) lowest energy content?
  (b) Which foods provide (i) most protein, (ii) least protein?
  (c) Which food is richest in lipids (oils and fats)?
  (d) Which vitamin is lacking from plant foods?
14 What is the difference between a vegetarian and a vegan diet?
Use Table 4.2 to answer questions 15 and 16 (NB: The amounts of energy are related to the body weight per kg.)
15 How does the energy requirement of a male change during his lifetime?
16 How do the energy needs of a female change during pregnancy and lactation? Explain why there is a change in energy requirement in each case.
17 Will the energy requirements of a tall, heavily built male labourer be the same as those of a short, lightly built female office worker? Explain your answer.
Use Tables 4.2 and 4.3 on page 91 to answer question 18.
18 Compare the daily energy and protein needs of a person throughout life. Explain any trends you identify.

Table 4.2 Average daily energy needs throughout life

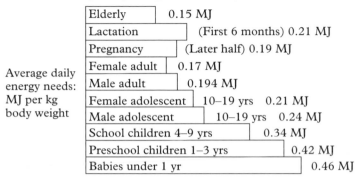

Average daily energy needs: MJ per kg body weight		
Elderly	0.15 MJ	
Lactation	(First 6 months) 0.21 MJ	
Pregnancy	(Later half) 0.19 MJ	
Female adult	0.17 MJ	
Male adult	0.194 MJ	
Female adolescent	10–19 yrs	0.21 MJ
Male adolescent	10–19 yrs	0.24 MJ
School children 4–9 yrs	0.34 MJ	
Preschool children 1–3 yrs	0.42 MJ	
Babies under 1 yr	0.46 MJ	

Table 4.3 Daily protein requirements throughout life

Average daily protein needs: grams per kg body weight		
Elderly	0.47 g	
Lactation	(First 6 months) 0.83 g	
Pregnancy	(Later half) 0.69 g	
Adult woman	0.53 g	
Adult man	0.57 g	
Female adolescents	10–19 yrs	0.64 g
Male adolescents	10–19 yrs	0.7 g
School children	4–9 yrs	0.93 g
Preschool children	1–3 yrs	1.2 g
Babies under 1 yr	1.92 g	

4.4 The stages in food processing

There are *five* stages in the processing of food in protoctists and animals (see classification, pages 49 and 55).
1 **Ingestion:** the intake of complex food molecules into the body.
2 **Digestion:** the breakdown of large molecules of mainly insoluble food substances into smaller molecules of soluble food nutrients. Digestion takes place by means of **hydrolytic enzymes** which are made of protein.
3 **Absorption:** the uptake of food nutrients into living cells.
4 **Assimilation:** the usage of food nutrients within the living organism. It is a process which follows the **transport** of food nutrients to cells.
5 **Egestion:** the elimination of indigestible food waste from the organism.

┌─ **Question** ─────────────────────────────────

19 What are the differences between
    (a) ingestion and digestion?
    (b) absorption and assimilation?
    (c) ingestion and egestion?

Ingestion in fluid feeders

Mosquito and greenfly

The mosquito and the greenfly (aphid) are insects with mouthparts which are modified for **piercing** and **sucking** fluid into the gut.

Figure 4.2 The piercing and sucking mouthparts of a female mosquito, *Anopheles* sp.

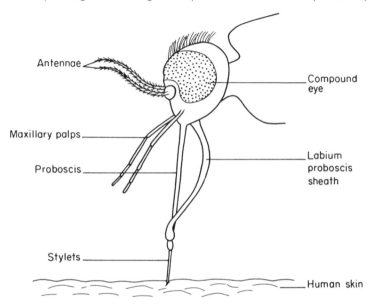

The female mosquito pierces human skin with needle-like **stylets**, and inserts a separate tubular **proboscis** which injects an **anticoagulant** substance, preventing the blood from clotting and blocking the stylets. Male mosquitos are without stylets and do not feed on blood.

Greenfly pierce the epidermis of leaves using stylets and suck up food in solution (sap) via the tubular proboscis in contact with the leaf vein or mesophyll cells.

Housefly

The housefly feeds mainly on dead and decaying material. It places its tubular mouthpart directly on solid or liquid material and then secretes a fluid, **saliva**, visible as 'fly spots', which *externally* digests the food and dissolves it to make a solution. This solution then enters the tubular proboscis partly by **capillary** action and by **suction**, passing into the gut. Alternatively, soluble foods are drawn into the gut directly, such as when the fly feeds on foul water or milk.

Butterfly

The butterfly feeds on a liquid called **nectar**, a solution of sucrose, produced in insect-pollinated flowers. The tubular proboscis reaches into the flower and the nectar is sucked into the gut. When not in use, the proboscis is coiled under the head.

Figure 4.3 The sucking mouthparts of a housefly, *Musca domestica*

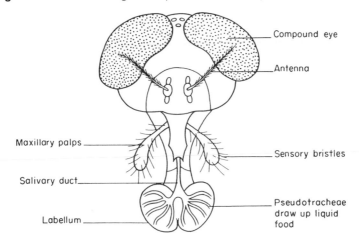

Compound eye

Antenna

Maxillary palps

Sensory bristles

Salivary duct

Labellum

Pseudotracheae
draw up liquid
food

Figure 4.4 The sucking mouthparts of a butterfly, *Pieris* sp.

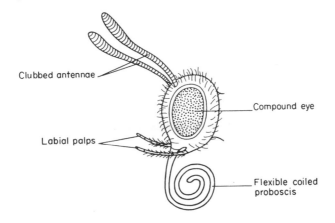

Clubbed antennae

Compound eye

Labial palps

Flexible coiled
proboscis

Questions

20 (a) Why is only the female mosquito able to pierce the skin?
  (b) Why would a mosquito starve without anticoagulant?
21 Which of the four insects described digests its food outside its body?
22 (a) Which insect has the longest mouthparts?
  (b) How are these mouthparts carried by the insect in flight?

Protoctists such as the malarial parasite, tapeworms and most bacteria are fluid-feeders. Fungi secrete digestive juices externally on to food and absorb the soluble nutrients.

Ingestion in filter-feeders

Filter-feeding animals are all aquatic, feeding on relatively *small-sized* food particles present in water as **plankton** (protoctists and tiny animal larvae). They include sponges, cockles, mussels, certain marine worms, herrings and the largest animal – the whalebone or baleen whale.

Filter feeders feed continuously, drawing a current of plankton-laden water into various filtering devices which strain off the small organisms. Cockles and mussels use **gills** as sieves. Herrings use **gill rakers** or the gill **bars** to collect the plankton, chiefly protoctists, tiny crustaceans and mollusc larvae.

Figure 4.5 Gill structure of a bony fish with operculum removed

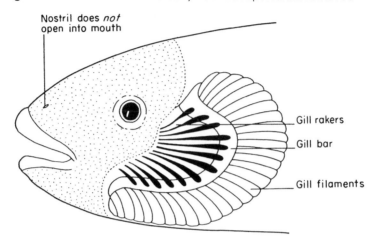

The baleen whale has comb-like whalebone or **baleen** hanging down from the upper jaw which strains the zooplankton or **krill** from the sea water.

Questions

23 Where would you find filter-feeding animals?
24 Name three kinds of filtering device.
25 Why do filter-feeders have to feed continuously?

Large-particle feeders

Fish, amphibia, reptiles and birds swallow their food whole. They have small, peg-like **teeth**, or **beaks** but no teeth as in birds. Mammals, with the exception of ant-eaters and baleen whales, have teeth which are used to hold and break up food.

Mammalian tooth structure

The main parts are the **crown** above the gum, and **root** below the gum in the jaw bone or **mandible**.

Enamel

The tooth crown is covered with extremely hard **enamel,** a substance composed of keratin and hardened with minerals such as calcium and magnesium compounds. The **ridges** seen in herbivore teeth are due to the enamel being worn away in the crosswise chewing action (see Figure 4.12).

Figure 4.6 Vertical section through a molar tooth showing internal structure

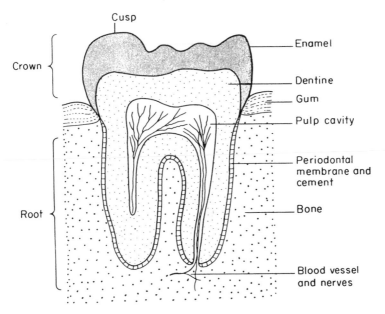

Dentine

The **dentine** is beneath the enamel layer and is a substance similar to but harder than bone. Fine canals connect it with the substance of the pulp cavity. Dentine is exposed in ridged, herbivore teeth (see Figure 4.12).

Pulp cavity

The **pulp cavity** is the tooth centre containing blood vessels and nerve endings which are sensitive to heat, cold and pain.

Cement

The root is held in the tooth socket by **cement** and the **periodontal** membrane. The fibres of this membrane allow the tooth to move slightly in the socket.

Roots

Open roots are seen in continuously growing teeth. These are present in rodent and herbivore dentition, as the teeth are being continuously worn away. **Closed** roots do not allow a good blood supply and are seen in teeth which grow to a certain size, as in carnivore and omnivore dentition.

4.5 How humans process food

Ingestion

The food is taken into the mouth and the teeth act upon it in the following ways:
• The **incisor** teeth are chisel-shaped and cut and bite the food into smaller pieces.

- The **canine** teeth are pointed. In many mammals they are large and used for tearing and grasping. In humans they are similar in size to incisors but their sharp points tear rather than cut the food.
- The **premolars** and **molars** have flattened or cusped crowns. They are used for grinding and mixing the food.

There are two sets of teeth in humans. The first set are the temporary or **deciduous** 'milk' teeth and are in position beneath the gums at birth. They comprise only 20 teeth with no molars. The second set of **permanent** teeth take from birth to age 21 to develop by calcification, using calcium, fluoride, phosphates, proteins and vitamin D from the diet. There are 32 teeth in this set.

Figure 4.7 Adult human dentition

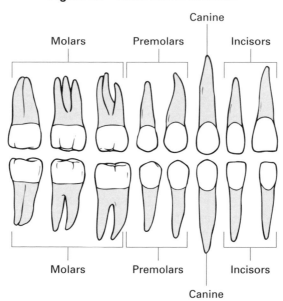

Oral and dental hygiene

The two threats to a healthy mouth are:
1 Gum disease, **gingivitis**, and loose teeth or **periodontitis**. These are caused by dental **plaque** which consists of bacteria growing in a soft substance formed from saliva and food. The plaque coats the teeth and fills the crevices between the gums and teeth. The bacteria infect the gum and attack the periodontal fibres, causing teeth to loosen in their sockets or form pus-filled abscesses.
2 **Dental caries** is caused by the conversion of sucrose into acids by bacteria. The acid causes demineralisation or loss of calcium and phosphates, with the formation of a **cavity** or hole in the tooth dentine, leading to rapid tooth decay and toothache.

Dental care

Gum disease and dental caries can be prevented by the following:
- **Brushing** teeth and gums correctly after every meal with fluoride toothpaste. This removes plaque and remineralises the teeth with calcium, phosphate and fluoride.
- Sugary foods must be restricted and a nutritious diet including hard foods such as fruit, raw vegetables, nuts and cheese substituted for sweets, chocolates, cakes and biscuits.

- Fluoride increases the resistance of teeth to decay and this mineral ion can be present in natural drinking water or taken as drops or tablets by children up to 5 years of age while the teeth are calcifying. The addition of fluoride to water supplies remains a controversial issue. Fluoride in small amounts prevents dental caries but in large amounts causes a discolouring of the enamel.

Questions

26 How does the shape of an incisor tooth differ from (a) a canine, (b) a molar?
27 How is the function of an incisor different from (a) a canine, (b) a premolar?
28 Where are the nerve endings in a tooth?
29 How do bacteria cause teeth to (a) become loose, (b) decay?
30 How can a change of diet improve dental health?
31 Why is the amount of fluoride in a toothpaste important to the appearance of the teeth?

Digestion

Digestion is the conversion of **large**, mainly **insoluble** molecules of food into **smaller**, **soluble** molecules of food nutrients by the combined processes of **physical** digestion and **chemical** digestion.

Physical digestion

In physical digestion, parts of the body such as the teeth, tongue, cheeks and wall of the gut act upon the food to make it easier to digest chemically. These actions include **breaking down** the food into smaller pieces to *increase* the food surface area on which the chemicals can work, **emulsifying** lipid substances to make them into small droplets on which chemicals can act, **lubrication** of the food so than it moves along the gut easily.

Chemical digestion

In chemical digestion extracellular enzymes break down the food molecules mainly by **hydrolysis** into molecules which can pass through the wall of the gut.

The human alimentary canal

The alimentary canal is an **organ system**. Organs such as the salivary glands, oesophagus, stomach, liver, pancreas and small intestine work together to digest food.

The gut wall from the oesophagus to the rectum is composed of two main layers:
- a **mucous coat** or **membrane** lined with epithelium, continuously moistened with mucus and with numerous folds.
- a **muscle coat** of inner **circular** and outer **longitudinal** involuntary smooth muscle.

A mesenteric membrane is wrapped around the gut, connecting it to the wall of the abdominal cavity.

The food is transported along the gut by **peristalsis** (see Figure 4.10). This is a muscular movement which takes the form of waves of contraction and relaxation around a tubular organ like the gut. It also occurs in other parts of the body, e.g. in the ureter and the uterus.

Figure 4.8 The human alimentary canal

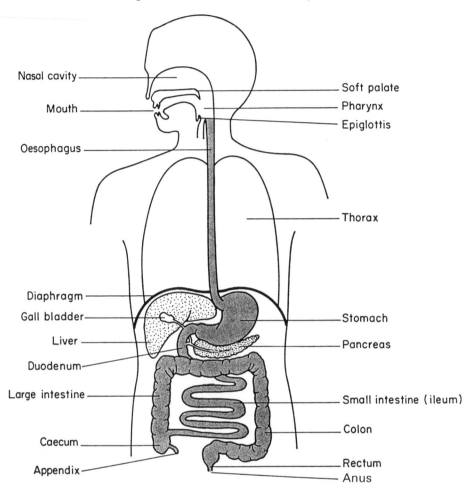

Nasal cavity

Mouth

Oesophagus

Soft palate

Pharynx

Epiglottis

Thorax

Diaphragm

Gall bladder

Liver

Duodenum

Large intestine

Caecum

Appendix

Stomach

Pancreas

Small intestine (ileum)

Colon

Rectum

Anus

Figure 4.9 Transverse section through the gut

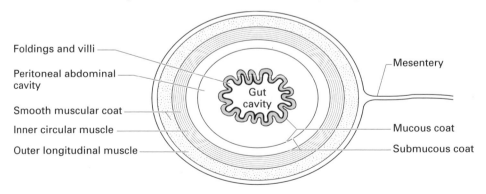

Foldings and villi

Peritoneal abdominal cavity

Smooth muscular coat

Inner circular muscle

Outer longitudinal muscle

Gut cavity

Mesentery

Mucous coat

Submucous coat

Transverse section of gut

Figure 4.10 Peristalsis

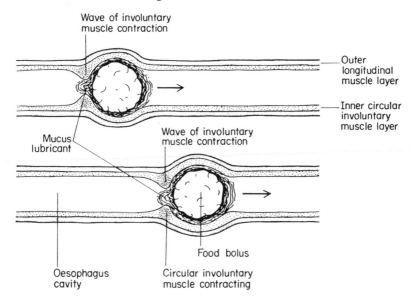

Digestion in the mouth

Between 1 and 2 litres of **saliva** are produced daily by the three pairs of salivary exocrine glands located beneath the tongue, jaw and near the ears. This solution, consisting of 99% water, dissolves soluble food nutrients. Saliva contains a lubricant called mucin which helps dry food to be swallowed easily.

The main enzyme present in saliva is **amylase** which digests cooked starch containing dextrins, producing **maltose** and a small amount of **glucose**. Raw, uncooked starch is changed by amylase into dextrins. The pH of saliva is 6–7, almost neutral. A round-shaped piece of food called a **bolus** is formed by the action of the tongue and cheeks. After being swallowed it is passed down the oesophagus by peristalsis (see Figure 4.10).

Digestion in the stomach

Between 2 to 4 litres of **gastric juice** are secreted from the gastric exocrine gland cells lining the stomach. The gastric juice is blended and churned with the food by peristalsis of the stomach wall, producing a creamy **chyme**.

Gastric juice is strongly acid, with a pH between 1 and 2. It consists of 98% water, 0.5% **hydrochloric acid**, the lubricant, **mucin** and the protease, **pepsin**. Hydrochloric acid provides the necessary pH for enzyme action. It also stops the action of swallowed saliva. Nucleoproteins, present in cell nuclei, are split into protein and nucleic acids. The acid also destroys bacteria on the food. The protease pepsin begins the breakdown of protein molecules into smaller **polypeptide** molecules.

In babies, the stomach also produces **rennin** which changes soluble **caseinogen** of milk into insoluble curds of the protein **casein**. In young children the stomach also produces **lipase** which acts upon small amounts of lipids to produce fatty (alkanoic) acids and glycerol (propanetriol).

Digestion in the small intestine

Periodically, chyme is allowed to leave the stomach through the **pyloric sphincter** muscle which guards the entry to the small intestine. The small intestine is 6 metres long and makes up about 65% of the total gut length.

The **duodenum** is the first 30 cm of the small intestine and it receives three main secretions: **bile juice** from the liver, **pancreatic juice** from the pancreas and **intestinal juice** from the wall of the duodenum itself.

Bile juice

Between 700 cm^3 to 1.2 l of bile juice is produced daily by the liver. It is stored in the **gall bladder** before it is released into the duodenum. Bile is a yellowish green liquid with an alkaline pH of about 8, which neutralises the acid from the stomach. Bile consists of 98% water and contains bile acids and salts which are able to change lipids physically into **emulsions**. The emulsions consist of tiny lipid droplets with a large surface area.

Excretory products such as bile pigments and cholesterol produced by the liver are present in bile juice. The bile juice does not contain any digestive enzymes.

Pancreatic juice

About 700 cm^3 of pancreatic juice is produced daily by the pancreas exocrine gland. The juice has an almost neutral pH value of 7 to 8. It contains the following enzymes:
- **amylase** to complete the conversion of starch to maltose;
- **lipase** which converts most of the lipid in a meal into fatty (alkanoic) acids and glycerol (propanetriol);
- **trypsin** which breaks down protein into polypeptides.

Intestinal juice

About 200 cm^3 of intestinal juice is produced daily by the walls of the duodenum and the small intestine. In the duodenum it has an acid pH of 4.5 to 6. In other parts of the small intestine it has a pH of 6 to 7. It contains carbohydrase, lipase and protease enzymes which complete the chemical digestion of the food.

Questions

32 How are physical digestion processes different from chemical digestion processes?

33 How is food moved along the gut?

34 What is the purpose of mucus?

35 What enzyme is present in the mouth and what does it do?

36 What is a bolus?

37 What does hydrochloric acid do in the stomach?

38 How does bile juice help lipid, fat and oil digestion?

Use Table 4.4 to answer these questions:

39 Trace the digestion of starch to glucose from the mouth to the small intestine.

40 Where is sucrose (cane sugar) digested?

41 What are proteins broken down into?

42 What are lipids, oils and fats broken down into?

Table 4.4 Summary of products of carbohydrate, protein and lipid digestion

REGION	ENZYME	SUBSTRATE	PRODUCTS
CARBOHYDRATE DIGESTION:			
Mouth	Salivary amylases	Cooked starch	Maltose (plus small amount glucose)
Stomach	NIL	NIL	NIL
Small Intestine	Pancreatic amylases. Intestinal juice: Carbohydrases:	Starch, glycogen	Maltose
		Lactose	Galactose/Glucose
		Sucrose	Fructose/Glucose
		Maltose	Glucose/Glucose

DIETARY FIBRE: Cellulose, gums, lignin and pectins are NOT digested and remain undigested in faeces.

REGION	ENZYME	SUBSTRATE	PRODUCTS
PROTEIN DIGESTION:			
Mouth	NIL	NIL	NIL
Stomach	Protease Rennin (children only)	Protein Milk Caseinogen	Peptides Casein
Small Intestine	Pancreatic proteases Intestinal juice proteases	Proteins and peptides	Amino acids
LIPID DIGESTION			
Mouth	NIL	NIL	NIL
Stomach	Gastric lipase (children only)	Milk lipids	Fatty acids and Glycerol
Small Intestine	Pancreatic lipase	Lipids	Fatty acids and Glycerol

Questions on enzyme experiments

A solution of amylase was mixed with a soluble starch solution and samples were taken every 30 seconds and tested with iodine solution. The early samples in the experiment went blue-black straight away but later samples failed to change the colour of the iodine.

43 Why was there a change in the reaction of the iodine to the samples?

44 Did the amount of starch in the samples (i) increase, (ii) decrease, (iii) stay the same as the experiment progressed? Explain your answer.

When the experiment was repeated at a range of temperatures below 38°C more samples turned blue-black at the the lower temperatures than at the higher ones.

45 What did the experiments at different temperatures show? Did the reaction (i) speed up as the temperature increased, (ii) slow down as the temperature increased or (iii) stay the same as the temperature increased?

When the experiment was repeated at a range of temperatures above 38°C more samples turned blue-black at the higher temperatures than at the lower ones.

46 Did heating the enzyme above 38°C (i) speed up its reaction with starch, (ii) slow down its reaction with starch or (iii) not affect its reaction with starch?

47 Enzymes are made of protein which coagulate and become denatured when they are boiled. If boiled amylase was added to soluble starch solution what would the colours of the samples be? (i) all brown, (ii) all blue-black, (iii) early ones blue-black; later ones brown?

48 What would the colours of the samples be if amylase and soluble starch solution were kept at 30°C and hydrochloric acid was added to the mixture? Would they be (i) all brown, (ii) all blue-black, (iii) early ones blue-black; later ones brown?

Absorption

The products and food nutrients for absorption

At the end of digestion, proteins have been broken down into amino acids by proteases, carbohydrates have been broken down into simple sugars by carbohydrases and lipids, oils and fats have been broken down into fatty acids and glycerol by lipases. These products and other food nutrients which do not require digestion – water, mineral ions and vitamins – are absorbed into the body.

The regions of absorption

The stomach

Water, glucose, ethanol, certain minerals, vitamins and certain drugs may be absorbed into the cells lining the stomach by osmosis, diffusion and active transport.

The small intestine

This is the main region of absorption of the products of digestion – simple sugars, amino acids, fatty acids and glycerol. Rapid absorption of the food occurs because the internal surface is covered with tiny finger-like projections called **villi** and thrown into folds to present a *very large* surface area through which the small molecules can pass.

Figure 4.11   Villi in the wall of the small intestine

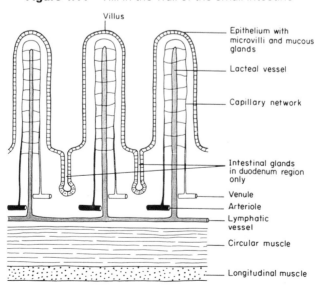

The surface is also very thin so the molecules can pass through quickly, and it has a large number of vessels beneath it which take the molecules away. The **lacteal vessels** are lymph vessels that drain the villi and transport fatty acids and glycerol together with lipid-soluble vitamins A and D. The **blood capillary** network takes away glucose, amino acids, and water-soluble vitamins such as vitamin C together with water-soluble minerals. The nutrients are quickly moved away from the absorbing surface by the blood. This creates a region of *low* concentration of nutrients in the blood vessels. The small intestine has a *high* concentration of nutrients so they move quickly by diffusion along the diffusion gradient into the blood vessels.

The large intestine

Most of the water present in the original saliva, gastric, bile and pancreatic juices is recovered by absorption in the **colon**. Certain mineral ions are also taken up by active transport. The **faeces** are all that remains. They are formed from indigestible dietary fibre, bacteria called coliform bacteria and other aerobic and anaerobic bacteria which may be harmful or pathogenic, excretory products from the liver and dead epithelial cells removed from the gut lining by the food pushing along its surface.

Mucus containing mucin is secreted from the lining of the large intestine to lubricate the faeces in preparation for egestion. (NB: Water is absorbed throughout the gut from mouth to anus but most absorption occus in the small and large intestines.)

Questions

49 What are the products of digestion of (a) proteins, (b) carbohydrates, (c) lipids (oils and fats)?

50 Which food nutrients do not need to be digested?

51 How is the small intestine adapted for the rapid absorption of nutrients?

52 Into which part of the villus: lacteal or blood capillary, are the following absorbed?
(a) amino acids, (b) vitamin A, (c) vitamin C, (d) fatty acids, (e) glucose, (f) glycerol, (g) vitamin A.

Assimilation and transport

The soluble food nutrients are absorbed into the blood capillaries and transported to the liver. Food nutrients in the lacteals pass through the lymphatic system (see page 135) and enter the blood near the heart. The **regulation** and fate of the nutrients within the body cells are described in Chapter 7.

Egestion

Faeces remain in the large intestine until a short time before the act of egestion or **defaecation**, when they move into the rectum which is normally empty. The presence of faeces in the rectum causes involuntary contraction of the rectum wall and relaxation of the anus sphincter muscle. Voluntary contractions may also occur in the diaphragm and abdominal muscles to help eject the faeces.

4.6 Digestion in other mammals

Teeth and jaws

Herbivores

Herbivores, such as the sheep and the horse, have sharp, chisel-like incisor teeth for cutting. They do not have canines or if present they are small. There is a space called the **diastema** between the front teeth and the premolars. The premolars and molars are similar and are ridged and serrated for grinding and chewing. The action of the jaws allows up and down movement and prolonged chewing action *across* the ridges for maximum grinding action.

Figure 4.12 Lower jaw and teeth of a herbivore, sheep

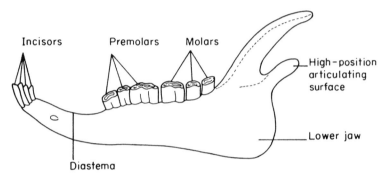

Carnivores

Carnivores, such as the cat and the dog, have small, pointed incisors for picking up food or for stripping flesh. They have sharp, pointed canines for tearing, stabbing and piercing. The premolars and molars are pointed for shearing, crushing and cracking. The jaw action is of a much shorter duration than that of the herbivores and is a strong, up and down, scissor action.

Figure 4.13 Lower jaw and teeth of a carnivore, dog

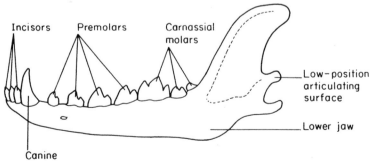

The gut

Food of herbivore mammals, such as rabbits, horses or cows, consists mainly of the carbohydrate, **cellulose**, chemically composed of glucose units. Cellulose can be changed into **organic acids** such as ethanoic (acetic) acid by bacterial action. Digestion occurs by means of bacteria which inhabit the gut. These bacteria secrete the **extracellular** enzyme **cellulase**.

Prolonged chewing, using the ridged herbivore teeth, breaks down the cellulose walls, releasing the protein, starch, nucleoprotein, vitamin and mineral cell contents.

Ruminant herbivores such as sheep, goats and cows have a four-chambered stomach. Food enters the **rumen** and **reticulum** chambers to undergo digestion of cellulose by bacteria. Periodically, portions of food are regurgitated into the mouth as `cud' and re-chewed and re-swallowed to enter the **psalterium** for digestion by gastric juice. In addition to food, the bacteria are also digested at this point.

Non-ruminant herbivores such as rabbits and horses have a very large **caecum** in the large intestine, and also a very large **appendix** (see Figure 4.14). Food enters the normal single-chambered stomach to undergo cellulose digestion by the bacteria in the large intestine.

Figure 4.14   The main regions of the gut of rabbit, a non-ruminant herbivore

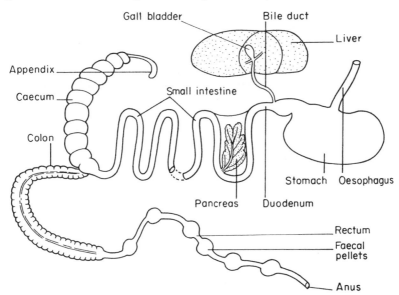

The rabbit has a large rectum full of faecal pellets undergoing dehydration. A feature of rabbit nutrition is that partly digested faeces are egested, ingested a second time and make another journey through the digestive system. This process of repeated digestion in the gut is known as **refection**.

The gut of a carnivore, such as a cat or dog, is much shorter than that of a herbivore. The carnivore tears off large pieces of food and swallows them whole. The food is stored and digested in the carnivore's large, single-chambered stomach.

Questions

53 How are the teeth of herbivores different from those of carnivores?
54 How are the actions of the jaws of herbivores different from those of carnivores?
55 Cellulose makes up the walls of plant cells. How is it broken down by a herbivore such as a sheep to release the nutrients in the cells?
56 What are the substances in a cell which can be used as nutrients by the sheep?
57 How does the method of digestion in the rabbit differ from that of the sheep?
58 How is the carnivore's method of feeding different from that of a herbivore?
59 How is the stomach of a cat different from that of a sheep?

4.7 Feeding relationships

There are several ways in which organisms are related through feeding. Three examples are **symbiosis** or **mutualism**, **parasitism** and **saprophytism** or **saprobiosis**. These relationships are considered in Chapter 14, Sections 14.3 and 14.4.

4.8 Decomposers

Decomposers are also called **saprotrophs**. They are mainly fungal yeasts, moulds or saprotrophic bacteria which are found in the soil or within organisms. Decomposers secrete enzymes onto food and digestion occurs **externally**. The food is broken down into simple substances and these are absorbed into the body of the fungus or bacteria.

The role of decomposers is in the **decay** of dead plant and animal remains, although many decomposers are useful in biotechnology. Decay is a process which proceeds best under the following conditions:
• Warmth: around 37°C for bacteria, 22–25°C for fungal moulds;
• Moisture;
• Alkaline pH to acid pH 5: moulds favour acidity whilst bacteria favour alkalinity;
• Oxygen is required by many decomposers but some can function without it.
Saprotrophic organisms are important in three ways:
• **Soil fertility** is improved by the conversion of plant and animal remains into humus through the process of decay or decomposition.
• **Antibiotics** are formed by many different kinds of bacteria and fungal saprotrophs. Examples of antibiotics are penicillin and neomycin.
• **Disease-causing** saprotrophs include the fungal saprotrophs which are responsible for ringworm in the skin and hair and bacteria which are responsible for the conversion of sucrose into acids which cause tooth decay.

4.9 Food preservation

Human food is also food for other consumers or heterotrophs such as insects, and decomposers such as fungi and bacteria. They are attracted to the food and are described as food pests, food spoilers and food decomposers accordingly.

Food preservation is the process of keeping food for long periods away from the influence of food decomposers. Food decomposition is prevented in two ways:

- Growth of microorganisms is prevented by depriving the microorganisms of water (by drying), warmth (by freezing), oxygen (by using a vacuum to remove air) or a neutral pH (by increasing the acidity).
- Microorganisms are destroyed by chemical preservatives, high temperatures and radiation with X and gamma rays or ultraviolet rays.

Table 4.5 shows the main food preservation methods.

Table 4.5 Food preservation methods

Principle	Methods	Application
Heating – denatures microorganism proteins (spores are heat-resistant)	*Pasteurisation* – temperatures of 61–80°C for various times destroy most pathogenic bacteria	Milk, eggs, ice cream, fruit
	Cooking by boiling, roasting, grilling and baking is a form of heat sterilisation	Most high-temperature cooking methods
	Canning and *bottling* involve pre-treatment by peeling, blanching and addition of liquor, followed by sterilisation involving cooking at 115–125°C in pressure-cooking vessels. Foods are then sealed into containers entirely free of air and oxygen	Most foods. Shelf-life 1–4 years
Low temperature cooling – below –20°C food poisoning and food spoilage microorganisms are dormant. Cell enzyme activity is reduced	*Domestic deep-freezing* – food frozen and stored at –20°C for up to 3 months	Mainly meat, fish, poultry, vegetables, fruits and pre-cooked convenience foods
	Commercial deep-freezing – food frozen by different methods at –30°C to –40°C. Foods stored for very long periods	
Dehydration – (a) water removal from food	*Sun-drying* – food liable to contamination	Fruits, vegetables and fish
	Commercial dehydration by fluidised bed, spray-, roller-, vacuum- or accelerated freeze-drying	Fruits, vegetables, eggs, milk, coffee, soup, potatoes and meat
(b) water removal from microorganisms by osmosis	*Sugar* – (syrups and preserves, jams and sugar crystallisation	Fruits
	Salt (sodium chloride) – salt and sweet-curing (high levels of sugar and salt in foods should be indicated on labels)	Vegetables, fish, bacon and ham
Acidity (pH) – strong acid solutions above pH 4 destroy most but not all microorganisms	*Pickling* – using vinegar (ethanoic/acetic acid)	Pickles and sauces
	Acid preservatives – sulphur dioxide gas; benzene carboxylic (benzoic) acid; propanoic (propionic) acid	Beer, dried fruits, dried potatoes and peas, jams, pickles, sausage, fruit juices and cordials; artificial creams, fruit juices, pickles, sauces; bread and cakes

4.10 Food additives

Foods are processed on a large scale by food manufacturers. This processing can cause losses in natural food colour, nutrient content, flavour and physical condition. Various substances are added to counteract the effect of these losses. The substances include:
- **Colourings:** natural or synthetic dyes;
- **Flavours:** natural flavours provided by herbs, spices and essences or synthetic flavours and sweeteners;
- **Conditioning additives:** e.g. moisturisers and emulsifiers;
- **Nutritive additives:** e.g. Vitamin C in fruit drinks, A and D in margarine, vitamins of the B group in breakfast cereals;
- **Preservatives:** including vinegar, salt or sugar (sucrose).

Salt

Salt, sodium chloride, is added either by the consumer and/or the food processor. It is considered to be a factor or suspected cause of heart disease. The recommended daily intake should be about 5 grams from all sources.

Sugar

Sugar (sucrose) is added to foods during processing and/or by the consumer. Sugar is known to cause:
- tooth decay
- obesity and overweight.

It is a suspected factor in heart disease. Sugar intake can be reduced and replaced by fruits, vegetables, bread or potatoes as a source of energy.

Summary

- Nutrients are needed for forming genetic material, growth, formation of body structures, metabolism and as energy sources. (▶ 83)
- The nutrient groups are carbohydrates, lipids (oils and fats), proteins, minerals and vitamins. Dietary fibre aids the movement of food along the gut. (▶ 84)
- The chemical reactions of cell processes take place in water. (▶ 85)
- Simple food tests can be carried out for starch, sugar, lipids and protein. (▶ 86)
- A balanced diet contains all the nutrients in their correct amounts. Malnutrition exists in two major forms: undernutrition and overnutrition. (▶ 87)
- Protoctists and animals process foods in five stages: ingestion, digestion, absorption, assimilation and egestion. (▶ 91)
- Some animals obtain food by filter-feeding. In this process water is filtered and small food particles left behind are eaten. (▶ 93)
- Gums and teeth can be destroyed by gingivitis and dental caries. Teeth can be preserved by brushing, attention to diet and the use of fluoride. (▶ 96)
- Food is digested in the human alimentary canal, principally in the mouth, stomach and small intestine. Food is absorbed mainly in the small intestine. (▶ 97)
- Sheep and rabbits are herbivores; dogs and cats are carnivores. Their teeth and digestive systems are adapted to the food they eat. (▶ 104)

- Decomposers break down the dead bodies of plants and animals. (▶ 106)
- Methods of food preservation either deprive microorganisms of the conditions they need to breed or kill them. (▶ 106)
- Processed foods contain food additives to improve their appearance, taste and nutritive value. (▶ 108)

5 How living things transport materials

Objectives

When you have completed this chapter you should be able to:
- understand the processes of diffusion, osmosis and active transport
- describe the transport of water and minerals through flowering plants.
- understand the process of transpiration in flowering plants
- describe the process of translocation in flowering plants
- describe the components of mammalian blood
- describe the functions of blood
- describe the structure of the circulatory system
- explain the action of the heart
- understand the differences between blood plasma, tissue fluid and lymph

5.1 Cell transport

All living things need to move or transport materials *within* their cells or *throughout* their bodies. Cell fluids consist of water, mineral ions, glucose, fatty acids, glycerol, amino acids and dissolved gases, oxygen and carbon dioxide. The composition of the cell fluid is kept constant within certain limits by the following regulatory processes which allow substances to move in and out of the cell. These are **homeostatic** processes (see Chapter 7).

Biological membranes are either:
- cell surface membranes surrounding a living cell or
- organelle membranes, e.g. surrounding chloroplasts or mitochondria.

They are **selectively permeable**, **living** membranes.

Examples of **physical** membranes are those composed of cellulose, cellophane or visking tubing. They are **non-living** membranes which are **selectively permeable**. Cell walls are porous, permeable structures.

Substances pass through biological membranes by two kinds of transport:
- **Passive transport**, which does not use energy from the cell. Examples of passive transport mechanisms are absorption, diffusion and osmosis.
- **Active transport**, which uses energy from the cell.

Passive transport

Absorption

This process is also called **imbibition**. Many dry or dehydrated substances can attract water. For example when dried seeds or dried fruits are soaked in water they take some of it in and swell up. It is an important process in the germination of seeds (see page 283).

The substances which can absorb water in this way are components of cells and include sugars, starches, cellulose and similar cellulose compounds called pectins and hemicellulose that are found in plant cell walls. Other cell substances which absorb water are proteins and mucilages.

Dried fruits such as raisins and prunes, and dried gelatine can absorb water from solutions of salt or sugar or from water without solutes. Dietary fibre consists mainly of cellulose and gums. It absorbs water in the gut and swells up.

Diffusion

Diffusion is the passive movement of substances from a region of *high* concentration to a region of *low* concentration. The difference in concentration between the two regions forms a **concentration gradient**. When the difference in concentration between the two regions is great, the concentration gradient is steep and the rate of diffusion is fast. As the difference in concentration between the two regions becomes less, the concentration gradient becomes flatter and the rate of diffusion slows down. Diffusion stops when the concentration of particles in the two regions is the same and the concentration gradient no longer exists.

Diffusion occurs in both liquids and gases and can take place over long distances. The smell of cooking, for example, can diffuse out of the kitchen to other parts of the home and even outside. Diffusion in gases is much faster than in liquids. The rate of diffusion of oxygen in air is 100 000 times the diffusion rate of oxygen in water. Diffusion can also take place through a permeable membrane. This is a membrane which lets all diffusing substances pass through it.

Diffusion of solutes through cell surface membranes

Dissolved sugar and ions of sodium and potassium can pass through cell surface membranes by diffusion. The cell can take in dissolved foods such as sugar molecules from its surrounding fluids when its stock of sugar is low. Waste products such as urea also pass out of the cell by diffusion.

Osmosis

Osmosis is the passive diffusion of water through a selectively or partially permeable membrane. The water moves from a region of a *high* water concentration (a dilute solution) to a region of a *lower* water concentration (a concentrated solution). The partially or selectively permeable membrane, which can be living or non-living, allows water molecules to pass through it but does not allow any dissolved substances to pass through.

Osmosis is a slow process in which only water is transported. It only takes place over a short distance but the cell does not have to use up any energy for it to take place.

The demonstration of osmosis illustrated in Figure 5.1 uses an artificial, non-living membrane called Visking or cellophane.

Osmosis in plant and animal cells

Figure 5.2 shows how a plant cell and a red blood cell change when they are put in a solution which is (a) more dilute and (b) more concentrated than their cytoplasm.

Figure 5.1 Demonstration of osmosis

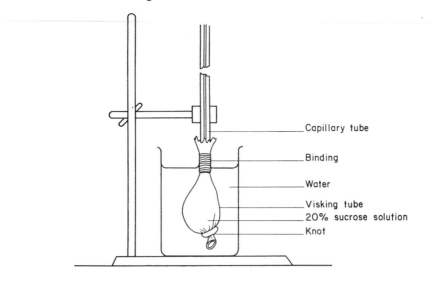

Capillary tube

Binding

Water

Visking tube
20% sucrose solution

Knot

Figure 5.2 Osmosis in plant and annual cells

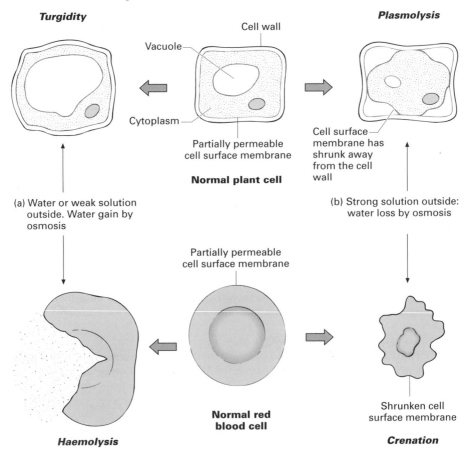

Turgidity

Plasmolysis

Cell wall

Vacuole

Cytoplasm

Partially permeable
cell surface membrane

Normal plant cell

Cell surface
membrane has
shrunk away
from the cell
wall

(a) Water or weak solution
outside. Water gain by
osmosis

(b) Strong solution outside:
water loss by osmosis

Partially permeable
cell surface membrane

**Normal red
blood cell**

Shrunken cell
surface membrane

Haemolysis

Crenation

In the plant cell, there are two cell barriers:
- wholly permeable cell wall and
- partially permeable cell surface membrane.

In (a), the entry of water into the vacuole by osmosis causes an *increase* in cell volume and **turgor pressure** which is opposed by the **cell wall** pressure. In (b), the loss of water from the vacuole by osmosis causes a *decrease* in cell volume and withdrawal of the cell surface membrane from the cell wall. This is referred to as **plasmolysis**.

In the animal cell there is only one cell barrier – the cell surface membrane. In (a), the entry of water into the red blood cell by osmosis causes an *increase* in cell volume resulting in bursting of the cell membrane (**haemolysis**) with the escape of haemoglobin. In (b), the loss of water from the red cell by osmosis causes a *decrease* in cell volume and shrinkage of the cell surface (**crenation**).

In plant cells, the wholly permeable cell wall acts as a resistance against the partially permeable cell surface membrane. Animal cells do not have a cell wall and therefore their cell surface membranes are not restricted.

Active transport

This method of transport uses **energy** from the cell. Certain selected solute ions, glucose, sucrose and amino acids, are transported rapidly through a short distance from a region of *low* concentration to a region of *high* concentration. This movement of substances takes place through a living, partially permeable membrane – the cell surface membrane. About a third of all the energy released in cell respiration is used to draw vital substances into the cell *against* their concentration gradients.

Questions

1 What is the difference between passive and active transport systems?
2 Why does dietary fibre swell up in the gut?
3 Which way do particles move when they diffuse?
4 What is a permeable membrane?
5 What substances pass through cell surface membranes by diffusion?
6 How is osmosis different from diffusion?
7 How is osmosis similar to diffusion?
8 What happens to plant and animals cells when they are put in
   (a) a more dilute solution than their cell sap?
   (b) a more concentrated solution than their cell sap?
9 How is active transport different from osmosis?

5.2 Transport of materials through flowering land plants

Vascular tissues

Green flowering plants have tissues which transport water and dissolved substances through their bodies. The tissues are called **vascular tissues**. There are two kinds of vascular tissue – **xylem** and **phloem**. The arrangement of these tissues is shown in cross-sections of a root (Figure 5.3) and shoot (Figure 5.4) of the buttercup.

The tissues are made of tubes. The tubes in the xylem tissue are called vessels. Figure 5.5 shows a xylem vessel. Figure 5.6 shows a sieve tube in phloem tissue.

Figure 5.3   Root of buttercup, *Ranunculus* sp., in transverse section, showing arrangement of tissues

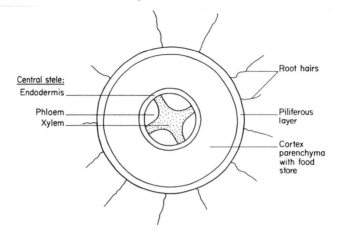

Central stele:
Endodermis
Phloem
Xylem

Root hairs

Piliferous layer

Cortex parenchyma with food store

Figure 5.4   Stem of buttercup, *Ranunculus* sp., in transverse section, showing arrangement of tissues

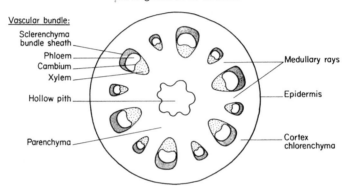

Vascular bundle:
Sclerenchyma bundle sheath
Phloem
Cambium
Xylem

Hollow pith

Parenchyma

Medullary rays

Epidermis

Cortex chlorenchyma

Figure 5.5   Xylem vessel

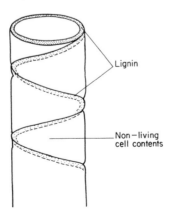

Spiral thickening

Lignin

Non–living cell contents

Figure 5.6 Phloem sieve tube

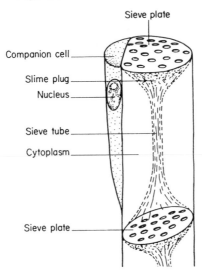

The tissues perform different tasks. Xylem tissue transports water and dissolved mineral ions from the root through the shoot, while phloem tissue transports food substances, mainly sucrose manufactured in the leaves, to other parts of the plant.

Questions

10 How are xylem and phloem tissues different in their structure?
11 How are xylem and phloem tissues different in their functions?

Uptake of water and mineral ions

The roots take up water and mineral ions from the soil and conduct them to the stem. The root hair cells through which the water and mineral ions pass have a *large* surface area which is almost twelve times the root surface area.

Question

12 If a plant lost a large number of its root hairs what would happen to
   (a) its absorbing surface,
   (b) the amount of water and mineral ions absorbed?

Mechanism of water absorption

Soil water is a solution of low concentration, whilst the sap in the vacuole of the root hair is of a higher concentration. Water moves by osmosis (see page 111) from the soil across the root hair cell wall and cell surface membrane, where it dilutes the root hair cell sap. It is moved on through the plant by the transpiration stream (see page 119).

It is believed that water is also able to move through the cell wall by diffusion through the tiny spaces *between* the cellulose fibres and also through the intercellular air

Figure 5.7 The pathway of water from the soil into the xylem

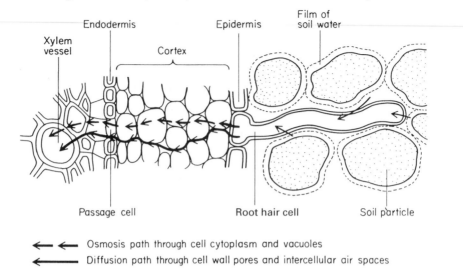

Osmosis path through cell cytoplasm and vacuoles

Diffusion path through cell wall pores and intercellular air spaces

spaces. Water in the root hair then, is transported across the root cortex by three routes. It finally passes from the cortex cells into the xylem vessels. If the root is in well-watered soil, it takes in the water and pushes it along the xylem vessels up into the stem. This pushing force is called **root pressure** and can be demonstrated by the experimental set-up shown in Figure 5.8. If a branch is cut off a grape vine or a tree in the springtime, the root pressure can be strong enough to force water out of the wound on the branch.

Figure 5.8 Demonstration of root pressure

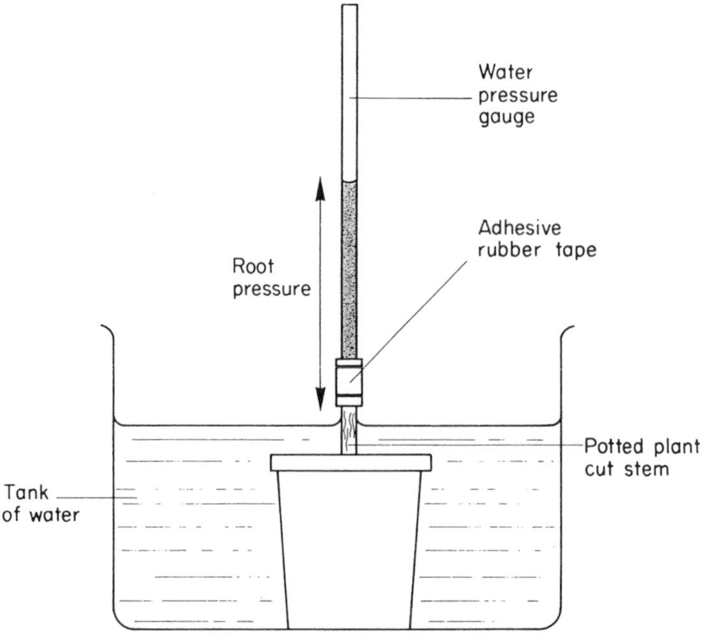

Mechanism of ion absorption

The mineral ions in the soil water are at a lower concentration than the mineral ions inside the root hair cells. For the plant to thrive, the ions must be taken in *against* the concentration gradient. They cannot passively diffuse into the cell but are brought in by **active transport**. This is a process which uses energy. The energy is released in aerobic respiration (see page 143) using oxygen from the air spaces in the soil and food made by the plant.

___ Questions _____

13 How is the uptake of water and mineral ions different?
14 Which one uses the most energy?

5.3   Transpiration in flowering plants

Transpiration is the loss of water vapour from the surface of land plants. It occurs mainly through the open stomata (Figure 3.9) of leaves by **evaporation**, which is a passive, physical process. A smaller loss of water occurs through the stem surface.

___ Measuring the rate of water loss using a potometer _____

Figure 5.9   A weight or mass potometer

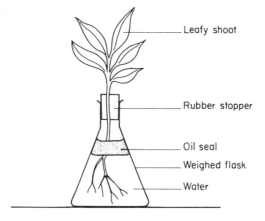

- Leafy shoot
- Rubber stopper
- Oil seal
- Weighed flask
- Water

When the potometer is set up, it is weighed before being exposed to the environmental conditions to be tested. At the end of the experiment it is weighed again to find the mass of water lost.

Figure 5.10 A water potometer for measuring rate of water uptake by a cut shoot

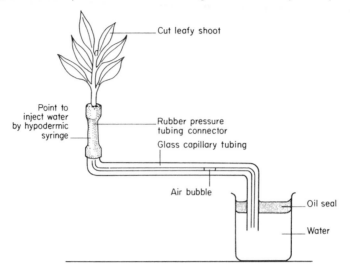

The rate of water uptake is normally closely related to the rate of water lost during transpiration but some water may be used up in photosynthesis (see page 69). The movement of the air bubble along the capillary tube is measured over a certain amount of time to find the rate at which the shoot takes up water. The bubble can be repositioned by using a hypodermic syringe to inject water through the rubber tubing.

Mechanism of transpiration

Water evaporates from the surface of the spongy mesophyll cells of the leaf into the leaf air spaces (see Figure 5.12). The cell vacuole sap becomes more concentrated. Water vapour exits the leaf by diffusion through the open stomata. The evaporation of water uses heat energy provided mainly by the external environment.

Demonstrating water loss from a leaf

Figure 5.11 To show water loss from a leaf

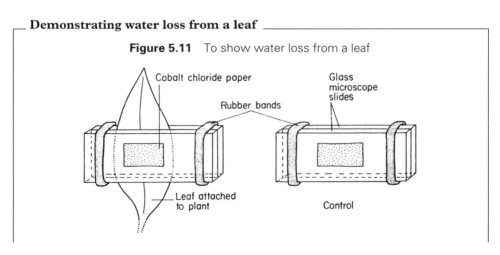

The water loss from the leaf of a dicotyledon plant (see page 54) can be investigated by attaching anhydrous cobalt chloride paper to each surface as shown in Figure 5.11. The paper indicates the presence of water by turning from *blue* to *pink*.

Question

15 A dicotyledon plant leaf has more stomata on one side than the other. How could you find which side has more stomata?

Factors affecting the rate of transpiration

Factors which affect evaporation also affect transpiration, but transpiration is in turn controlled by stomata which are able to open and close.
- **Light intensity**: In daylight, the stomata of most plants are open. They close in darkness. The light intensity consequently affects the transpiration rate.
- **Wind**: Moving air removes water vapour from the leaf surface, allowing more air to diffuse out of the leaf. Still air allows the water vapour concentration around the leaf to build up.
- **Humidity**: The concentration of water vapour in the air may vary from low concentration in dry air to a higher concentration in humid air. Water vapour will only diffuse along the concentration gradient from a high concentration in the leaf to a lower concentration in the surrounding air.
- **Temperature**: The concentration of water in air is related to air temperature. An increase in air temperature allows more water to evaporate; a decrease in air temperature slows down the evaporation rate. Warm air holds more water than cold air.
- **The number of stomata**: Leaves with a large number of stomata have a higher transpiration rate than leaves with a small number of stomata.

Questions

16 What would be the environmental conditions which would produce a very high transpiration rate?
17 What would be the environmental conditions that would produce a very low transpiration rate?

The transpiration stream

The transpiration stream is the **upward flow** of water and mineral ions from the roots to the leaves by the way of the xylem tissue. This flow is the result of:
- water loss by evaporation from the leaves;
- pull on the water in the xylem vessels in the leaf vein and stem, due to the withdrawal of water into the mesophyll cells.
- root pressure (see page 116).

Transpiration pull can be demonstrated by the experimental set-up shown in Figure 5.13 on page 121. **NB:** Mercury vapour is poisonous so care should be taken when setting up this apparatus.

Figure 5.12 The transpiration stream

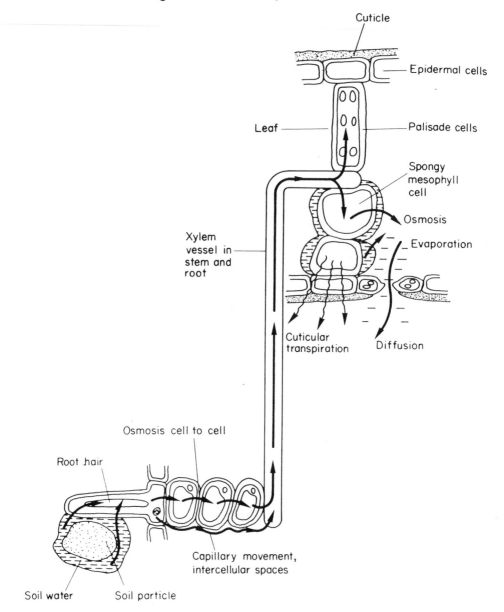

Wilting and the transpiration stream

When a plant has drooping leaves and a floppy stem it is said to be **wilting**. This is caused when the rate of transpiration from the leaves *exceeds* the rate of water absorption by the roots. It can also be expressed by saying that the water output is greater than the water input. Plants can reduce the risk of wilting by closing their stomata when the water input is low.

The size of a stoma (pore) is controlled by the guard cells that surround it. When they take in water from surrounding cells they swell up a little but because they have a

Figure 5.13 Apparatus to demonstrate transpiration pull

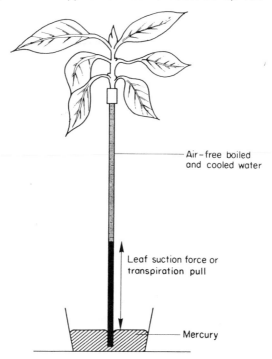

Air-free boiled and cooled water

Leaf suction force or transpiration pull

Mercury

thicker wall on the stoma side this makes them bend, causing the stoma to open. When they lose water to surrounding cells there is less pressure on their cell walls and they straighten up, causing the stoma to close (see Figure 3.9).

Plants wilt when they are short of water because water supports the plant cells. When there is plenty of water, it is drawn into the cell sap by osmosis. The water inside the cell then pushes outwards on the cellulose cell wall. It exerts a pressure on the walls called **turgor pressure** which gives strength to the cell walls. Turgid cells make a strong tissue which can support the body of the plant. If the plant is short of water the plant cells lose water to try and maintain the transpiration stream. The pressure of the water inside the cells falls and the cells become weak, making the plant wilt.

The plant tries to maintain the transpiration stream because it provides water for photosynthesis, mineral ions for food production and the evaporation of water from the leaf helps to prevent the leaf cells overheating and dying in hot weather.

--- **Questions** ---

18 Which plant wilts: plant A where transpiration rate is higher then the rate of water absorption, or plant B where transpiration rate is less than the rate of water absorption?
19 How can stomata reduce the rate of transpiration?
20 What makes cells turgid?
21 The stem of a seedling is supported by cells full of water. What happens to the stem if the seedlings are not watered on a hot sunny day?
22 Why is it important for a plant to maintain the transpiration stream for as long as possible?

Preventing water loss

Flowering plants which live in soil where there is an adequate water supply are called **mesophytes**. They have a thin layer of wax on their leaf surfaces. It is called the **cuticle** and prevents water escaping from the epidermis into the air. It ensures that water loss only occurs through the stomata.

Flowering plants which grow in dry conditions are called **xerophytes**. The dry conditions which result in a lack of soil water may be due to:
- **heat** in deserts,
- **wind** on mountains and moorlands.

Xerophytes have adapted in various ways to surviving in dry conditions:
- some have small or needle-like leaves to reduce their surface area, e.g. heather conifers, cacti.
- some have a thick cuticle, e.g. conifers, succulents
- some have stomata sunken below the leaf surface, e.g. conifers, marram grass.
- some have a store of water, e.g. succulent plants such as cacti.

Questions

23 Why does a needle-like leaf lose less water than a broad, flat leaf?
24 Still air around a stoma has a high concentration of water vapour.
   (a) How will this affect the rate of diffusion of water vapour from inside the leaf?
   (b) What happens to the rate of diffusion of water vapour if the wind blows across the leaf surface?
   (c) If the stoma is sunk into a pit in the leaf surface, how will this affect the rate of diffusion of water vapour? Explain your answer.

5.4 Translocation

Translocation is the process of movement of **soluble** materials, mainly sugars and amino acids, from one part of the plant to another. It occurs in phloem tubes which take food, mainly sucrose, from the leaves down to the growing points of the roots or into storage organs. The phloem tubes also take food upwards to the growing points of the shoot and to developing flowers, fruits and seeds. The materials are moved by active transport using cell energy.

Some translocation also occurs in xylem vessels from roots upwards to the leaves, flowers and fruits. No cell energy is used in this passive transport process.

Investigating translocation routes

Ringing

This is done by removing the bark and the underlying phloem tissues from two stems in a complete ring. In stem A, the xylem is blocked with wax. In stem B, the phloem is blocked.

The results show:

Figure 5.14 Stem-ringing experiment

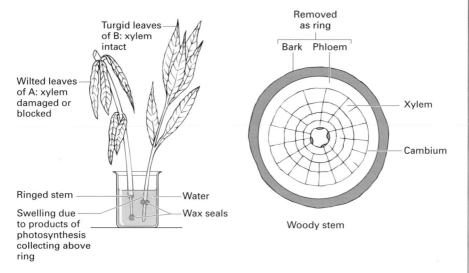

A Transpiration has ceased because the xylem is blocked, causing the leaves to wilt.
B Transpiration is normal from the leaves because water is reaching them through unaffected xylem. The products of photosynthesis, e.g. sugars, collect above the ring and are not translocated below the ring, because the phloem has been disturbed.

Radioactive tracers in translocation routes

Sucrose (cane sugar) is the carbohydrate which is normally translocated in flowering plants. Sucrose containing radioactive ^{14}C (carbon) can be used as a tracer to follow the path of sugar translocation in tomato plants. Radioactive ^{14}C sucrose is applied to the tomato plant leaves after removing some of the cuticle.

Autoradiographs (made by exposing photographic film to the plant parts) are taken at different times and they trace the path of sugar out of the leaf, up and down the stem to collect finally in the tomato fruits. Autoradiographs of transverse sections of the stem show ^{14}C sucrose in the phloem tissues which demonstrate they are the regions of translocation of the sugar.

Systemic pesticides are chemicals which are taken up by the plant and transported through its body. They poison any insects which feed on the plant. **Systemic herbicides** are absorbed through the plant surface and translocated by the phloem tissues, eventually killing the plant.

Questions

25 How are the contents of xylem vessels different from the contents of phloem sieve tubes?
26 Which substance is found in both xylem vessels and phloem sieve tubes?
27 In what way is a systemic pesticide less harmful to the environment than one used as a spray?

5.5 Transport in mammals

The mammalian circulatory system

The mammalian circulatory system is composed of the **heart** (**organ**) and **blood vessels**. **Blood** is the **tissue** which circulates through the system. This system uses a huge amount of cell energy compared to the small amount of cell energy used in flowering plant transport methods. This cell energy is used for **active transport** in the contraction of the heart muscle to pump the blood around the system. **Passive transport,** by means of osmosis, which does not use cell energy, returns fluid into the blood and is used to a far smaller extent than in plants.

Compared to plants, the mammalian circulatory system is almost unaffected by the external environment, apart from changes in air pressure with altitude.

5.6 Human blood

The amount of blood in a healthy human adult is about 5 litres, forming about 8% of the body weight. The blood is composed of 45% solid components, the **blood cells** and 55% liquid component, the **blood plasma**.

Blood plasma

This is 92% water and contains the following soluble components:
- **proteins**: 7% hormones, albumins, globulin antibodies, and 0.3% fibrinogen. **Serum** is blood plasma without fibrinogen and blood clotting substances. It is a yellow, watery fluid.
- **food nutrients**: glucose, amino acids, fatty acids and glycerol (propanetriol).
- **metabolic materials**: enzymes, vitamins, hormones, lactic acid, carbamide (urea) and uric acid.
- **inorganic ions** of sodium, potassium, calcium, iron, chloride, sulphate and phosphate; hydrogen carbonate radicals.

About 3 litres of plasma are present in adult human blood.

Figure 5.15   Mammalian blood cells

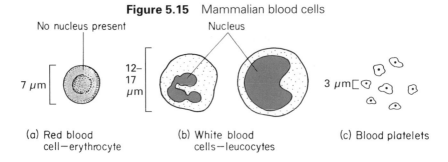

No nucleus present   Nucleus

7 µm

12–17 µm

3 µm

(a) Red blood cell—erythrocyte

(b) White blood cells—leucocytes

(c) Blood platelets

Red blood cells

Red blood cells, or **erythrocytes**, are formed in the red bone marrow of the ribs, sternum and vertebrae. 1.2 million cells are formed every second. Half a tonne of cells is produced in a human life time. There are 5 million in 1 mm^3 of blood. Each cell is a biconcave disc which presents a large surface area for taking in and releasing oxygen. The erythrocytes in the blood present a total surface area of 4500 m^2 through which oxygen can pass. There is no nucleus in a red blood cell. The cytoplasm is filled with the red blood pigment **haemoglobin,** which combines with oxygen. This pigment contains iron which is essential for carrying oxygen.

Haemoglobin is a functional protein and combines with oxygen to form **oxyhaemoglobin**. This occurs in regions of high oxygen concentration in the lungs. Oxyhaemoglobin is unstable. In regions of low oxygen concentration around the body cells it decomposes to oxygen and haemoglobin again. The oxygen then enters the cells (see Chapter 6).

Red cells last for about 120 days in the bloodstream. After this time they are destroyed in the liver and the spleen. Part of the iron and protein material from the cells is stored in the liver for further use and part is excreted in bile juice as the pigment bilirubin.

White blood cells

There are two kinds of white blood cell: **phagocytes** and **lymphocytes**.

Three kinds of phagocyte can be found; all are made in the **red bone marrow**. There are 10 000 in 1 mm^3 of blood. They have a lobed nucleus and granular cytoplasm. They protect the body from bacteria by **phagocytosis** (see page 126). Phagocyctes last in the blood for about nine days before they are replaced.

Lymphocytes are formed in the **lymphatic system** (see page 135). There are 2000 in a mm^3 of blood. They have a bean-shaped nucleus and protect the body by forming **antibodies** (see page 126). They last in the body for 100–200 days before they are replaced.

Platelets

Platelets are fragments of cells formed in the red bone marrow. There are 250 000 platelets in 1 mm^3 of blood. They help to clot the blood (see page 126) and last for 8–14 days.

Questions

28 Why does the circulatory system use more cell energy than that used in the plant transport system?

29 How is plasma different from serum?

30 How is a red blood cell adapted for carrying oxygen?

31 Describe how haemoglobin transports oxygen from the lung to the cells.

32 Which cell lasts for (a) the longest time, (b) the shortest time in the blood?

33 Which cell is the (a) most numerous (b) least numerous?

34 What are the functions of the solid components of blood?

The functions of the blood

The functions of the blood can be divided into two categories: transportation and defence.

Transportation

Cells receive oxygen, water, cell nutrients, heat and chemical energy from the blood. The oxygen is carried as oxyhaemoglobin in the red blood cells. The heat is brought from the liver cells and exercising muscles. The chemical energy is in the form of blood sugar – glucose.

The blood takes away carbon dioxide, urea (carbamide), heat and cell secretions such as hormones and enzymes. Carbon dioxide is removed from cells mainly (80%) in the form of **hydrogen carbonate ions** which are soluble in blood plasma. Small amounts are dissolved as **gas** in plasma or **combined** with the red blood cell.

Body defence

Bacteria and foreign bodies can cause disease. The body defends itself by means of white blood cells. Phagocytes ingest bacteria and antibodies or defensive proteins are formed by certain lymphocytes. These are able to destroy bacteria (see Section 5.10).

Blood clotting is a process to prevent blood loss and bacterial entry when a blood vessel is cut. Calcium ions, the plasma protein fibrinogen, platelets and the presence of air or oxygen are needed. The platelets form an enzyme which changes the soluble **fibrinogen** into microscopic threads of insoluble **fibrin**. These spread across the wound and stop the blood escaping from the body and bacteria from entering it.

Human blood groups

The four main human blood groups are called **A**, **B**, **AB** and **O**. They differ because their red blood cells have different proteins called **antigens** on their cell surface membranes and different **antibodies** in the plasma. (See also Section 12.8 on page 316.)

A blood **transfusion** takes place between a **donor** or giver of blood and a **recipient** or receiver of blood. A successful blood transfusion can only take place when the donor blood is of the same type or **compatible** with the recipient's blood.

Incompatible blood will cause:
• red blood cell breakdown or haemolysis (see Figure 5.2)
• red blood cell clumping or **agglutination** which blocks the circulatory system and can cause death.

Table 5.1   Blood group frequency in the UK

Group	Frequency/%
O	47
A	42
B	9
AB	2

Blood compatibility in blood transfusions

Table 5.2 summarises blood group donors and acceptors.

Table 5.2   Compatibility of blood groups

Blood group	DONORS Can act as donors to	RECIPIENTS Can accept blood from
AB	AB	all groups UNIVERSAL RECIPIENTS
A	A and AB	A and O only
B	B and AB	B and O only
O	all groups UNIVERSAL DONORS	O only

Questions

35 How is the transportation of oxygen in the blood different from the transportation of carbon dioxide?
36 How does the blood defend the body against bacteria?
37 How are threads made across a wound and what do the threads do?
38 Which of the following will lead to a successful blood transfusion?

Donor	Recipient
AB	A
A	AB
B	B
O	A
AB	O

5.7 Blood vessels

There are three kinds of blood vessel in the circulatory system: **arteries, veins** and **capillaries**. The blood vessels and the heart are lined with a single layer of cells called the **endothelium**.

Figure 5.16 Internal structure of blood vessels

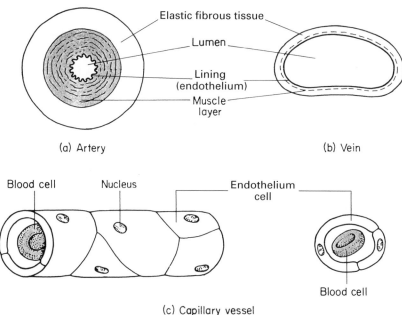

(a) Artery   (b) Vein

(c) Capillary vessel

Arteries

Arteries carry **oxygenated blood** under a **high** pressure *from* the heart. An exception is the pulmonary artery which carries deoxygenated blood to the lungs. Artery walls are thick and muscular to withstand the high blood pressure surges which are detectable as a **pulse**. The pulse is a pressure wave that corresponds to the heart beat (see page 130). The resting pulse in adults is between 60–80 beats per minute. Muscle fibres in the artery walls can contract and relax, causing either constriction or dilation of the vessel.

Veins

Veins carry **deoxygenated** blood *towards* the heart under much **lower** blood pressure than in arteries. An exception is the pulmonary vein which carries oxygenated blood from the lungs. The walls are thinner, have less muscle fibres and are more elastic than arteries. Small **pocket valves** are located at intervals inside the veins to prevent the backflow of blood.

Venous blood returns to the heart by means of the 'muscle pump'. This is due to the effect of the skeletal muscles contracting around the veins as occurs in exercise. The action of the muscles compresses the vein and drives the blood towards the heart.

Capillaries

These are small blood vessels with walls one cell in thickness; they receive blood from **arterioles** and deliver it to **venules**. Capillaries form extensive networks in organs such as the skin, lungs, liver, alimentary canal and muscles.

A comparison of the three kinds of blood vessel can be made using Table 5.3.

--- Questions ---

39 How is an artery different from a vein?
40 How is blood in the veins made to travel in only one direction?
41 Why does blood move more quickly through the veins in the legs when a person is walking than when the person is just standing?
42 How is a capillary different from arteries and veins?
43 How does the blood change as it moves from an artery and passes through a capillary?
44 How does the blood change as it moves from a capillary and passes into a vein?
45 Which vessels carry mainly oxygenated blood?

5.8 The adult human heart

The heart is located between the lungs in the chest or thoracic cavity. The adult human heart weighs about 300 g (male) and 250 g (female). The walls of the heart are made of **cardiac muscle**. It has a different structure to skeletal muscle and works continuously and automatically without experiencing fatigue.

The mammalian heart has *four* chambers: two atria and two ventricles. The left and right sides of the heart are separate and do not communicate.

The heart functions as two force pumps:
• the right atrium and ventricle pump deoxygenated blood to the lungs;
• the left atrium and ventricle pump oxygenated blood to the body tissues.
Blood passes *twice* through the heart, once through the right side and once through the left side, before being circulated to the body.

Table 5.3 Comparison of mammal blood vessels

	Arteries	Veins	Capillaries
Structure			
1	Artery walls have a *thick* muscle and elastic tissue layer	Vein walls have *thin* muscle and elastic tissue layer	Capillary walls are *one* cell thick. *No* muscle or elastic tissue. 'Pores' may be present
2	Valves are *not* present	Valves present to prevent backflow	No valves
3	Cross-section is circular	Cross-section is oval	Cross-section is circular
4	Fluid and white blood cells *cannot* pass through artery wall	Fluid and white blood cells *cannot* pass through vein wall	Fluid *without* proteins *can* pass through wall. White blood cells pass out *between* the cells
5	Muscle in walls can *relax* and artery dilates. Muscle *contraction* causes narrowing or *constriction*	Dilation and constriction are limited	Capillaries can dilate or contract by cell wall changing shape
Blood composition and flow			
1	Flow is *away* from heart	Flow is *towards* the heart	Flow is *from* artery *to* vein
2	Oxygenated blood (except in pulmonary artery)	Deoxygenated blood (except in pulmonary vein)	Mixed oxygenated and deoxygenated blood
3	Rapid flow	Slow flow	*Very* slow flow
4	High pressure, 11 to 16 kPa	Low pressure, 1 to 2 kPa	*Very* low pressure
5	Pulse strong	No pulse	No pulse

Figure 5.17 Human heart – external surface

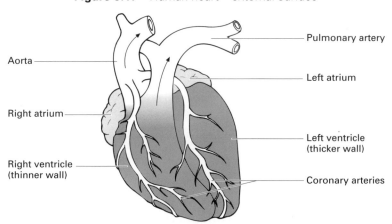

Pulmonary artery

Aorta

Left atrium

Right atrium

Left ventricle (thicker wall)

Right ventricle (thinner wall)

Coronary arteries

The internal structure of the heart

Figure 5.18 Human heart – internal structure

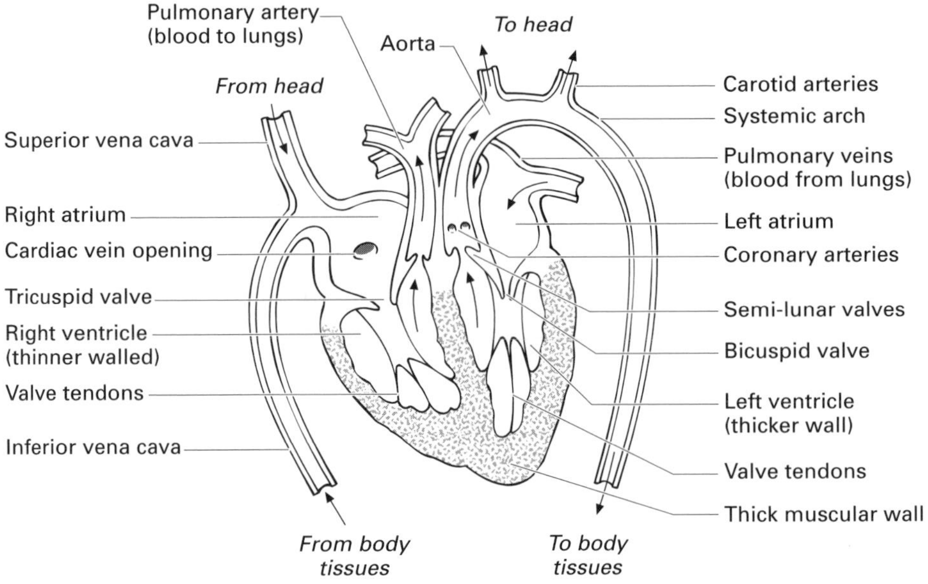

The heart valves

The heart valves are positioned at the outlets from each atrium and ventricle. The valve between the right atrium and the right ventricle is called the **tricuspid** valve. The valve between the left atrium and the left ventricle is called the **bicuspid** valve. On the outlets of the ventricles into the aorta and the pulmonary artery are the **semi-lunar valves**.

The heart valves prevent the backflow of blood when the atria and ventricles are contracting.

The heart chambers

The **right atrium** receives **deoxygenated** blood from the head and body. The **left atrium** receives **oxygenated** blood from the lungs. The **right ventricle** forces **deoxygenated** blood to the lungs. The **left ventricle** forces **oxygenated** blood to the head and body. The muscular wall of the left ventricle is *thicker* than that of the right ventricle because it needs to pump the blood for a greater distance.

The heart or cardiac cycle

The heart muscles in the atria and ventricles contract and relax simultaneously. Contraction of the heart muscle is called **systole** and relaxation of the heart muscle is called **diastole**. When the atria are contracting or are in systole, the ventricles are relaxing or are in diastole.

Atrial systole occurs when:
- atrial muscle *contracts*;
- main vein valves *close*;
- blood enters the ventricles past *open* tricuspid and bicuspid valves.

Ventricular systole occurs when:
- ventricle muscle *contracts*;
- tricuspid and bicuspid valves *close*;
- blood passes into the aorta and pulmonary vein;
- atria draw in blood through vein openings.

The heart beat is the 'lub dub' sound heard through a stethoscope applied to the chest. This is the sound of the tricuspid and bicuspid valves closing, followed by the semi-lunar valves closing. The heart beat of a resting adult is about 70 beats per minute.

The heart beat slows down during sleep, shock, fainting, or if the person takes digitalis type drugs. The heart beat speeds up during excitement, stress or fright, pain, vigorous exercise, tobacco smoking and if the person takes certain drugs. The secretion of the hormones thyroxine and adrenalin also speed up the heart beat.

Questions

46 How is the action of the heart muscle different from skeletal muscle?
47 (a) How is the blood in the right side of the heart different from the blood in the left side of the heart?
     (b) Why is there a difference in the blood in the two halves of the heart?
48 What is happening in the heart when the (a) lub heart sound is made and (b) the dub heart sound is made?

5.9 The mammalian circulatory system

The mammalian circulatory system is composed of a closed circuit – the blood is enclosed within the heart and blood vessels. The circuit is a **double circulation**; the **pulmonary** circulation to the **lungs** and the **systemic** circulation to the **body**.

Figure 5.19   Double circulation and the lymphatic system

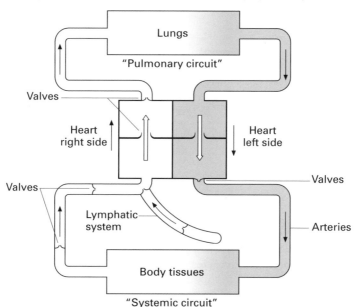

- The pulmonary circulation carries deoxygenated blood via the pulmonary artery to the lungs and returns oxygenated blood from the lungs to the heart via the pulmonary veins.
- The systemic circulation circulates 70% of the blood through parallel arranged vessels in organ system sub-circuits as shown in Figure 5.20.

Figure 5.20   The heart and circulatory system of a mammal

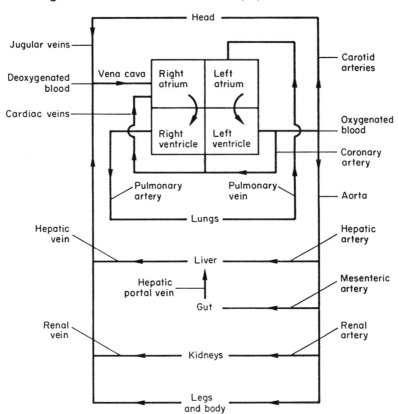

The **arterial system** of this circuit contains the aorta, one of the largest arteries in the body. This gives off other branch arteries to the arms, the head (the carotid arteries) and the chest. Other branches supply the liver (hepatic artery), intestine (mesenteric artery), the kidneys (renal artery) and finally the legs.

The **venous system** consists of the great veins or vena cavae, delivering deoxygenated blood to the heart from the kidney (renal vein), liver (hepatic vein) and legs. The head (jugular vein) and arms link with the vena cava into the right atrium.

Blood from the kidneys is minus **excretory wastes**, blood from the liver contains **nutrients**, whilst other venous blood is loaded with **carbon dioxide**. Between the gut and liver is the **hepatic portal vein** carrying venous blood from one capillary network to another. Portal blood vessels begin and end in capillaries.

The heart or **coronary circuit** is a sub-circuit of the systemic circulation which serves the heart. The **coronary arteries** supply the heart muscle with energy, oxygen and nutrients to power the heart pump. The **coronary veins** drain the heart and run parallel to the coronary arteries. The coronary vessels can become blocked during heart disease and can be replaced by means of **bypass** surgery.

49 Describe the path of blood from the vena cava to the aorta.
50 Where may the blood go from the aorta before it returns to the right atrium?
51 How does the blood change as it passes through the capillaries of (a) the small intestine, (b) the kidney, (c) a muscle?
52 How is a portal vein different from other veins?

Blood pressure

Blood pressure can be measured *directly* by inserting a connection into the artery or vein and recording pressures on a manometer gauge. Blood pressure is measured *indirectly* with a sphygmomanometer, an inflatable cuff connected to a mercury manometer gauge.
• Diastolic blood pressure is 1 kPa in veins and 11 kPa (80 mm of mercury) in arteries.
• Systolic blood pressure is 2 kPa in veins and 16 kPa (120 mm of mercury) in arteries.
The surrounding air has an atmospheric pressure of 101 kPa, acting all over the body surface. The heart must produce a blood pressure in excess of this in order to circulate the blood. The absolute blood pressure will be air pressure plus gauge pressure, or absolute systolic blood pressure will be 101 kPa plus 16 kPa, a total of 117 kPa.

Coronary heart disease

Large numbers of people in developed countries die from coronary heart disease every year. Studies have been made on people who suffer from the disease and on people who do not suffer from the disease. From the results of this work, risk factors which can lead to an increased chance of suffering from heart disease have been identified. The risk factors are divided into two groups – unavoidable risk factors and avoidable risk factors.

Unavoidable risk factors

The studies have shown that:
• more men die from coronary heart disease than women. Males are therefore at greater risk than females;
• some families have more family members dying of coronary heart disease than other families. The tendency to develop the disease is inherited and passed on by certain genes;
• older people develop the disease more often than younger people. The disease is associated with the ageing processes of the body.

Avoidable risk factors

The studies have shown that:
• people with higher cholesterol levels in their blood may have a greater chance of developing the disease. Cholesterol forms a coating inside the blood vessels which narrows the passage for blood movement. The heart must pump harder to maintain the blood supply and the blood pressure rises;
• excess salt raises the blood pressure;
• excess strain on the heart occurs in people who are overweight or obese;
• smoking causes a narrowing of the blood vessels (vasoconstriction) and an increase in blood pressure;
• stressful life style increases blood pressure by vasoconstriction;
• a lack of exercise weakens the heart muscle.

Prevention of coronary heart disease

While nothing can be done about the unavoidable risks, the chances of anyone developing the disease can be reduced by adopting a life style which reduces the effect of the avoidable risk factors.

Cholesterol is a component of fatty food from animals, e.g. hard fats (butter and lard), cheese and fatty meat, so the amounts of high cholesterol food in the diet should be reduced. Less salt should be put on food. A balanced diet of soft and hard lipids (see page 84) should be eaten.

Soft lipids are vegetable oils. They are unsaturated and cholestrol is low or absent. Hard lipids are animal fats. They are saturated and they have a high cholesterol content.

There must be no smoking. A lifestyle must be adopted which aims to reduce stress and provides time for recreation. Frequent physical exercise must be taken as this strengthens the heart muscle and helps to reduce stress.

Questions

53 How is a smoker who eats plenty of fatty foods of animal origin, takes little exercise and leads a stressful life, at risk of developing heart disease?

54 A young female and an older male both have lifestyles described in question 53. Which one is at greater risk and why?

The exchange of substances at the cells

Capillary networks are found in the lungs, skin, gut, liver and placenta. The networks have an enormous surface area which is essential for the rapid exchange of materials such as nutrients, water, gases, wastes and the exchange of heat. The blood capillary vessels are in close contact with the cells. The blood pressure forces fluid through the pores and spaces between the capillary endothelium cells. This filtered liquid forms **tissue fluid** around the cells. White blood cells are able to squeeze through the spaces between adjacent capillary cell walls and enter the tissue fluid.

Figure 5.21 The formation of tissue fluid

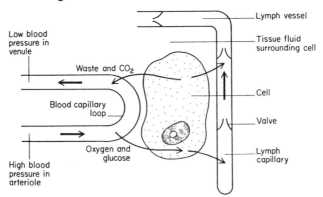

After bathing the cells and supplying and removing various substances essential to cell homeostasis (see Chapter 7), the fluid returns to the capillary vessels at the end close to the venules. The difference in concentration of the plasma and the tissue fluid due to the presence of plasma proteins and the absence of proteins in tissue fluid, regulates the return of fluid back through the capillary wall by osmosis.

Lymph and the lymphatic system

A small amount of tissue fluid (10%) does not return to the blood capillary system but is drained into **lymph capillaries**. These are blind-ended, fluid-filled tubes between the cells. These lymph capillaries empty into large **lymph vessels** and then into main **lymph ducts** which deliver the lymph fluid into the innominate vein and superior vena cava near the heart.

Table 5.4   Comparison of blood plasma, tissue fluid and lymph

	Blood plasma	Tissue fluid	Lymph
Location	Inside the blood vessels	Bathing living cells as *interstitial* or *intercellular* fluid	Inside the lymphatic vessels
Composition	Water, protein (6–8%) glucose, amino acids, salts, enzymes and hormones. Oxygen is present	Little protein (1–2%) otherwise similar. Oxygen is present	More protein than tissue fluid, but less than blood plasma (4–5%). More lipids, otherwise similar. No oxygen
Cells	Red, white and platelets	White (emerge between capillary cell walls)	Lymphocytes. No platelets
Transport	Blood pressure forces fluid through capillary at artery end. Osmosis returns fluid at vein end of capillary	From capillary under pressure and return by osmosis to capillary. 90% returns to capillary, 10% drains to lymph	From tissue fluid by drainage under pressure and diffusion. Lymph vessels empty into veins

Lymph capillaries are highly **permeable**, composed of thin-walled tubes of endothelium. They allow large molecules and bacteria to pass through. These are unable to pass into blood capillary vessels as they are composed of thicker walled tubes of endothelium.

Lymph **nodes** or **glands** are swellings found at certain points in the lymphatic system such as in the neck, armpits, groin and intestine wall. The nodes produce **lymphocytes**, which are white blood cells and can engulf bacteria. The tonsils and adenoids are lymphoid tissues with similar functions to the lymph nodes.

The movement of lymph in the lymphatic system is by means of the muscle pump mechanism similar to that operating on veins (see page 128). The lymphatic system defends the body by producing lymphocytes. It also absorbs lipids through the intestinal villi and returns proteins to the blood circulation.

Questions

55 Why do capillaries form large networks?
56 (a) *How* and (b) *why* is tissue fluid different from plasma?
57 How do tonsils and adenoids help to fight disease?

5.10 Immunity

Immunity is a means of defending the individual and the community. It is a natural resistance to infection, due to the presence of specific antidotes called **antibodies** which react with **antigens** made by pathogenic microbes.

Naturally acquired immunity

White blood cells produce the antibodies which destroy or neutralise the antigens of a pathogen. The immunity is acquired after an attack of most diseases of childhood, namely measles, German measles, mumps and chickenpox. **Active** immunity normally gives immunity for life – most epidemic diseases give active immunity.

A **fetus** acquires short-term, **passive** immunity from antibodies passed to it from the mother across the placenta or in the mother's milk if breast-fed after birth.

Artificially acquired immunity

Vaccination or **immunisation** is the production of immunity by artifical means other than by an attack of the disease itself. Edward Jenner (1749–1823), a doctor, was the first to discover the vaccination process.

Vaccination can be done by injection of **weakened** live microbes, or **dead** microbes. These cause the white blood cells to produce antibodies, giving **active** immunity. **Passive** immunity is obtained by injecting ready-made vaccines of antibodies made in another animal, e.g. from the blood serum of a horse, as for tetanus and diphtheria vaccination.

- **Weakened** living microbes or **attenuated** vaccines are used against smallpox, tuberculosis, rabies and poliomyelitis.
- **Dead** microbes are used in vaccines against cholera, influenza, typhoid and whooping cough.

Summary

- Passive transport (diffusion and osmosis) and active transport are used to transport materials in living things. (▶ 110 and 111)
- Flowering land plants take up water by osmosis and mineral ions by active transport. (▶ 115)
- The transpiration stream is the upward transport of water through a plant. (▶ 119)
- Water, mineral ions and manufactured food substances are transported within the plant via conducting xylem and phloem tissues by translocation. (▶ 122)
- Blood contains proteins, food nutrients, metabolic materials, inorganic ions, red cells, white cells and platelets. (▶ 124)
- The blood provides a transport medium and a body defence mechanism. (▶ 125)
- The circulatory system is composed of the heart, arteries, veins and capillaries. (▶ 129)
- The beating of the heart occurs in a series of stages which ensure that the blood moves through the circulatory system under pressure. (▶ 130)
- Plasma, tissue fluid and lymph have different compositions and functions. (▶ 135)

Respiration

Objectives

When you have finished this chapter you should be able to:
- distinguish between internal and external respiration
- understand the process of anaerobic respiration
- explain some applications of anaerobic respiration
- understand aerobic respiration
- describe metabolism and metabolic rate
- explain external respiration in humans
- explain external respiration in bony fish and insects
- understand external respiration in flowering plants

6.1 Introduction

Respiration is a characteristic process of living organisms (see Introduction, page 2). It involves the release of **energy** by breaking down high-energy organic chemical compounds. The main energy-providing chemical compounds are the organic chemical nutrients, carbohydrates, lipids and proteins (see Section 4.1).

All living organisms need energy in order to make genetic material and carry out metabolic processes. There are two main kinds of respiration:
- **Internal respiration**: a biochemical process which takes place *inside* living cells in either the cytoplasm or the **mitochondria** (see Section 2.1). This process is also called **cell respiration**. It may or may not use oxygen.
- **External respiration**: a physical process. It is also called **ventilation** or **breathing** and is concerned with gas exchange in air for land organisms and in water for aquatic organisms.

(**NB:** Non-living materials, e.g. carbon, sugar and fuels, release energy by chemical action called **oxidation** or **combustion**.)

6.2 Energy release in living cells

Glucose is the main respiratory substrate which is broken down or respired in living organisms to yield energy. Energy can exist in many different forms but the main forms of energy in the respiring cell are **heat** and **chemical** energy.

Figure 6.1 Diagram of a mitochondrion, the main site of internal respiration inside the cell

In the living cell a series of enzyme-controlled chemical changes slowly release the potential energy from glucose. Approximately 35% of the energy released is in the form of heat whilst about 65% is chemical energy in the form of an energy-rich compound called **adenosine triphosphate** or **ATP**. ATP can be moved about inside the cell. It is a form of stored energy that can be released rapidly as required to a living cell. The energy is used in different life maintenance processes.

Energy usage

The energy released by cell respiration is used for:
- **genetic material** formation and its subsequent duplication. Growth involves making new cells and the synthesis of cell material;
- **nutrition** – energy is needed for ingestion, digestion and absorption by active transfer and for moving food through the gut by peristalsis;
- **heat energy** – essential for all the cellular chemical processes of metabolism;
- **movement** – muscle contraction in respiration, transport, peristalsis, body movements and locomotion;
- **irritability** – as nerve impulses or messages which travel along nerves in the form of electro-chemical energy;
- **reproduction**.

Experiment to show that respiration produces heat

Figure 6.2   To show that heat is produced by respiring seeds

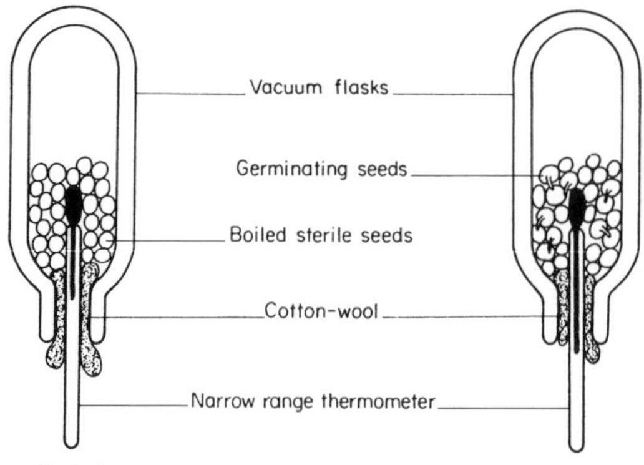

Vacuum flasks

Germinating seeds

Boiled sterile seeds

Cotton-wool

Narrow range thermometer

Control

Experiment

Procedure

1 Prepare pea seeds, or wheat or barley grains by soaking in water for 24 hours. Sterilise them externally with sodium hypochlorite.
2 Prepare seed or grain for the control by boiling for 5 minutes.
3 Pack one vacuum flask with live peas or grain and pack a second *control* vacuum flask with dead peas or grain. Plug the flask mouths with cotton-wool surrounding a narrow range thermometer.

Respiration and combustion

Dry, powdered glucose, a non-living material, can be burnt in the laboratory in a chemical process called **combustion**. Living cells cannot respire glucose in the same way. Chemical combustion would be far to rapid and uncontrolled a process, giving out a large amount of heat energy too quickly for the cell to cope with. Energy is released slowly in enzyme-controlled stages in living cells.

$$C_6H_{12}O_6 + 6O_2 \rightarrow 6CO_2 + 6H_2O + 2900 \text{ kJ Energy}$$

glucose   oxygen   carbon   water
          (air)   dioxide

Questions

1 State two differences between internal and external respiration.
2 In what form is most energy released in a cell?
3 What is the chemical energy store which links the energy released from glucose to a life maintenance process?
4 State nine ways in which energy is used in the body.
5 In the experiment with respiring seeds, (a) How can you tell there is a difference in energy released between the contents of the two flasks?
   (b) How is the energy being used by the embryo plants inside the seeds?
6 How is the way energy is released in cells different from energy released when a material burns?

6.3 Respiration without oxygen

The early Earth was without oxygen. Consequently the first living organisms had to obtain energy without using oxygen. **Anaerobic respiration** is the process of partly breaking down glucose within living cells *without* using oxygen. Certain bacteria and parasites such as the tapeworm can live continuously without oxygen and are called **complete anaerobes**. Oxygen is poisonous to bacteria which are complete anaerobes; hence fresh air and oxygen-rich chemicals are used to combat certain bacterial diseases.

Green plants with the exception of Canadian pond weed (*Elodea*), certain yeasts, mudworms and diving mammals such as whales and seals, are **partial anaerobes**. They are able to live for a short while completely without oxygen. Mammalian skeletal muscle cells can also respire anaerobically for short periods.

Anaerobic respiration in yeast – a fungus

When yeast respires anaerobically the glucose is broken down to produce two chemical substances. They are ethanol (ethyl alcohol) and carbon dioxide gas. The reaction can be summarised as follows:

$$C_6H_{12}O_6 \rightarrow 2C_2H_5OH + 2CO_2 + 210 \text{ kJ Energy}$$
glucose   ethanol   carbon
dioxide

Experiment to show anaerobic respiration in yeast

Figure 6.3 To show anaerobic respiration in yeast, *Saccharomyces* sp.

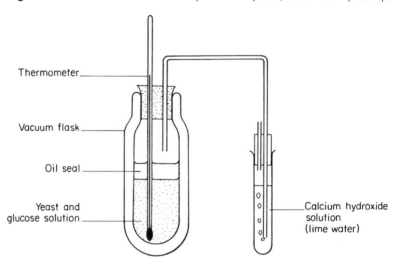

Thermometer

Vacuum flask

Oil seal

Yeast and glucose solution

Calcium hydroxide solution (lime water)

Procedure

1 Place a boiled and cooled solution of 10% glucose in a vacuum flask. Stir 3 g of baker's yeast into a thin cream with a little water and add to the glucose solution. Add liquid paraffin oil to form an air-excluding layer.
2 Insert a stopper supporting a narrow-range thermometer dipping into the yeast–glucose solution, and a delivery tube leading to a test tube of hydrogen carbonate indicator or clear calcium hydroxide solution.
3 Set up a control made from a boiled and cooled suspension of yeast in 10% glucose solution.

Observations

A rise in temperature is observed in the experiment flask, showing a release of heat. Carbon dioxide gas is evolved causing the calcium hydroxide solution to turn cloudy or the hydrogen carbonate indicator to turn from red to yellow.

Further work and observations

If the yeast and glucose solution was filtered, a clear liquid would be produced. Distillation of part of this liquid would produce ethanol solution which is flammable and burns, producing heat and light energy.

Features of anaerobic respiration in yeast

- **Occurrence:** Anaerobic respiration occurs with glucose dissolved in solution in the cell cytoplasm.
- **Mechanism:** The reaction occurs slowly, and is controlled by intracellular enzymes which are destroyed by heat as shown by the 'control' experiment. Oxygen is not required.
- **Energy:** Almost 50% of the energy produced is in the form of heat detected by the thermometer and 50% of the total energy produced is for cell use as ATP. Energy release is slow and controlled.
- **Products:** The carbon dioxide causes foaming. It is used to raise flour dough in bread-making when the gas expands in the proving and baking process. The gas causes foaming in beers and certain wines. Ethanol becomes poisonous to respiring yeast cells when the concentration exceeds 15%. This means that green plants and yeasts are partial anaerobes and must respire at some time using oxygen to prevent a high concentration of ethanol forming. Alcoholic drinks are made using different species of yeasts in solutions of sugars from different sources such as barley, potatoes and grape juice.

Questions

7 (a) In the experiment on page 140, how does the boiling then cooling of the glucose solution help to develop anaerobic conditions?
  (b) What other liquid helps to create anaerobic conditions for the yeast?
  (c) Why does the yeast survive in the anaerobic conditions?
  (d) When would yeast die in these conditions?
8 How are alcoholic drinks containing up to 40% alcohol made?

Anaerobic respiration in animals

When animal cells, particularly skeletal muscle cells, respire anaerobically, the glucose is broken down to produce a chemical product called **lactic acid**. No carbon dioxide is produced.

The chemical process can be summarised as follows:

$$C_6H_{12}O_6 \rightarrow 2CH_3 \cdot CH \cdot OH \cdot COOH + 150 \text{ kJ Energy}$$

glucose   lactic acid

Features of anaerobic respiration in animals

- **Occurence:** Anaerobic respiration occurs with glucose in solution in the cell cytoplasm.
- **Mechanism:** It is a slow process controlled by intracellular enzymes. It is important to note that the equation above shows only the reactants and the products. There are many intermediate compounds which make the reaction complex. The reaction cannot continue for long as oxygen is eventually needed because the products have a toxic effect on the cells.
- **Energy:** Over 60% of the total energy released is heat energy, whilst less than 40% is available for cell use as chemical energy in ATP.
- **Product:** Lactic acid increases in concentration in muscle tissue and causes poisoning. This is experienced as aches and fatigue.

Oxygen debt

An **oxygen debt** builds up when a human being is involved in very strenuous athletic exercise such as taking part in a 100 m sprint. During this time anaerobic respiration occurs in the muscles because the body cannot breathe fast enough to provide all the oxygen needed for releasing energy in aerobic respiration. Anaerobic respiration produces lactic acid. After exercise, rapid panting occurs for some minutes in order to provide enough oxygen for the aerobic respiration of the lactic acid. It is converted into carbon dioxide, water and further energy. The extra oxygen taken in to restore the cells after exercise is said to *pay off* the oxygen debt. The relationship between the concentration of lactic acid in the blood, exercise time and the period of time to repay the oxygen debt is shown in Figure 6.4.

Figure 6.4 Graph to show the increase in lactic acid concentration due to insufficient oxygen. This shortfall is repaid by increased intake of oxygen during the oxygen debt period

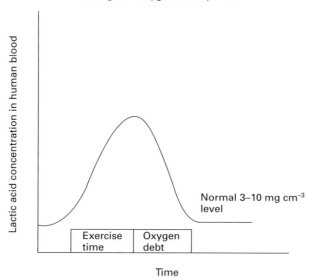

Questions

9 How does anerobic respiration in animals compare with anaerobic respiration in plants?
10 Why do animal cells eventually need oxygen?
11 (a) When and why is an oxygen debt built up?
    (b) How is an oxygen debt repaid?
12 How does the concentration of lactic acid in the blood change during a sprint and the rest period afterwards.

6.4 Respiration with oxygen

The amount of oxygen in the air has gradually increased over millions of years mainly due to photosynthesis which produces oxygen as a by-product in green organisms (producers). Today, approximately 20% of the volume of the air is oxygen (see Tables 6.2 and 14.2). As a result living organisms have developed aerobic respiration or the use of oxygen to break down glucose completely into carbon dioxide and water. The chemical reaction is summarised on page 139.

Features of aerobic respiration

Occurence

Glucose dissolved in solution is brought to the **mitochondria** of plant and animal cells.

Mechanism

The process of energy release is slow and passes through many intermediate stages. The first stages are particularly similar to anaerobic respiration and do not require oxygen. The later stages require a continual supply of oxygen and pass through a complex repeated cycle known as the **Krebs cycle**. All stages are controlled by intra-cellular enzymes. Most or 95% of aerobic respiration takes place in the mitochondria and a little, 5%, in the cytoplasm.

Energy

The energy released in aerobic respiration is approximately 35% in the form of heat energy and 65% as ATP for cell use. Almost 19 times more energy is released from glucose by aerobic respiration than by anaerobic respiration. In birds and mammals the heat released is used to maintain a constant body temperature.

Products

- **Water:** The water produced by respiration is called **metabolic water** because it is produced as a result of metabolism (see Section 6.5).
- **Carbon dioxide:** Carbon dioxide is less toxic than ethanol and lactic acid produced by anaerobic respiration. However, it is still a poisonous substance if allowed to accumulate in living cells. It is therefore removed by efficient external respiration methods described in the later parts of this chapter.

Experiment to show that oxygen is taken up during aerobic respiration

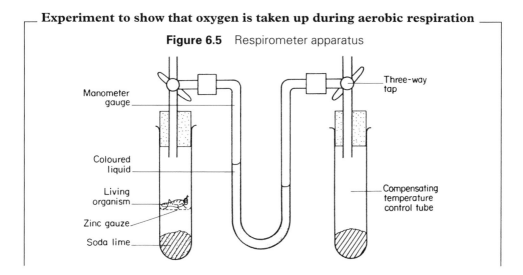

Figure 6.5   Respirometer apparatus

Three-way tap

Manometer gauge

Coloured liquid

Living organism

Zinc gauze

Soda lime

Compensating temperature control tube

Procedure

1 The apparatus in Figure 6.5 is a respirometer. Soda lime is a chemical mixture which absorbs carbon dioxide, leaving nitrogen and oxygen gas in the air.
2 Respiring organisms such as sterile germinating seedlings are placed in one tube supported by the zinc gauze platform.
3 Liquid levels in the manometer gauge are equalised by means of opening the three-way taps.
4 A separate *control* apparatus is set up containing non-living items such as glass marbles or sterile dead seedlings.

Observations

A change in the air volume inside the apparatus is shown by the manometer gauge. This indicates that the gas oxygen is consumed; nitrogen is inert and takes no part in cell respiration. Table 6.1 shows the oxygen consumption of various animals.

Note: The oxygen consumption rate is equal to the rate of respiration, and this in turn is a measure of the metabolic rate.

Table 6.1   Oxygen consumption of different animals

Earthworm	$0.06 \text{ cm}^3\text{g}^{-1}\text{h}^{-1}$
Butterfly (resting)	$0.6 \text{ cm}^3\text{g}^{-1}\text{h}^{-1}$
Maggot	$1.3 \text{ cm}^3\text{g}^{-1}\text{h}^{-1}$
Frog	$1.5 \text{ cm}^3\text{g}^{-1}\text{h}^{-1}$
Mouse (resting)	$2.5 \text{ cm}^3\text{g}^{-1}\text{h}^{-1}$

A respirometer can be used to determine the amount of oxygen measured in cm^3 taken up for each gram of body weight per hour.

Questions

13 In what ways are the chemical reactions of aerobic respiration and anaerobic respiration in muscles (a) similar, (b) different?
14 How else are aerobic and anaerobic respiration (a) similar, (b) different?
15 (a) Describe the fate of carbon and hydrogen atoms in a glucose molecule.
   (b) What happens to the energy that the glucose molecule contained?
16 (a) In the respirometer apparatus, why does the volume of gas in the left-hand respirometer get less?
   (b) If the right hand tube was warmer than the left hand tube how would the levels of the liquid in the manometer be affected?
   (c) In Table 6.1, why is the oxygen consumption of the different animals not the same?

6.5 Metabolism

Metabolism is the word used to describe the chemical processes which take place within a living thing. The main chemical metabolic reactions occuring in cells are either **synthesis** or the building up of a compound from simpler substances, which is also called **anabolism,** or **decomposition** or breakdown of chemical compounds, also called **catabolism**.

Metabolic rate is the rate at which a body uses energy, and can be measured as either oxygen uptake or carbon dioxide output using a respirometer or spirometer (see page 143). It is found that the rate of respiration or metabolic rate of plants is very much less than animals. Also the rate differs among resting animals.

The rate of metabolism or respiration also differs for different tissues found in a human being; the cells with the greatest number of mitochondria having the highest rate. Liver cells have a higher number of mitochondria than blood cells and blood cells have a higher number of mitochondria than fat cells.

Basal metabolic rate, BMR, is the rate at which the human body uses energy when at complete rest, 12 hours after a meal in a moderately warm room. In such resting conditions energy is used for heart beat, external respiration and cell metabolism. Basal metabolism excludes all physical skeletal muscle activity.

Factors affecting the rate of respiration

The rate of respiration in animals is affected by physical activity or exercise, also body size and weight, pregnancy, lactation and age. This is related to the daily needs described in Chapter 4.

- **Hormones**, which regulate metabolic processes in animals, affect the rate of respiration. Growth substances in plants affect the rate of growth. Mammals and humans are affected by the secretion of the thyroid gland hormones.
- **Temperature** increases up to 45°C generally cause an increase in rate of respiration and growth seen in plants, above 45°C the rate falls rapidly due to damage and destruction of respiratory enzymes. Human beings in cold climates have a *higher* rate of metabolism than human beings in hot tropical climates.
- **Age:** BMR is highest in babies and decreases with age.
- **Sex:** females have a lower BMR than males.
- **Stimulant drugs**, such as nicotine increase BMR.

Questions

17 What is the difference between a catabolic process and an anabolic process?
18 Which kind of process is (a) the respiration of glucose, (b) the building up of proteins from amino acids?

6.6 External respiration

Aerobic respiration requires a continuous flow of oxygen into cells with a rapid removal of carbon dioxide from the cells. This is called external respiration and may involve two processes depending on the type of organism:
- **ventilation**, or 'breathing' – the process of moving oxygen and carbon dioxide to and from the respiratory surface of the skin, gills or lungs.
- **gaseous exchange** – the process by which oxygen and carbon dioxide pass through the respiratory membrane into and out of the cell.

Conditions needed for external respiration

- **Oxygen** is available dissolved in sea or fresh water or in its gaseous form in the air. There is 25 times more oxygen in air than is dissolved in an equal volume of sea water (see Table 6.2). Aquatic animals must transport great volumes of water to their respiratory surfaces.

Table 6.2 Summary of dissolved substances in fresh water and sea water compared with air (see also Table 14.2)

Component	Sea water (%)	Fresh water (%)	Air (%)
Oxygen gas	0.85	1.03	20
Nitrogen gas	1.45	1.90	79
Salts, mainly sodium chloride	3.5	traces	nil
Water	94	97	varies to about 6%

- **Diffusion** is the movement of substances in air or water. The rate of diffusion of oxygen in the air is 100 000 times the diffusion rate of oxygen in water. Oxygen and carbon dioxide will diffuse very slowly in cell fluid, compared to rapid diffusion in the air spaces of the lungs.
- **Moisture**, in the form of water or mucus on the respiratory surface, is needed to dissolve the gases before they can diffuse through the membrane.
- **Permeable thin membranes** or skins allow rapid diffusion compared to non-permeable scaly skins.
- The **surface area** of the respiratory membrane (see also Figure 1.21) must be as large as possible to allow rapid diffusion.

Body size in respiration

Amoeba, a unicellular organism, has a surface area of 2.5 mm^2 and a body volume of 0.1 mm^3 whilst an adult man, a multicellular organism, has a surface area of 1.8 m^2 and a body volume of 80 litres.

Surface area/volume ratios are shown as follows:

$$Amoeba \ \frac{2.5 \ mm^2}{0.1 \ mm^3} = 25$$

$$\text{Human adult} \ \frac{18 \ 000 \ cm^2}{80 \ 000 \ cm^3} = 0.225$$

Comparing these ratios it is seen that *Amoeba* has 25/0.225 or 111 times more surface area/ body volume than a human adult.

As organisms increase in size, the surface area for gas exchange purposes relative to the body volume becomes less. As a result, larger organisms develop special **respiratory surfaces** or organs. In unicellular organisms and small, multicellular organisms, e.g. cnidarians like *Hydra* (Figure 6.14), the diffusing oxygen and carbon dioxide need only travel a short distance to reach the centre of the unicellular organism or the innermost cell of a small multicellular organism.

Respiration in air and water

- **Aquatic organisms** have delicate respiratory organs such as gills, supported and kept moist by the surrounding water. Since large volumes of viscous water must reach the respiratory organs, it does so in a one-way flow.
- **Air-breathing organisms** have the delicate respiratory organs protected inside the body, and kept moist by mucus which prevents the respiratory surface from drying out by evaporation. Since air is 100 times less viscous than water and contains more oxygen per unit volume, a two-way flow of air operates in the respiratory organs of air-breathing organisms.

Questions

19 Why do aquatic animals have to move large volumes of water over their respiratory surfaces?
20 Why is it advantageous to an organism to have
   (a) a large respiratory surface?
   (b) a moist respiratory surface?
   (c) a thin respiratory surface?
21 How do the respiratory systems of aquatic and air-breathing organisms compare?

6.7   External respiration in mammals

Respiratory system

The respiratory system (Figure 6.6) consists of a pair of **lungs** contained in the thorax which is formed from the ribcage and the dome-shaped diaphragm muscle. Double-layered **pleural membranes** containing **pleural fluid** surround the lungs. The fluid acts as a lubricant and allows easy movement of the membranes during ventilation.

The respiratory route

The respiratory route or airway is contained within the respiratory organ system. It is composed of tubular airways, which connect the nasal cavity to the lungs. Air enters and leaves the respiratory organ system along the following route:

Nasal passages → Larynx → Trachea → Bronchi → Bronchioles → Alveoli

Air enters the nasal cavity where it is warmed and cleaned, and passes the **pharynx** to reach the **trachea,** guarded by the flap valve or **epiglottis** (see Figure 6.7).

The trachea divides into two tubes called **bronchi**. One bronchus goes to each lung. These tubes are kept open by C-shaped cartilage rings and are lined by a ciliated epithelium which wafts mucus and foreign particles, such as dust, upwards to the throat. Similar epithelium lines the nasal passage.

Each bronchus divides into numerous **bronchioles** which terminate in a group of **alveoli**. Each alveolus has a wall one cell in thickness in contact with a blood capillary network (Figure 6.8).

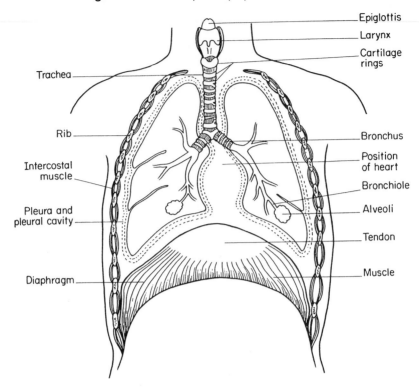

Figure 6.6 The respiratory system of humans

Epiglottis

Larynx

Cartilage rings

Trachea

Rib

Intercostal muscle

Pleura and pleural cavity

Diaphragm

Bronchus

Position of heart

Bronchiole

Alveoli

Tendon

Muscle

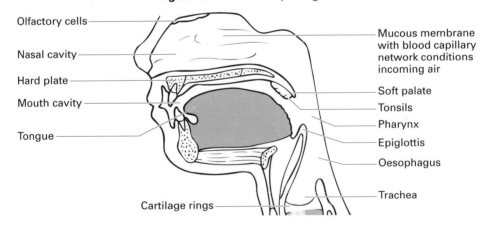

Figure 6.7 The nasal passages

Olfactory cells

Nasal cavity

Hard plate

Mouth cavity

Tongue

Cartilage rings

Mucous membrane with blood capillary network conditions incoming air

Soft palate

Tonsils

Pharynx

Epiglottis

Oesophagus

Trachea

Questions

22 What are the functions of (a) the pleural fluid, (b) the nasal cavity (c) the C-shaped cartilage, (d) the ciliated epithelium?

23 What is the path taken by carbon dioxide as it leaves the surface of an alveolus?

Figure 6.8 The blood supply to mammalian alveoli

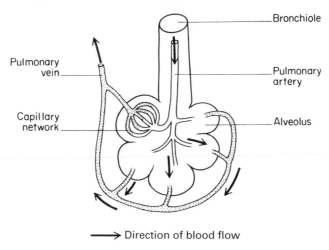

→ Direction of blood flow

Ventilation

Ventilation is achieved by a change in volume of the thorax, brought about by movement of the **diaphragm** and ribcage **intercostal muscles** (Figure 6.9). A summary of ventilation movements in inspiration and expiration is given in Table 6.3.

Table 6.3 Summary of respiratory movements in humans

	Part of respiratory tract	Inspiration	Expiration
1	Diaphragm	Contracts and flattens downwards	Relaxes and moves upwards to dome shape
2	Intercostal muscles	Muscles contract	Muscles relax
3	Ribcage and sternum	Move upwards and outwards	Move downwards and inwards
4	Thorax volume	Increases	Decreases
5	Air pressure	*Decrease* in pressure inside thorax and lung	*Increase* in pressure inside thorax and lung
6	Air movement	External air pressure drives air *into* lungs at *low* pressure	Air forced *out* of lungs by thorax compression and elastic recoil of lungs
7	Pleural cavity	Pleural fluid lubricates pleural membranes	Pleural fluid lubricates pleural membranes

The volume of air moving into and out of a human lung is shown in Table 6.4 on page 151, from which it is seen that some air always remains in the lung as residual air, whilst a small portion of the total lung capacity passes in and out in normal unforced breathing called **tidal volume**.

Figure 6.9 Ventilation movements in mammals

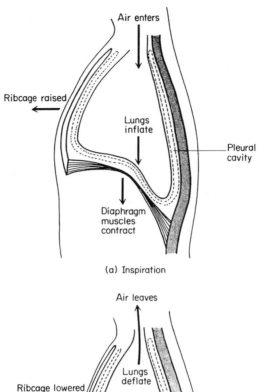

(a) Inspiration

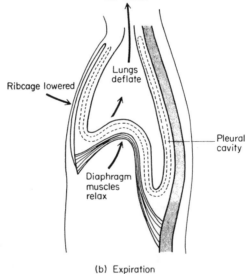

(b) Expiration

Questions

24 Write a paragraph describing the changes that take place during (a) inspiration, (b) expiration.

25 From Table 6.4, what is the difference between:
(a) the total lung capacity and the vital capacity?
(b) the residual volume and the tidal volume?

26 What is the pulmonary ventilation/minute volume of someone breathing in and out eight times a minute with a tidal volume of 0.85 litres?

Table 6.4 Normal lung volumes in resting adults

Lung volume	Range	Meaning of term
Total lung capacity	4.2–5.9 litres	Total volume of *fully inflated* lungs
Vital capacity	3.1–4.7 litres	Volume of air inspired or expired in *forced* breathing
Residual volume	1.10–1.19 litres	Volume of air which *cannot* be expelled by forced breathing. Remains in alveoli
Tidal volume	0.76–0.90 litres	Volume of air inspired and expired in *normal unforced* breathing
Pulmonary ventilation/ minute volume	6–10 litres per minute	Volume of air inspired in one minute, given by rate of breathing × tidal volume

Gas exchange

Gaseous exchange occurs in the millions of alveoli which provide a respiratory surface area of between 60 and 100 m^2. By comparison, the skin surface area of a human adult is 2 m^2. Figure 6.10 shows the gas exchange in one alveolus. Figure 6.8 on page 149 shows the blood supply to a group of alveoli.

Figure 6.10 Gas exchange at the alveolus

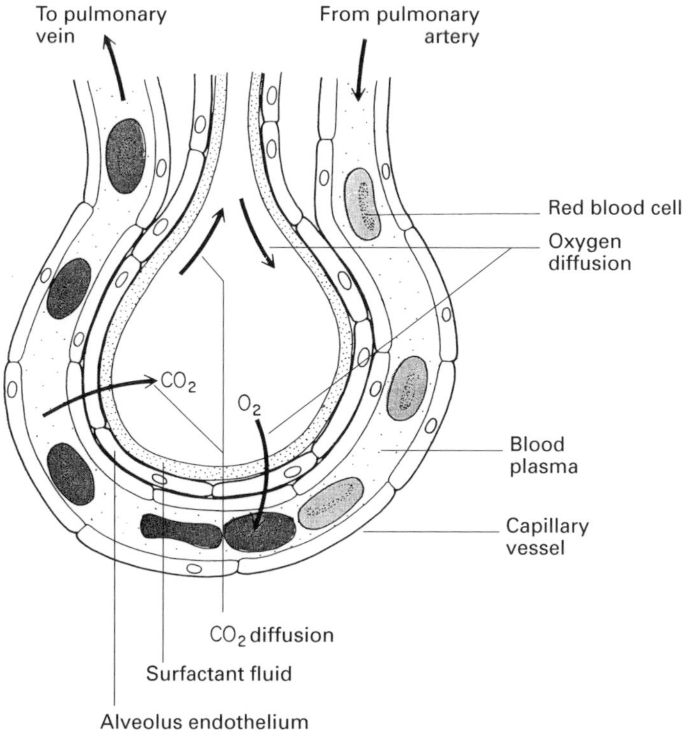

Inside the alveolus is a **surfactant solution** which serves to reduce surface tension and prevents the delicate air sacs from collapsing when air is expelled from the lungs. It increases lung elasticity.

Oxygen transport

Oxygen transport occurs rapidly (within one second) by passive diffusion through the following layers or barriers:
- surfactant solution dissolves oxygen
- alveolar endothelium permits oxygen diffusion
- capillary endothelium allows oxygen to dissolve in blood plasma
- red blood cell surface membrane allows oxygen to diffuse into the cell and combine chemically with the functional protein, haemoglobin, to form bright red **oxyhaemoglobin** seen in oxygenated blood.

$$Hb \quad + O_2 \quad \rightarrow HbO_2$$
$$\text{haemoglobin} + \text{oxygen} \rightarrow \text{oxyhaemoglobin}$$

Oxyhaemoglobin readily decomposes into oxygen and haemoglobin in regions of low oxygen concentration to form deep red-purple deoxygenated blood.

$$HbO_2 \qquad \rightarrow Hb \qquad + O_2$$
$$\text{oxyhaemoglobin} \rightarrow \text{haemoglobin} + \text{oxygen}$$

Carbon dioxide transport

Carbon dioxide transport occurs by means of the blood plasma or red blood cells. Carbon dioxide is mainly transported chemically combined in the form of **hydrogen carbonate** ions, 65% in the blood plasma and 25% in the red blood cells. About 5% of carbon dioxide can dissolve in blood plasma to form a physical solution. When the blood reaches the alveoli the hydrogen carbonate ions break down to water and carbon dioxide. Carbon dioxide leaves the blood and passes into the alveolar air by diffusion.

The essential **diffusion gradient** is maintained by:
- **lung ventilation** transporting respiratory gases to and from the alveolus;
- blood flow from the **pulmonary artery** bringing blood with a *low* concentration of oxygen and a *high* concentration of carbon dioxide to the alveolus;
- blood flow to the **pulmonary vein** taking away blood with a *high* concentration of oxygen and a *low* concentration of carbon dioxide;
- **haemoglobin** rapidly combining chemically with oxygen;
- **carbon dioxide** being released from the blood plasma.

Table 6.5 shows the varying composition of inspired, expired and alveolar air.

Table 6.5 Composition of inspired and expired air

Air component	Inspired air	Expired air	Alveolar air
Oxygen	21%	17%	14%
Nitrogen	79%	79%	80%
Carbon dioxide	0.03%	4%	6%
Water vapour	variable	saturated	saturated
Temperature	prevailing atmospheric temperature	38°C	+40°C
Bacteria	low count	high count	high count

27 How many times larger is the lung area than the skin area of a human adult?
28 Describe the path taken by oxygen from just outside the nose to the inside of a red blood cell.
29 In what forms is carbon dioxide found in the blood and in which parts of the blood does it travel?
30 How does ventilation maintain a high concentration of oxygen for diffusion into the blood?
31 How does the circulatory system maintain a low concentration of oxygen for diffusion into the blood?
32 How would the diffusion of the respiratory gases be affected if the blood flowed more slowly?
33 How does expired air differ from inspired air?

Experiment to demonstrate the presence of carbon dioxide in exhaled air

Figure 6.11 To demonstrate CO_2 in exhaled air

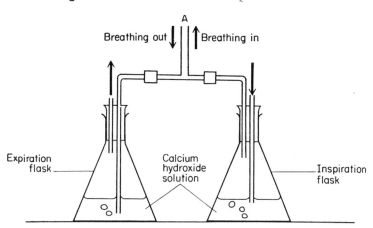

If a person breathes in gently at A for 20 seconds, air will be sucked through the inspiration flask and into the lungs. When the person breathes out gently for 20 seconds the air will pass out through the expiration flask. Calcium hydroxide solution (lime water) turns milky in contact with carbon dioxide gas.

Observation

The expired air contains over 130 times the amount of carbon dioxide as inspired air.

Question

34 In the experiment above
   (a) What will happen if the person breathes in and out too strongly?
   (b) What would you expect to happen to the lime water in each flask?

Control of respiration in mammals

The normal breathing rate for a resting adult is 16 to 20 respiratory cycles per minute, one respiratory cycle being composed of inhalation followed by expiration.

The respiratory centre of the brain is in the part called the **medulla oblongata** (see page 214). This controls the diaphragm and intercostal muscles. The activity of the respiratory centre is **homeostatically** controlled by chemical and by nervous activity (see Chapter 7). An *increase* in carbon dioxide in the blood stimulates the respiratory centre to produce faster and deeper ventilation. This occurs in strenuous exercise.

Conscious or **voluntary control** over breathing is exerted in speaking, singing or playing wind instruments.

Insufficient oxygen or oxygen starvation for more than 2 minutes causes brain cell damage and finally asphyxiation. Hence the need for speedy artificial respiration by the 'kiss of life' method. *Excess* oxygen or oxygen poisoning causes damage to body cells through excessive oxidation, leading to death. This can occur by breathing 100% oxygen under high pressure.

Breathing and health

Clean, unpolluted fresh air, together with frequent physical exercise, is essential for the development of a healthy respiratory system. Physical exercise strengthens and develops respiratory muscles and increases lung capacity. Cigarette tobacco smoke contains the alkaloid drug nicotine, carbon monoxide, carcinogenic tars and lung irritants. These are inhaled by both smokers and non-smokers.

- **Nicotine** causes bronchiole constriction, destruction of the protective cilia of the ciliated epithelium, increase in blood pressure and pulse rates. It is also responsible for addiction.
- **Tar** causes lung cancer.
- **Irritant** components cause 'smokers cough'.
- **Carbon monoxide** forms carboxyhaemoglobin causing oxygen shortage.

The four main diseases related to cigarette smoke are:
- **bronchitis**, which occurs when the lining of the bronchi are damaged and become inflamed;
- **emphysema**, which occurs when some of the alveoli are destroyed;
- **cancer** of the lung and bronchi;
- **coronary heart disease** (see Section 5.9).

Large amounts of money are spent in hospital care for tobacco smokers. For good health all people need clean, unpolluted fresh air, frequent and regular exercise and well-ventilated homes, buildings and public places.

Questions

35 How many breaths may you take at rest in an hour?

36 Why do people who take part in physical exercise recover faster after running fast than those who are not fit?

37 Why are mucus and dust not removed from a smoker's respiratory system as efficiently as they are in a non-smoker's respiratory system?

38 How does nicotine affect the circulatory system?

39 What happens to the area of the respiratory surface in emphysema? How may this affect the life of the emphysema sufferer?

6.8 External respiration in bony fish

Respiratory system

The respiratory system of a bony fish, e.g. cod, whiting or herring, consists of **gills** in an opercular cavity protected by an **operculum** or gill cover. The gill slits connect the buccal cavity with the opercular cavity.

A gill is made up of the **gill arch** which supports the two rows of **gill filaments**. Each gill filament has rows of **gill flaps** with a blood capillary network.

Figure 6.12 Ventilation movements in a bony fish

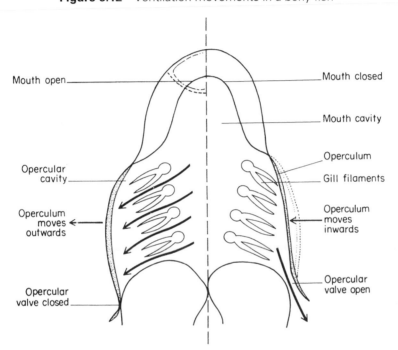

Ventilation

Ventilation movements in a bony fish are shown in Figure 6.12. Inspiration begins by water being drawn into the buccal cavity as a result of lowering the mouth floor. The outward movement of the operculum sucks water into the operculum cavity; this is aided by the raising of the mouth floor. Surrounding water pressure keeps the opercular valves closed.

Expiration occurs by the inward movement of the operculum and opening of the opercular valves. This causes the buccal cavity to contract in volume and drive water out through the open opercular valves.

Gas exchange

Gaseous exchange takes place on the gill flaps or secondary lamellae which provide a respiratory surface 10 times the fish's body surface area. The gill flap walls are one cell in thickness and separated from the blood by the capillary cell wall thickness.

Water passing over the gill flaps does so in the opposite direction to the blood flow in the capillary vessels. This is a very efficient **counterflow** system extracting more than 80% of the available oxygen in the water.

Haemoglobin in the blood transports oxygen to all parts of the body. This, together with the counterflow at the gills, maintains a diffusion gradient of oxygen in the blood.

___ Questions ___

40 What provides the large surface area for the uptake of oxygen?
41 Describe the actions of the fish that cause water to be drawn over the gills.
42 Describe the path taken by oxygen dissolved in the water in the mouth of the fish to the time it is transported by the fish's circulatory system.

6.9 External respiration in insects

Respiratory system

The respiratory system is a **tracheal system** or network of interconnecting air-fillled tubes called **tracheae** delivering air directly to the body tissue cells. Tracheae are lined with cuticle which is continuous with the outer epidermis as a waterproof, non-permeable outer layer (see Figure 6.13).

The tracheae have spiral rings called **taenidia** which serve to keep the tubes open. Air enters the tracheal system via **spiracles** (pores), a pair usually being located on the sides of each segment of the thorax and abdomen. A spiracle may have a pit or an atrium with hairs acting as filters, together with a muscular valve, regulating the spiracle opening and closing. This controls water loss from the internal tissues.

Ventilation

Ventilation is achieved by contraction and relaxation of abdominal muscles; this results in a rhythmic 'pumping' of air into and out of the tracheae. Some insects have flexible **air sacs** connected to the tracheae and these may help in air flow. Cockroaches, beetles and locusts show a flattening of the abdomen in their ventilation movements, whilst bees, flies and wasps show a telescoping of the abdomen.

Gas exchange

Gas exchange takes place in the unthickened tracheoles which are permeable to gases and are filled with a fluid at the ends in contact with the body tissues. The fluid disappears when an insect is active, thus allowing respiratory gases to enter and leave the tissues rapidly. When the insect is inactive the fluid returns. Gases are *not* transported in the blood of insects.

___ Questions ___

43 Describe how oxygen in the air outside an insect reaches its tissues.
44 How is the external respiration of an insect different from that of a human?

Figure 6.13 The respiratory system of an insect

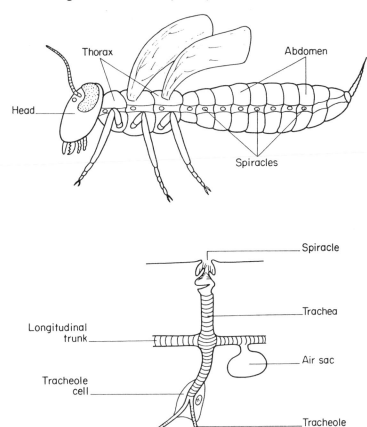

6.10 External respiration in small organisms

The cell surface in *Amoeba*, a Protoctist, and the body surface of *Hydra*, a cnidarian function as the respiratory surface. The high surface area/volume ratio of these small organisms allows the movement of the respiratory gases to occur by simple diffusion. No special system is required.

Earthworms have an epidermis layer one cell in thickness, kept moist by mucus secretion. The respiratory gases diffuse over a short distance to enter the blood system. Due to its very low solubility, oxygen combines chemically with the blood haemoglobin.

Questions

45 Why do the cells in the body wall of *Hydra* not need a special respiratory system?
46 How is the external respiration of an earthworm similar to that of a human?

Figure 6.14 (a) *Amoeba* and (b) *Hydra*, small organisms that do not have a special respiratory system

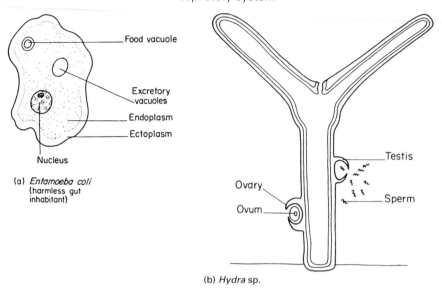

Food vacuole

Excretory vacuoles

Endoplasm

Ectoplasm

Nucleus

(a) *Entamoeba coli* (harmless gut inhabitant)

Testis

Ovary

Sperm

Ovum

(b) *Hydra* sp.

6.11 External respiration in green plants

Flowering and green, non-flowering plants, in contrast to animals, have an additional source of oxygen apart from air. They use the oxygen produced during photosynthesis.

Oxygen produced during photosynthesis can diffuse over very short distances into neighbouring cells, or from chloroplasts to mitchondria in the same cell. In daylight, green plants will therefore use much less oxygen from the air compared to animals, who rely entirely on air for their oxygen supply.

Tissues involved

Respiratory systems and organs are not required or developed even in the largest plants such as trees. The leaves and roots provide an extensive surface area for entry and exit of gases through the following:

- **Stomata** are present in great numbers as pores in leaves in the epidermis layer which is mainly non-permeable due to the cuticular layer (see Section 3.5).
- **Lenticels** are raised breathing pores in the surface of woody stems. They contain loosely packed cells allowing passage of gases (Figure 6.15). The remaining bark of stems has a non-permeable layer of cork cells.
- **Root hairs** are without a waterproofing cuticle and the root hair cell walls are permeable and permit diffusion. Cultivated soil is aerated by digging and ploughing. This soil air dissolves in the film of soil water surrounding the soil particles (see Figure 5.7).

Figure 6.15 Gas exchange in the stem lenticel of a flowering plant

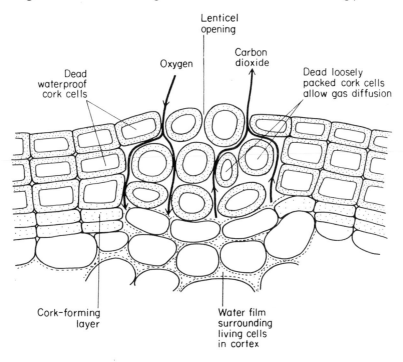

Ventilation and gas exchange

Ventilation movements do not occur in plants since they are less active than animals. **Gaseous exchange** occurs after entry of oxygen into the numerous air spaces present between cells in the stem, leaf and root.

Few cells are far from an oxygen supply from the air or photosynthetic tissue. Internal cells have a film of water coating their outer surfaces; this is essential to dissolve the respiratory gases prior to diffusion across the cell wall. The **cell walls** are therefore the respiratory surfaces of the plant.

Table 6.6 compares plant and animal respiration.

Questions

47 Where do plants get their supply of oxygen from?
48 How may oxygen enter a woody plant?
49 In what ways is the respiration of plants and animals (a) similar; (b) different?

Photosynthesis and respiration

Photosynthesis in flowering plants is dependent on light intensity. In daylight, both photosynthesis and respiration occur. Photosynthesis provides oxygen for respiration,

Table 6.6 Comparison of respiration in plants and animals

Plant respiration	Animal respiration
1 No external ventilation movements	External ventilation movements in all except small animals, *Amoeba* and *Hydra*
2 Gaseous exchange is on cell wall of unicellular plants, and in air spaces between cells in leaf or stem of multicellular plants	Gaseous exchange is on cell membrane of unicellular animals, or in lungs, skin, gills, or by air tubes in multicellular animals
3 No special transport system	Blood systems transport oxygen
4 Cellular respiration produces energy stored as ATP in mitochondria	Cellular respiration produces energy stored as ATP in mitochondria
5 Aerobic respiration forms $CO_2 + H_2O$	Aerobic respiration forms $CO_2 + H_2O$
6 Anaerobic respiration forms ethanol and carbon dioxide	Anaerobic respiration forms lactic acid[*] only, and no CO_2
7 Green plants produce little detectable heat	Animals produce considerable detectable heat
8 Aerobic plants have additional oxygen source from photosynthesis	Aerobic animals have one oxygen source, the air
9 Respiration rate low	Respiration rate high

[*] Lactic acid = 2-hydroxypropanoic acid.

Figure 6.16 Graph showing effect of light intensity on photosynthesis

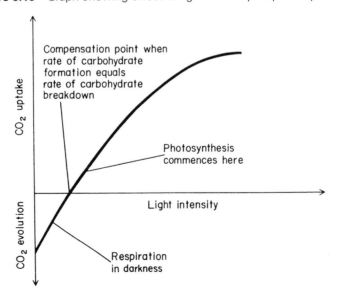

CO_2 uptake

Compensation point when rate of carbohydrate formation equals rate of carbohydrate breakdown

Photosynthesis commences here

Light intensity

CO_2 evolution

Respiration in darkness

whilst respiration provides carbon dioxide for photosynthesis. In darkness, only respiration occurs, using oxygen from the outside air.

In bright light the rate of photosynthesis is 10 to 20 times the rate of respiration, and thus provides an ample oxygen source. With decreasing light, the photosynthesis rate drops until a point is reached when the rate of respiration *equals* the rate of photosynthesis (see Figure 6.16). This is called the **compensation point**. In very dim or no light at dawn or dusk the rate of photosynthesis will become *less than* the respiratory rate.

Table 6.7 compares photosynthesis with respiration.

Table 6.7 Comparison of photosynthesis and respiration

	Photosynthesis	Respiration
1	A process of *anabolism* or *synthesis* of organic materials	A process of *catabolism* or *breaking down* of organic materials
2	Energy *required* from sunlight	Energy *released* and stored as ATP
3	Occurs in cells containing *chloroplasts*	Occurs in all cells containing *mitochondria*
4	Light essential	Occurs at all times in light or dark
5	Oxygen produced	Oxygen needed for *aerobic* respiration
6	Carbon dioxide required	Carbon dioxide produced by plant and animal aerobic respiration, and plant anaerobic respiration
7	Typical of all green plants and certain bacteria	Typical of all living organisms – aerobes and anaerobes

Questions

50 Why is there oxygen left over for animals and protoctists when plants photosynthesize?

51 What is the compensation point and when does it occur?

52 How does the amount of carbon dioxide in the air around a plant change from mid day to midnight?

53 Look at Table 6.7. In how many ways is photosynthesis different from respiration?

Summary

1 There are two main kinds of respiration – internal and external respiration. (▶ 137)

2 In cells, enzyme-controlled chemical changes release energy from glucose for transfer to ATP. (▶ 138)

3 Energy is used in nutrition, growth, to provide heat, for muscular contraction and the transport of impulses along the nervous system. (▶ 138)

4 Cell respiration can take place without oxygen in a process called anaerobic respiration. Different products are produced in plants and animals. (▶ 139)

5 The products of anaerobic respiration in humans are removed by paying off the oxygen debt. (▶ 142)
6 Cell respiration takes place with oxygen in a process called aerobic respiration. (▶ 142)
7 The metabolic rate is linked to the rate of respiration. (▶ 144)
8 External respiration may involve ventilation and gaseous exchange. (▶ 149)
9 External respiration in humans (▶ 148) differs significantly from external respiration in fish (▶ 155), insects (▶ 156) and small animals. (▶ 157).
10 Flowering plants and non-flowering green plants use oxygen from the air and from the process of photosynthesis when they respire. (▶ 159)
11 The processes of respiration and photosynthesis are widely different. (▶ 160)

Homeostasis – the steady state

When you have finished this chapter you should be able to:
- understand the principles of homeostasis in protoctists and animals
- explain how the liver functions
- explain how the blood sugar is regulated
- understand excretion
- understand osmoregulation
- explain how the kidney works
- understand ways in which kidney failure can be treated
- describe the structure of the skin
- describe the variety of functions of the skin
- explain how temperature is regulated in a variety of animals including humans
- understand how life-supporting conditions are provided for the development of the fetus

7.1 Principles of homeostasis

The environment or surroundings of a living organism is of two main kinds:
- The **external environment** outside the living organism's body. The features of this environment are water, soil, light, heat and other living organisms (see Chapter 14).
- The **internal environment** or the 'inside' of a living organism's body which features the body fluids in multicellular animals.

The body fluids in mammals and human beings include:
- **blood plasma** (see Section 5.6);
- **tissue fluid**, which bathes and surrounds the cells of multicellular animals and is also called the intercellular fluid (see Section 7.5.)
- **intracellular fluid** which is the fluid inside the cell and amounts to over 25 litres in an adult human body or over 60% of the total body fluid.

Living things do not like severe changes in their internal or external surroundings – it can harm their genetic material. **Homeostasis** is the maintenance of a steady state in living organisms, mainly protoctists and animals, by control of the internal environment, despite changes in the external environment. Plants have growth substances which regulate and control responses to changes in their external environment.

Tissue fluid has chemical, biological and physical components which must be kept within a narrow quantitative range called 'norms ' or 'set points' in order to maintain a steady state. Any changes from the 'norm' will be automatically corrected by homeostatic control mechanisms. Table 7.1 summarises the norm set values of the main components of tissue fluid.

Table 7.1   Summary of norm values for tissue fluid components

Component	Norm range	Mean or average
Water	97–99%	98%
Glucose	0.08–0.13%	0.1%
Proteins	1–2%	1%
Amino acids	0.03–0.05%	0.04%
Lipids	0.2–0.5%	0.30%
Carbamide (urea)	0.02–0.04%	0.03%
Sodium ions	0.3–0.36%	0.30%
Acidity or alkalinity (pH)	7.34–7.43	7.4
Temperature	37–39°C	38°C

Biological components include antibodies and white blood cells.

The blood composition will vary in different body tissues. For example the blood leaving the gut may have more water, amino acids and glucose in it than blood in other parts of the body. The kidney tissue fluids may contain more urea (carbamide) than the tissue fluids in other organs and the liver tissue fluids may be at a higher temperature than other body tissues.

The homeostatic control process

Homeostatic controls in living organisms, protoctists and animals, are **automatic** and **self-adjusting** and function in a similar manner to controls in machines, for example the thermostat in a water bath or an aquaruim tank for tropical fish. The study of control mechanisms in machines and living organisms is called **cybernetics**.

Any change in the norm of a tissue fluid component is detected by **sensory receptors** which are part of the nervous system (see Section 8.5). These send information by **feedback** to the **central control system** – usually the brain. It sends information on to an **effector** (see Section 8.6).

The central control causes corrective mechanisms to operate by means of **nerves** or **hormones**. These mechanisms either *increase* or *decrease* the particular feature and bring it back to the norm. The control of a feature in this way is called **negative feedback**. If the central control is defective, the norm is not restored but continues to increase or decrease and is difficult to reverse. This is called **positive feedback**. The relationships between the processes and components of a homeostatic control mechanism are shown in Figure 7.1. Homeostatic balance is the relationship between cell

input and cell **output** and cell material usage or cell **throughput**. This can be summarised as:

$$\text{INPUT} \xrightarrow[\text{USAGE}]{\text{CELL}} \text{OUTPUT}$$

Figure 7.1 How an altered norm is affected by positive and negative feedback

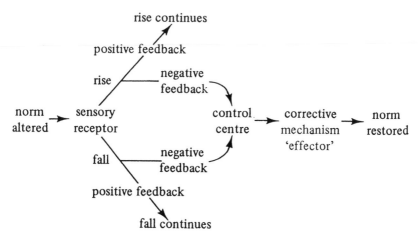

Questions

1 What is the difference between intercellular fluid and intracellular fluid?
2 What do you think would be the effect on an animal's body if the levels of the following tissue components fell? (a) water, (b) glucose, (c) temperature.
3 What is the sequence of events that takes place when an altered norm is restored?
4 What is the effect of (a) positive feedback (b) negative feedback on the body?

7.2 The mammalian liver as a homeostatic organ

The human liver weighs about 1.5 kg and is located beneath the diaphragm, above the stomach in the abdominal cavity. Two blood vessels supply it with blood at a rate of 1.5 litres per minute (Figure 7.2). The hepatic portal vein provides 80% of the blood and is rich in food nutrients from the small intestine and hormones from the pancreas. The remaining 20% is provided by the hepatic artery as mainly oxygenated blood. The blood then returns to the heart via the hepatic vein.

The cells of the liver tissue are in direct contact with the blood because the blood passes through large pores in the capillary walls. These cells contain many energy-providing mitochondria.

Liver functions

The liver has many functions which are either concerned with the blood or with food.

Figure 7.2 Blood supply to the mammalian liver

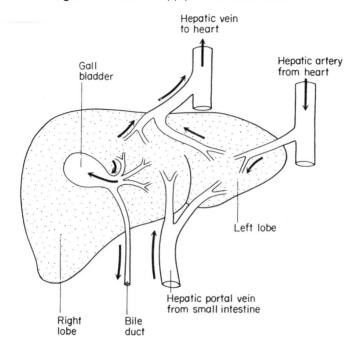

Blood functions

- Red blood cell formation occurs here in embryos. In adults the liver destroys old red blood cells.
- Blood plasma proteins, fibrinogen, prothrombin and albumen are formed.
- Up to 1.5 litres of blood are stored in the liver capillaries.
- Heat is produced. The liver supplies up to 12% of the body's heat.
- Detoxification occurs here. Harmful poisons, ethanol (alcohol) and drugs are converted into harmless substances. Drug abuse damages the liver.
- Deactivation of hormones occurs to prevent their continued activity.

Food functions

- Homeostatic regulation of glucose is achieved by adding or removing it from the blood. When glucose is removed from the blood it is stored as insoluble glycogen.
- Homeostatic regulation of lipids occurs here.
- Homeostatic regulation of proteins occurs. Amino acids which not required for making enzymes, parts for cells or for the repair of body tissues cannot be stored. These surplus amino acids are converted by deamination into ammonia and carbohydrate. The carbohydrate is used either for supplying cell energy or stored as glycogen. The ammonia is poisonous. It is combined with carbon dioxide to make urea (carbamide) which is non-toxic.
- Vitamins A, D and B12 and iron are stored in the liver.
- Bile juice is formed.

Questions

5 How many litres of blood flow through the liver in an hour?
6 How would the liver be affected if it only received blood from the hepatic portal vein?
7 How does the red blood cell composition change as it passes through the liver?
8 How does the temperature of the blood change as it passes through the liver?
9 What happens to the liver's glycogen store when the concentration of glucose in the blood falls below its norm?
10 What is the fate of surplus amino acids when they reach the liver?

7.3 Blood sugar regulation

Soluble blood sugar, **glucose**, is the source of energy for all living animal cells. The brain and nerve cells have no energy reserves and must have a continuous supply of glucose. If they are deprived of glucose the cells are harmed and the body may suffer convulsions and enter a coma. The norm for glucose concentration in tissue fluid is shown in Table 7.1 on page 164. Even though some meals have more carbohydrates than others, the blood plasma concentration of glucose must be maintained at the norm value of 0.1%.

The mechanism of regulation

The liver, pancreas (Figure 7.3) and adrenal glands function in a complex homeostatic mechanism of maintaining the level of blood sugar at about 0.1%.

Figure 7.3 The mammalian pancreas and its connection with the duodenum

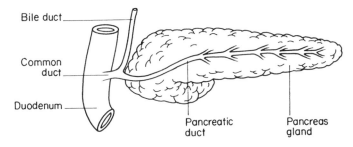

Glucose addition to blood – fasting

Glucose is added to the blood when its concentration falls below the norm of 0.1% in **hypoglycaemia**. Sensory centres in the brain detect a low glucose concentration and by negative feedback set off the following corrective mechanisms to restore the norm:
- Glycogen reserves in the liver are changed into glucose by the hormones **adrenalin** from the adrenal gland and **glucagon** from the pancreas. In this way sufficient glucose can be provided for the liver reserve for maintaining the norm for 12 hours.
- After all the liver glycogen reserves are consumed and there is no further intake of carbohydrate food, the non-carbohydrate food reserves are used. Lipids from the fat deposits are changed into glucose by adrenalin. This occurs in slimming. Muscle proteins are converted to glucose by another adrenal gland hormone. This causes muscle wastage seen in extreme starvation.

Glucose removal from the blood – after feeding

An increasing concentration of blood sugar glucose above the norm causes **hypergly-caemia**. Sensory centres in the pancreas and brain set off the following corrective mechanisms by negative feedback:

- **Insulin** is a hormone produced by the islets of Langerhans in the pancreas. It causes the liver to remove surplus glucose and change it into insoluble glycogen. Up to 1900 g of glycogen can be stored in the human liver. Its main function is to facilitate the uptake of glucose by skeletal muscle and liver cells. A form of insulin is now produced by certain bacteria in large-scale biotechnology (see Section 15.6).
- Glucose which cannot be stored in the liver accumulates in the skeletal muscles as glycogen. Up to 400 g of glycogen is stored in human muscle.
- Further surplus glucose is then stored as lipids in the fat deposits causing obesity.

Sugar diabetes

When the pancreas is defective it is unable to produce insulin. The blood sugar concentration increases above the norm, causing hyperglycaemia. Glucose then appears in the urine and the person shows signs of sugar diabetes in fatigue, weight loss, urine increase, thirst and hunger.

Injections of insulin directly into the blood restore the norm. If excessive amounts are used the blood sugar concentration falls below the norm, causing sweating and finally the coma symptoms of hypoglycaemia. As insulin is a protein, it cannot be taken by mouth as it would be digested like any other protein.

Questions

11 Why must a certain level of glucose be maintained in the blood?
12 What is the difference between hypoglycaemia and hyperglycaemia?
13 How do the actions of glucagon and insulin compare?
14 Where do the energy reserves come from as a person moves from a healthy balanced diet to a starvation diet?
15 How is the glucose stored as a person moves from a healthy balanced diet to one which is very high in carbohydrates?

7.4 Excretion

Excretion is the process of removing the waste products of metabolism from the body. The main excretory products are:

- respiratory waste products, **carbon dioxide** and **water** (this water is called metabolic water);
- **nitrogenous waste**, the main excretory product of animals unable to use and store amino acids. They are produced by deamination in the liver. The nitrogenous waste of invertebrates and bony fish is **ammonia**. In insects, molluscs, reptiles and birds it is **uric acid** and in mammals and adult amphibians it is **urea** (carbamide).

Certain plants produce nitrogen-containing materials called **alkaloids**, such as **nicotine** in the tobacco leaf and **quinine** in Chinchona bark. Plants are able to use the nitrogen they take in to make amino acids and so do not produce nitrogenous excretory products and have no specialised excretory organs such as the kidneys in mammals.

16 Name five excretory products.
17 What are the nitrogenous waste products of plants?

7.5 Osmoregulation

The concentration of water in the tissue fluid of mammals is maintained within the range 97–99% or a mean of 98%. Similarly osmoregulation regulates the amount of water in the blood at around 92%. If the norm is exceeded, water collects in the tissues causing **oedema** (dropsy); if it falls below the norm the tissues suffer dehydration. Movement of water into and out of living cells is by **osmosis** (see Section 5.1).

Osmoregulation is the control of the osmotic pressure or the amount of water and dissolved substances in protoctists and animals. The control of water and mineral ions (sodium, potassium, chloride and sulphate) in animals is combined in the same excretory organ. Excretion and osmoregulation are both processes concerned with blood plasma and tissue fluid homeostasis. They are closely linked and almost inseparable homeostatic mechanisms.

Osmoregulation in plants

There are no specific osmoregulation organs in higher plants. Control of water intake and loss is by means of those internal and external factors which affect the rate of transpiration.

Plants share with animals the problems of obtaining water and in disposing of the surplus. Certain plants develop methods of water conservation. **Xerophytes** are plants in dry habitats such as deserts which are able to withstand prolonged periods of water shortage. Succulent plants such as the cacti have water stored in large parenchyma tissues. Other plants have leaf modifications to reduce water loss, such as needle-shaped leaves, sunken stomata and thick, waxy cuticles as in the pine. The sand-dune marram grass has rolled leaves with stomata on the inner surface.

Excretion and osmoregulation in protoctists and animals

Amoeba

Amoeba has a **contractile vacuole**. This collects soluble excretory waste, possibly ammonia, from the cytoplasm intracellular fluid by diffusion and active transport. Many energy-providing mitochondria surround the contractile vacuole. Soluble substances are also able to diffuse along concentration gradients through the cell surface membrane from the cytoplasm into the surrounding water.

Pond water enters the *Amoeba* cell by osmosis, thus increasing the concentration of water above the norm for the cytoplasm. The contractile vacuole moves to the cell surface and discharges both water and excretory products into the external environment (Figure 7.4).

Freshwater fish

Bony fish, such as trout, take in water by osmosis through the gill semi-permeable membrane – they do not drink water. The nitrogenous waste is mainly **ammonia** which can diffuse out of the body by way of the gills, or pass out through the kidney in

Figure 7.4 The process of excretion and osmoregulation in *Amoeba* sp.

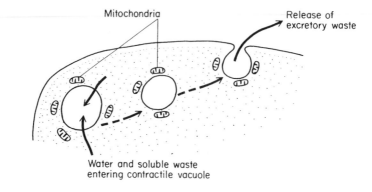

Mitochondria

Release of
excretory waste

Water and soluble waste
entering contractile vacuole

copious amounts of dilute urine. Freshwater fish therefore eliminate surplus water and take up mineral ions.

Sea-water fish

A bony fish such as a herring, loses water by osmosis through the gill semi-permeable membrane. Sea water is drunk in large quantities and water enters the body by osmosis through the gut wall. Mineral ions are eliminated through salt water glands in the gills.

The nitrogenous waste is mainly **urea** which leaves the kidney in scanty, concentrated urine. Sea-water fish therefore retain water, and eliminate excess mineral ions.

Land-living animals

Land animals take in water as part of their food and drink to replace the water continually lost in urine and by evaporation from the skin and respiratory surfaces. There is a need to conserve water by various means. This is achieved by waterproof body coverings and reabsorption of water from faeces in the large intestine and from urine in the kidney of mammals. Birds and reptiles produce a semi-solid paste-like excretory waste consisting mainly of insoluble **uric acid**. Mammals produce a liquid **urine** which is of a greater concentration than the tissue fluid.

Questions

18 What is the difference between excretion and osmoregulation?
19 How are the excretory and osmoregulatory processes of the herring different from those of the trout?
20 How do land plants and animals conserve water?

7.6   The human kidney as a homeostatic organ

The kidney, like the liver, is in close contact with the blood. The kidney is a homeostatic organ regulating the concentration of water by osmoregulation, removing urea by excretion and regulating the concentration of the blood mineral ions and also its pH.

Mammalian kidney structure

Two kidneys are located on the dorsal (back) wall of the abdominal cavity. Each receives oxygenated blood from the **renal artery** and blood is removed by the **renal vein**. Close to these blood vessels are the paired **adrenal glands** (Figure 7.5).

Figure 7.5 The urinary system of a human female

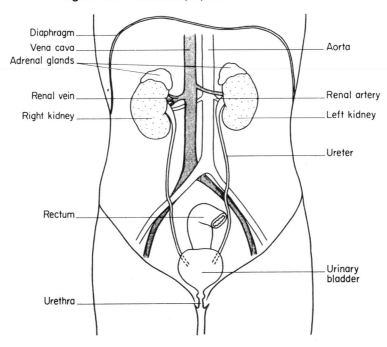

Tubular **ureters** convey urine by peristalsis from each kidney into the urinary **bladder** which is emptied by way of the **urethra** during micturition (urinating). The urethra is controlled by a ring-like sphincter muscle.

Internally, each kidney has two regions, the outer **cortex** and inner **medulla** with pyramids (seen only in humans) connecting with a collecting space called the **renal pelvis** (Figure 7.6).

The nephron

The **nephron** (Figure 7.7) is the basic unit of the kidney, there being over a million in each kidney. The renal artery supplies an **afferent arteriole** which connects with a bunch of porous blood capillaries called a **glomerulus** which projects inside a **Bowman's capsule**. Each Bowman's capsule is the blind end of a kidney tubule which is made up of a first **convoluted tubule**, descending and ascending tubules with a **loop of Henle** in between and a **second convoluted tubule** leading to a **collecting duct**. An **efferent arteriole**, which has a narrower bore than the afferent arteriole, removes blood from the glomerulus to supply a network of blood vessels to the tubule before returning the blood to the renal vein.

Figure 7.6 Vertical section to show the internal structure of a human kidney

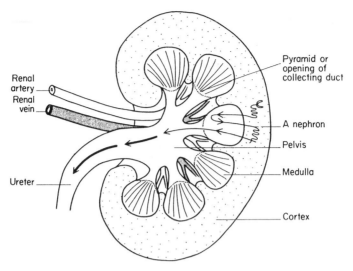

Figure 7.7 The structure of a nephron from a human kidney

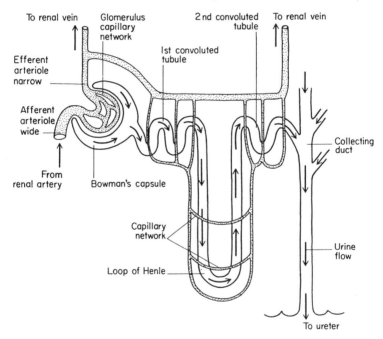

Questions

21 Look at Figure 7.5 and trace the path of water from the blood in the renal artery until it leaves the body in urine.

22 Look at Figures 7.6 and 7.7 and trace the path of water from the afferent arteriole (Figure 7.7) to the ureter (Figure 7.6).

Mechanism of kidney function

Two important processes occur in the nephron, filtration and reabsorption.

Filtration

The capillary walls of the glomerulus have pores allowing filtration. Blood pressure provides the force for blood plasma to be filtered through the thin, porous epithelium of the glomerulus and the neighbouring Bowman's capsule wall to produce the **glomerular filtrate** which enters the kidney tubule. The glomerular filtrate is blood plasma without blood cells and large-molecule proteins. Small molecule proteins may be present. All the body's blood passes through the kidneys 350 times a day at the rate of 1.2 litres per minute. This produces 125 cm^3 of glomerular filtrate per minute. When a kidney is defective, proteins pass into the glomerular filtrate and appear in the urine.

Reabsorption

Cells lining the kidney tubule have numerous energy-producing mitochondria, and take up glucose, amino acids and variable amounts of mineral ions by active transport using energy supplied by ATP. Almost 99% of the water in the glomerular filtrate enters the convoluted tubules and collecting ducts by osmosis.

Only 1% of the original glomerular filtrate becomes urine. Between 170 and 200 litres of glomerular filtrate pass into the tubule to be eliminated as 1.5–2.0 litres of urine daily. Urine in humans consists of 96% water, 2% urea (carbamide) 0.03% uric acid and 1.4% mineral ions (sodium, potassium, chloride and phosphate). Table 7.2 compares the composition of glomerular filtrate and urine.

Table 7.2 Comparison of glomerular filtrate and urine percentage composition

Component	Glomerular filtrate (%)	Urine (%)	Approx. change in urine over glomerular filtrate
Water	98	96	Variable
Carbamide (urea)	0.03	2	×65
Glucose	0.10	nil	–
Large molecule proteins[*]	nil	nil	–
Sodium chloride:			
sodium ion	0.32	0.32	×1
chloride ion	0.30	0.60	×2
Potassium	0.02	0.15	×8
sulphate ion	0.003	0.18	×60

[*]Small molecule proteins in glomerular filtrate are rapidly reabsorbed.

Questions

23 How many litres of blood pass through the kidney in an hour and how much glomerular filtrate is produced?

24 Which vital substances form part of the glomerular filtrate?

26 Look at Table 7.2 and answer these questions:
   (a) Which component of the blood does not enter the filtrate?
   (b) Which is the largest component of urine?
   (c) Which component becomes the most concentrated in the urine?
   (d) Which component is removed from the filtrate?

Homeostatic function of the kidney

The kidney has three main homeostatic functions in regulating the pH, mineral ion and water composition of the blood fluid.

pH regulation

The kidney tubule exchanges certain hydrogen or hydroxyl ions to maintain the blood plasma at a norm of pH 7.4. The exchanged ions in the urine cause it to be either acid at pH 5 or alkaline at pH 8. Urine pH can be shown by dipping a BDH Universal indicator paper onto a urine sample and matching the colour formed with the pH colour chart.

Mineral ion regulation

Sodium ions are controlled in a complex homeostatic process involving the adrenal gland hormone **aldosterone** which increases sodium ion absorption in the convoluted tubule.

Water balance regulation

The adult human body contains 40 litres of water which forms 60% of the body weight. This water is divided up as follows:
- inside the cells: 63% – 25 litres;
- tissue fluid: 30% – 12 litres;
- blood plasma: 7% – 3 litres.

The water concentration of the tissue fluid is maintained within the range 97–99% (or mean 98%), and in blood at 92%, by making the daily water intake equal the daily water loss.

Table 7.3 Daily water loss and gain in adult male humans (in good health, at rest, in a temperate climate)

	Loss (cm³)			Gain (cm³)	
1	By skin, sweat	500 (20%)	1	In drink	1500 (60%)
2	By lungs, exhalation	400 (16%)	2	In food	700 (28%)
3	In faeces	100 (4%)	3	Metabolic water, a product of respiration*	300 (12%)
4	In urine	1500 (60%)			
	Total output	= 2500 cm³		= Total input	= 2500 cm³

*Glucose + oxygen = carbon dioxide + metabolic water.
Lactating or breastfeeding mothers, and pregnant women require an extra daily intake of water amounting to over 750 cm³, for the purpose of forming milk or body fluids in the developing embryo or fetus.

Table 7.3 shows the daily water loss and gain, in a healthy male, at rest in a temperate climate. The **metabolic water** is the product of respiration of lipids, proteins and carbohydrates, 100 g of each producing 100 cm^3, 40 cm^3 and 60 cm^3 of water, respectively.

Homeostatic mechanisms operate when insufficient or excessive water intake occurs.

Insufficient water intake

- A shortage of water causes the water concentration to fall *below* the norm of 98%; this in turn causes the blood osmotic pressure to *rise*.
- Sensory receptors in the brain (hypothalamus) detect the change in osmotic pressure and signal the pituitary gland (Section 8.6) by negative feedback.
- The pituitary gland sets off the corrective mechanism by secreting **antidiuretic hormone** (ADH) or **vasopressin**. This causes the kidney tubule to absorb more water. The urine becomes **scanty** and **concentrated**. Thirst develops and water is taken into the blood from the intestine, diluting the blood and returning the water concentration to its norm of 98%.

Excessive water intake

- Excessive water intake *increases* the tissue fluid water concentration above the norm of 98%. This causes the blood osmotic pressure to *fall*.
- Sensory receptors in the brain detect the osmotic pressure change and signal the pituitary gland by negative feedback.
- The gland sets off a corrective mechanism which *stops* the gland secreting ADH or vasopressin. No more water is absorbed in the tubules and the urine becomes **copious** and **dilute**. The tissue fluid concentration *decreases* and returns to the norm of 98% and blood osmotic pressure *rises* to its norm.

Water diabetes is due to defective water control through positive feedback and lack of ADH or vasopressin secretion. This causes the production of large volumes of dilute urine and of thirst, leading to body dehydration. Oedema or dropsy is the accumulation of water in body tissues.

Questions

27 List the ways in which water is lost from the body, starting with the largest cause of water loss.
28 How much of the water gained by the body comes from the substances it takes in?
29 (a) How and where is metabolic water formed? (see also Section 6.4).
    (b) Which substance produces the most metabolic water?
30 List the stages that take place in the body to restore the concentration of the water in the blood when (a) insufficient water is taken in, (b) too much water is taken in.

Kidney failure

Kidney failure can be due to various causes and can be relieved by **dialysis** or cured by **kidney transplantation**.

Transplantation

Kidney transplantation involves the insertion of a **donor kidney** which is connected to the patient's blood and urinary systems. This can provide a cure, provided the patient's blood group and immune system (see Section 5.10) do not *reject* the transplant organ. The kidney may be supplied by someone who has died or donated by a close relative. If the kidney of a close relative is used, the tissues will be similar to the patient's tissue and this will reduce the chances of the transplant being rejected.

Dialysis

There are two methods of dialysis:

Arterial blood dialysis (haemodialysis)

This method uses an artificial kidney machine (see Figure 7.8).

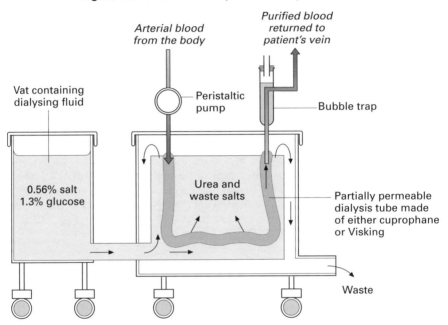

Figure 7.8 Artificial kidney or renal dialysis machine

This machine consists of a tank of sterile dialysis fluid surrounding a long length of dialysis tubing. Arterial blood from the person with kidney failure is pumped slowly through the dialysis tubing. Urea and other waste materials diffuse from the blood through the dialysis membrane into the dialysis fluid. Blood is withdrawn from and returned to the patient by a shunt fixed in a forearm vein.

A kidney machine is used by a patient two or three times a week. Removing the wastes with a kidney machine takes about five hours.

Body cavity (peritoneal) dialysis

Peritoneal dialysis is used in kidney failure, when a sterile dialysing fluid is pumped by a portable machine into the abdominal peritoneal cavity which is the space between the gut and the abdomen (see Figure 7.9). The gut capillary wall acts as a dialysing membrane. Urea and other waste materials diffuse into the dialysing fluid, which is drawn by another pump leading the fluid to waste.

Figure 7.9 Peritoneal dialysis

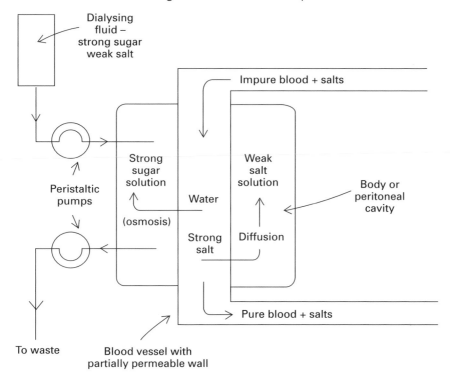

Dialysing fluid

Dialysing fluid is used at body temperature, and consists of a solution of glucose, amino acids and mineral ions. The solution is sterile – free from harmful microorganisms.

Because the dialysing solution does not contain urea (carbamide), the urea present in the blood plasma diffuses down the diffusion gradient into the dialysing fluid. Concentrations of glucose, amino acids and mineral ions are similar to that of normal blood plasma, in order to prevent loss of these nutrients from the blood.

Higher concentrations of glucose, amino acids and mineral ions are used in certain circumstances to serve as a nutrient supplement which enters the blood of a sick patient. Osmosis of water from the blood plasma will occur if a strong glucose solution is used in the dialysing fluid; the higher concentration of glucose will produce a solution with a low water potential. This in turn will remove water from the blood plasma as a high water potential.

If the dialysing fluid was pure water, without dissolved solutes, diffusion of all urea, amino acids, glucose and mineral ions would occur from the blood plasma. Osmosis of water into the blood plasma would also occur.

Questions

31 How are urea and wastes removed from the blood without the loss of nutrients and salts?

32 Why must the dialysing fluid be sterile?

33 How would the amounts of the components of the blood change if pure water was used instead of dialysing fluid?

34 How does a patient's life style alter when changing from using a kidney machine to having a transplant?

35 What are the risks involved in receiving a transplanted kidney, and how do they compare with the risk of not having a transplant?

7.7 The human skin as a homeostatic organ

Structure

Figure 7.10 A photomicrograph of mammalian skin, seen in vertical section, showing the main structures (Griffin Biological Laboratories)

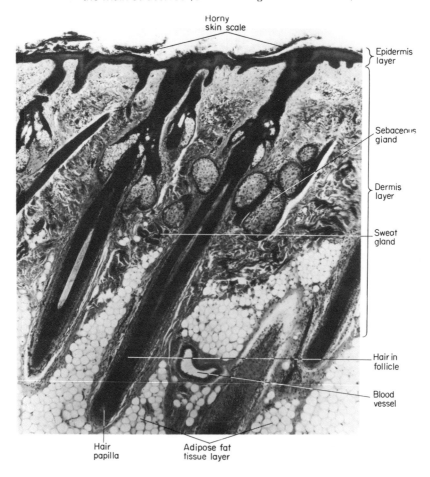

Skin (Figure 7.10) is one of the largest organs in the human body, having a surface area of between 1.5–2.0 m² in an adult man and providing a continuous covering to the body.

(see page 25 for the relationship between the size and surface area of a body). Mammalian skin is composed of two main layers, the **epidermis** and **dermis** (Figure 7.11).

Figure 7.11   A vertical section through human skin, showing its internal structure

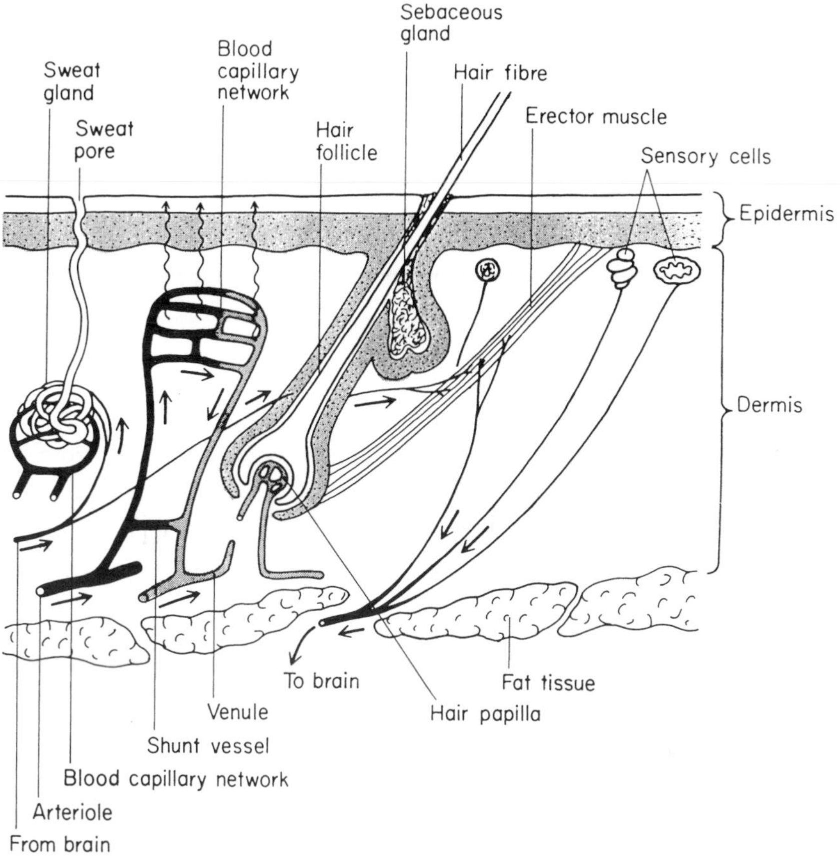

Epidermis

This is the surface layer without blood or lymph vessels, consisting of three epithelial cell layers:
- The **Malpighian layer** is composed of actively dividing cells with granules of melanin pigment. It forms the inner boundary with the dermis.
- The middle layers are called the **prickle, granular** and **clear** layers.
- The **horny layer** is composed of flaking scales of keratin, a protein substance which is also the component of hair and fingernails.

Dermis

This is the deep, inner skin layer with blood and lymph vessels, nerve endings, muscles, sweat glands and hair follicles. **Hair fibres** are in **hair follicles** with a **hair papilla** providing growth and pigment materials from a small capillary network. The **erector muscle** is attached between the papilla and epidermis. The muscle's contraction causes the hair fibre to be pulled upright and 'goose pimples' appear on the skin surface.

Sebaceous glands are exocrine glands which produce a secretion of **sebum,** a mixture of lipid and waxy substances, having skin-softening, lubricating, waterproofing and bactericidal properties.

Sweat glands are blind-ended exocrine glands. Each one has a long duct opening onto the skin surface by a pore. A blood capillary network supplies the components of sweat: 99% water, 0.1–0.4% sodium chloride, 0.015% urea (carbamide) and small amounts of vitamins B1 and C and lactic acid.

Blood capillaries are found beneath the epidermis, linked to an arteriole and a venule. An **arterial shunt vessel** may bypass the capillary network in the skin of ears, nose and finger-tips.

Subcutaneous fat deposits of lipid **adipose tissue** also insulate the body. Brown fat layers are seen in young newborn and hibernating mammals. These are heat generating tissues.

Sensory cells receive sensations of heat, cold, touch and pain. A network of nerve fibres surrounds the hair fibre, making 'whiskers' sensitive to touch. A nerve fibre supplies the erector muscles.

Questions

36 How is the structure of the epidermis different from that of the dermis?
37 How are the position and opening of the sebaceous gland different from those of the sweat gland?
38 How is the secretion of the sebaceous gland different from that of the sweat gland?

Skin functions

Skin has several functions:

Protection

Skin acts as a barrier between the internal and external environment in body defence described in Section 8.4.

Sensation

Sensory function makes the body aware of changes in the external environment through touch, pain, pressure, heat and cold sensory cells.

Storage

Skin acts as a storage centre for lipids and water, and a means of synthesis of vitamin D by the action of ultraviolet light (sunlight) on certain skin components.

Absorption

The horny layer of skin is generally non-permeable, but certain substances, oxygen, nitrogen and carbon dioxide, can diffuse into the epidermis in small amounts, and small amounts of water can evaporate through the skin at the rate of 120 to 200 cm^3 a day in an adult man.

Certain harmful organophosphorus insecticides are readily absorbed through the skin.

Excretion

The concentration of urea (carbamide) in sweat is 130 times less than its concentration in urine; sweating as a means of excretion is only a secondary function to temperature regulation (see Section 7.8).

Camouflage

The skin of many animals provides a means of protection through the different skin colour pigments, which may blend with the background, as in the brown fur of a rabbit.

Questions

39 What are the sensory and storage functions of the skin?

40 How does the skin offer protection from (a) microorganisms and (b) predators?

7.8 Temperature regulation

Living organisms can survive within a temperature range from 0 to 45°C. Above 45°C the enzymes are destroyed, proteins are denatured, cell surface membranes break down and cells suffer lack of oxygen. Below 0°C the cells burst due to the formation of needle-like ice crystals within and between the cells.

A constant body temperature of 37°C is favoured by most animals since this is the **optimum temperature** for efficient enzyme action within and outside cells.

Endotherms

Endotherms, also called **homoiotherms**, include birds and mammals, and are able to generate their heat internally and maintain their body temperature within a narrow range of 2°C (37–39°C) by homeostatic mechanisms and by good insulation.

Ectotherms

Ectotherms, also called **poikilotherms**, include all animals except birds and mammals, who are unable to generate sufficient body heat, but must gain it from the external environment. Their body temperature fluctuates and is usually slightly higher than the prevailing external temperature. They are unable to control body temperature by homeostatic mechanisms but gain and lose heat rapidly, having little insulation.

Questions

41 What happens inside the body of a living organism if the temperature rises above 45°C or falls below 0°C?

42 What are the differences between an exotherm and an endotherm?

Temperature regulation in ectotherms

Ectotherms have a *low* food intake. For example, a reptile needs only 10% of the food eaten by an equal-sized mammal. They have low rates of metabolism and produce very little heat by respiration. Since there is no insulation beneath the thin skin, heat is readily lost to and gained from the external environment.

Aquatic ectotherms, such as fish, gain and lose heat mainly by conduction between an external environment with a fairly steady temperature. **Terrestrial ectotherms**, such as insects and reptiles, are subject to rapidly changing external air temperatures between day and night. Heat loss and gain will be mainly by radiation, convection and conduction with evaporation playing a lesser role in order to conserve water.

Body temperature control is achieved by the animal's behaviour in moving *towards* or *away from* the heat source. This involves sunbathing, or lying on a warm rock or sand to gain heat. Heat loss is achieved by seeking shade, cooling winds, burrowing or hiding under stones.

Plants are also ectotherms with low metabolic rates, unable to produce much heat by respiration. They gain heat from the external environment as seen in heated greenhouse plant culture.

Questions

43 Why does a reptile not generate as much heat as a mammal?
44 Why does heat pass through the skin of ectotherms quickly?
45 How does a lizard raise its body temperature in the morning and lower it in the hot afternoon?

Temperature regulation in endotherms

Endotherms have a *high* level of food intake to provide heat and body energy by a *high* rate of metabolism. Half the energy produced in cell respiration is heat energy. Body heat comes mainly from cell respiration (50%), muscle activity (38%) and liver functions (12%).

Birds and mammals have good insulation from hair, feathers and subcutaneous fat. Under normal conditions the body temperature will be maintained at 37°C by self-regulating homeostatic mechanisms.

Endotherms gain the following advantages in having a steady body temperature:
• steady rate of chemical and enzyme activity;
• can be active at any time of day, compared to ectotherms who are inactive at night;
• can inhabit cold, polar climates, but are not so numerous in hot, desert climates where ectotherms thrive.

Heat regulation in overheating

Endotherms can control their body temperatures by several physiological and behavioural methods.

Physiological methods

• **Less insulation:** achieved by having *less* subcutaneous fat in warm climates. The hair or feathers lie flat on the skin surface by relaxation of the erector muscle; this *reduces* the layer of air insulating the body, allowing heat loss.
• **Vasodilation:** the expansion of the arterioles, providing *more* blood to the skin capillaries. This encourages heat loss by **radiation**.
• **Evaporation:** Evaporation of 1 cm^3 of water or sweat requires 2.4 kJ of heat. Sweat glands produce between 600 cm^3 of sweat in a body at rest and up to 8000 cm^3 of sweat in a body working in the tropics.

- **Panting:** encourages evaporation of water from the tongue and lung surfaces. In normal breathing 400 cm^3 of water evaporates a day which is equal to a heat loss of 950 kJ.
- **Metabolism and muscle activity:** these are slowed down resulting in a *lower* food intake and *less* heat from the muscle and liver tissues. Cold food and drink can also help to reduce overheating.

Behavioural methods

Overheated animals will seek shade, bathe or wallow in water. Humans can reduce the amount of clothing.

Questions

46 How can the level of insulation be reduced?
47 Why does vasodilation increase heat loss?
48 How much heat is lost by the evaporation of (a) 600 cm^3, (b) 8000 cm^3 of sweat?
49 (a) Why does resting cool an animal down?
   (b) How else could an animal cool itself down?

Heat regulation in over-cooling

Physiological methods

- **More insulation:** achieved by laying down *more* subcutaneous fat as blubber in whales and animals about to overwinter. Hair and feathers are raised by the contraction of erector muscles. This *increases* the insulating layer of air and retains *more* body heat.
- **Vasoconstriction:** the narrowing of skin arterioles, reducing blood flow to skin capillaries. The blood may bypass the skin surface capillaries by the arterial shunt in the skin of body extremities such as the nose and ears.
- **Evaporation:** evaporation by sweating is stopped by vasoconstriction of blood vessels to sweat glands.
- **Metabolism and muscle activity:** these are increased. There is an increased intake of food, leading to a *higher* metabolic rate in the liver and body tissues. Muscles produce heat, first by contracting and then by involuntary 'shivering'.

Behavioural methods

Over-cooled animals will huddle together, curl up into a ball, or retire to burrows and nests to keep warm in a static 'fug'. Migration and hibernation are also means of avoiding unfavourable changes of climate. Humans adjust to over-cooling by wearing more clothing or doing physical exercise.

Questions

50 How can the level of insulation be increased?
51 How does vasoconstriction reduce heat loss?
52 How can metabolism and muscles be used to prevent over-cooling?
53 How can an animal's behaviour prevent it from over-cooling?

The heat balance

The mammalian body has a heat balance which is composed of heat gains which are almost equal to the daily heat losses.

heat input gains = heat output losses

If a human being has a daily energy requirement of 10 MJ (see Section 4.3) the same person will have an energy input of 10 MJ and this will be converted into an equal daily heat output of 10 MJ.

energy intake 10 MJ = heat loss 10 MJ

The heat gains and heat losses are summarised in Table 7.4.

Table 7.4 Heat balance in mammals

Heat gains – **inputs**	=	Heat losses – **outputs**
1 Aerobic respiration		1 Panting or expiration
2 Muscle activity		2 Sweat evaporation
3 Liver metabolism		3 Urine, faeces
4 Heat transfer *from* environment		4 Heat transfer *to* environment
5 Hot food and drink ingestion		5 Cold food and drink ingestion

Body size in heat regulation

Small animals or body parts will have a *high* surface area/volume ratio, or have *more* skin compared to body volume. Consequently there will be *more* heat loss from small, exposed animals or exposed body parts such as ears and noses. Small endotherm animals such as the shrew have a *high* food intake and a *high* metabolic rate. A shrew consumes its own weight of food in 24 hours. In cold, unfavourable conditions, the shrew is unable to maintain the high metabolic rate and has a need to hibernate. Its lifespan is short, being 12–18 months.

Large endotherms, such as the elephant, have a *low* surface area/volume ratio or have *less* skin to body volume. Consequently they have *less* heat loss per unit volume. The nose and ears are large and are used to radiate heat from the body, whereas the shrew has smaller ears and nose close to the body. Elephants have a moderate food intake in comparison to that of the shrew. This supplies a *lower* metabolic rate. Elephants have ample lipid and energy reserves to face unfavourable conditions without hibernating. An elephant's life span is 50–80 years.

Newborn mammals and birds will experience heat loss from their high skin surface area/volume ratio and will need to gain heat from the external environment in the nursling stage. The brown fat present in some newborn mammals including humans, is believed to generate heat by metabolism in a manner resembling a heating blanket.

Experiment to compare heat loss from different sized surfaces

1 Set up 50 cm³ and 25 cm³ round-bottomed flasks on stands. Each flask represents an animal's body.
2 Fill each flask with boiling water and fit a 0–100°C thermometer and rubber stopper as shown in Figure 7.12.

Figure 7.12

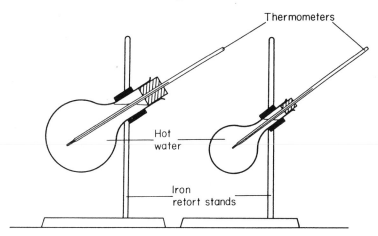

3 Record the temperature of water in each flask every 5 minutes and plot a graph of how the temperature falls in each flask. The smaller flask has a greater surface area/volume ratio and shows a more rapid fall.

Question

54 (a) How do the surface area/volume ratios of the shrew and the elephant compare?
  (b) How does the surface area/volume ratio affect the life of the (i) shrew, (ii) elephant?
  (c) Do the flasks in the experiment represent ectotherm or endotherm animals? Explain your answer.

Homeostatic temperature regulation in endotherm animals

The automatic self-regulating mechanism of heat control in birds and mammals operates through the nervous system and by hormones from the pituitary, thyroid and adrenal glands (see Section 8.6). Sensory detectors in the skin and inside the body relay information by **negative feedback** to the central temperature control located in the brain, called the **hypothalamus** (see Section 8.5).

The hypothalamus is similar to a thermostat and is itself sensitive to changes in the temperature of the blood which bathes it. It is responsible for setting off corrective mechanisms when the blood temperature rises or falls.

- When the blood temperature *rises*, the **adrenal gland** secretes various hormones, resulting in **vasodilation**, *increased* skin blood supply and **sweating**. The muscles relax and hair lies flat, resulting in heat loss and body temperature being restored to the norm of 37°C.
- When the blood temperature *falls*, the **thyroid gland** secretes **thyroxine** hormone. This *increases* the metabolic rate, muscles contract and twitch, hair is raised, **vasoconstriction** occurs with *decreased* skin blood supply and sweating stops.

Positive feedback occurs in defective homeostatic control. The blood temperature will continue to fall, with lowered metabolic rate, and death by **hypothermia** occurs at body temperatures below 26°C. Hypothermia occurs in elderly people and young babies in unheated rooms in winter-time. Hypothermia is also induced intentionally for open heart surgery.

Continuous exposure of the human body to low temperatures through immersion in sea water causes hypothermia through heat loss by conduction into the surrounding water. Similar hypothermia occurs when a body clothed in wet clothing is exposed to winds as on moors or mountains. The heat is lost through the cooling effect of evaporation, where every cm³ of water evaporated removes 2.4 kJ of heat from the body (see page 182).

A rising blood temperature may occur when the body is unable to correct a temperature over 41°C. Increased metabolic rates lead to higher body temperatures. Death by **hyperthermia** or **heat stroke** occurs at body temperatures above 43°C.

Questions

55 What changes take place in the skin when the temperature (a) rises, (b) falls?
56 What happens to the body temperature if positive feedback occurs instead of negative feedback?
57 What are (a) the differences and (b) the similarities between hypothermia and hyperthermia?

7.9 The embryonic environment

Fish, amphibia, reptiles and birds

The immature forms or embryos of most fish and amphibia develop *externally* in freshwater or sea water, independently of the female parent. The embryos of reptiles and birds develop *internally* within a watery environment provided by the egg white. Most reptile embryos develop independently of their parents while birds depend on their parents for incubation.

Figure 7.13   The human embryo inside the uterus

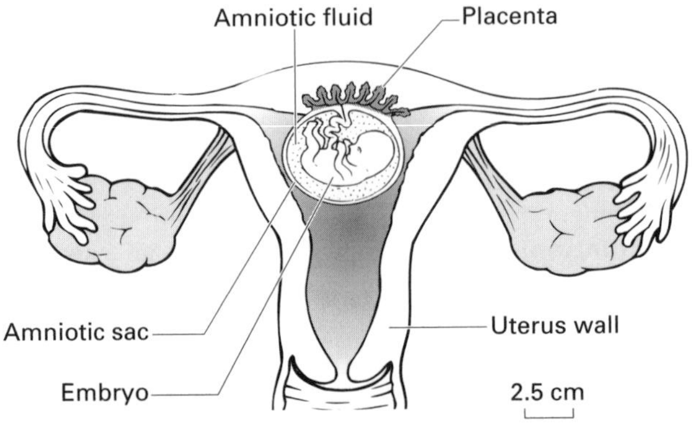

The human embryonic environment

A human embryo develops *internally* within the female parent's **uterus** (Figure 7.13). The embryo is surrounded by **amniotic fluid** and is continually dependent on the female parent for life support through the **placenta**. After 8 weeks the embryo is called a **fetus**.

The average composition of amniotic fluid is shown in Table 7.5. It is very similar to the composition of blood plasma (see Table 5.4). This composition varies, particularly after 4 months when the fetus excretes urine into the amniotic fluid.

Table 7.5   Average composition of amniotic fluid

Amniotic fluid composition	Approximate average value (%)
Solvent:	
Water	95
Solutes:	
Protein	4
Sugars	0.1
Lipids	0.03
Urea (carbamide)	0.03
Mineral ions:	
Sodium chloride	0.6
pH	7.15
Temperature	37–38°C
Cells:	
Fetus/Embryo skin cells	Variable

A sample of the amniotic fluid and its cells can be removed by a technique called **amniocentesis** (see page 321).

Functions of the amniotic fluid

The amniotic fluid provides an external environment for the fetus and six life protection features. They are:
- **Buoyancy:** The fluid supports the embryo body as it grows.
- **Mechanical protection:** The fluid protects the embryo from bumps and knocks.
- **Steady temperature:** The temperature is maintained at a steady 37/38°C.
- **Drinking water:** The fluid provides a source of water for the embryo.
- **Urine reservoir:** The urine produced by the fetus is collected in the amnion.
- **Lubrication:** The amniotic fluid provides lubrication for the fetus which helps it to pass along the birth canal during the birth process.

Functions of the placenta

The main function of the placenta is homeostatically to regulate the internal environment of the embryo/fetus. The placenta blood circulation is in close contact with the mother's blood, the mother providing life-support for the embryo/fetus.

The placenta provides:
- nutrients, sugars, amino acids, lipids and mineral ions for cell nutrition;
- oxygen for cell respiration.

The placenta removes:
- waste carbon dioxide and urea.

Questions

58 How would the development of the embryo/fetus be affected if amniotic fluid was not present?

59 How is the placenta's role in providing life-support to the embryo different from that of the amniotic fluid?

Summary

- Homeostasis provides steady conditions for life processes to take place. (▶ 163). **Hyper** = above 'normal'; **Hypo** = below 'normal', e.g. hypothermia and hyperthermia. (See also Figure 7.14 overpage.)
- The liver takes part in regulating conditions in the blood and the storage and processing of food. (▶ 165)
- The amount of blood sugar is regulated by the action of adrenalin, glucagon and insulin. (▶ 167)
- Excretion is the process of removing the waste products of metabolism and occurs in both animals and plants. (▶ 168)
- Osmoregulation is the process of regulating the amount of water in the body of an animal or plant. (▶ 169)
- The processes of excretion and osmoregulation are linked together in the life processes of all animals. (▶ 169)
- The kidney removes waste products and regulates the water content of the body without losing vital substances from the blood. (▶ 170)
- Kidney failure can be treated by transplantation or the use of a kidney machine. (▶ 175).
- The skin plays an important part in the regulation of body temperature. (▶ 178)
- Animals called **ectotherms** have fluctuating body temperatures. Animals called **endotherms** (e.g. mammals and birds) have a constant body temperature. (▶ 181)
- Body size plays an important role in heat loss. (▶ 184)
- Endotherms control their body temperature in a variety of ways including the path of the blood through the skin, sweat secretion, nerve and hormone action, the rate of metabolism, muscle action and the raising or lowering of fur and feathers. (▶ 185)
- The amniotic fluid and the placenta provide conditions for the safe growth and development of the embryo and fetus. (▶ 187)

Figure 7.14   Summary of homeostasis

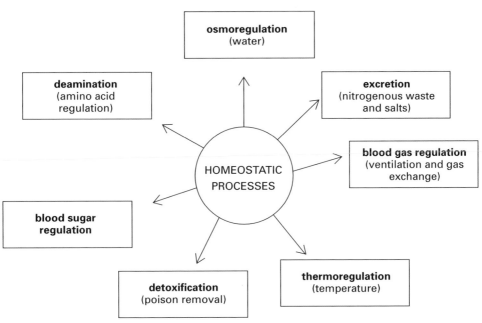

THEME III

Life Protection Processes

8 Irritability

The delicate genetic material in a living organism must be protected against harmful effects which could damage or destroy it. To prevent damage and destruction, living organisms show **irritability**. This means that they are able to **sense** changes to their **internal** and **external** surroundings and are able to **respond** to them. For example, changes which occur in the external surroundings of an organism such as increasing heat or cold cannot be altered by the organism so the organism reacts or responds mainly by moving *away* from the harmful conditions and *towards* more favourable ones.

In addition to irritability there is a second life-protecting process. This is **homeostasis** (see Chapter 7). It prevents damage to life processes despite extreme changes in the external environment.

8.1 The principles of irritability

Irritability is a characteristic of all living organisms. It has *three* main components: **sensitivity**, **coordination** and **responsivity**. The relationship between these components is shown below:

$$\underbrace{\text{stimulus} \rightarrow \begin{array}{c}\text{sensory}\\\text{receptors}\end{array}}_{\substack{\text{SENSITIVITY}\\\text{stimulus detection}}} \xrightarrow{\begin{array}{c}\text{CHEMICAL OR NERVOUS}\\\hline\text{COORDINATION}\end{array}} \underbrace{\text{effectors} \rightarrow \text{response}}_{\substack{\text{RESPONSIVITY}\\\text{stimulus reaction}}}$$

Sensitivity

Stimulus

A **stimulus** is any change in the organism's *external* or *internal* environment. It tends to disturb the steady state or homeostasis which all organisms need for maintaining life. Each environment has physical, chemical and biological components, all of which can provide stimuli:

- **Physical stimuli** include changes in temperature, pressure, light and sound in an environment.
- **Chemical stimuli** include changes in pH (acidity or alkalinity) or the concentrations or amount of water, ions or organic compounds in the body fluids, foods, water, oxygen and odours of an environment.
- **Biological stimuli** are caused by other living organisms such as predators, and harmful toxins from microorganimsms in body fluids, or the sexual attraction or aggression between living organisms, leading to reproduction, survival or death.

Receptor

A receptor is a structure which *detects* a stimulus. The receptors of plants and animals will be described in detail in Sections 8.2 and 8.3.

Coordination

Coordination is the working together of receptors and effectors through the transmission of messages by chemical hormones or nervous action.

Responsivity

A response is the **reaction** to the stimulus, due to the activity of the effectors. This is seen as movement or secretion.

Questions

1 'Without irritability there would not be life.' Is this true? Explain your answer.
2 If one component of irritability is unable to function, how are the other components affected and how does this affect the survival of the organism?
3 Draw up a table with two columns headed *internal stimuli* and *external stimuli* and classify the examples of the three types of stimuli under the two headings.
4 When a harmless beetle is placed on the palm of the hand it runs off back to the ground. Explain its behaviour in terms of the three components of irritability.

8.2 Irritability in flowering plants

Sensitivity

Stimuli

Stimuli affecting flowering plants include:

- **Physical stimuli:** changes in light intensity, gravity, temperature and factors responsible for the seasons, summer and winter.
- **Chemical stimuli:** changes in concentrations of water, mineral ions, carbon dioxide and oxygen.

- **Biological stimuli:** due to shading by other plants, or from harmful disease-causing fungi, viruses and bacteria, or from grazing animals.

Receptors

Receptors which receive and detect stimuli are located in the **apical meristem**. This is a tissue at the tip of a stem or a root of flowering plants. When removed from the plant, the stem or root is unable to detect and respond to stimuli.

It should also be noted that the plant cell surface membrane is sensitive to stimuli but the cell wall is not. The cell membrane forms a protective barrier for the cell contents against the external environment.

Coordination

Coordination in flowering plants is by means of chemical substances called **plant growth substances**. These are made in the apical meristems when the cells in these tissues receive a stimulus. (It is important to note that it is incorrect to call plant growth substances plant hormones because chemically they are not like animal hormones which are composed of or derived mainly from proteins.)

Transmission of plant growth substances is by diffusion between cells, or translocation in the xylem and phloem. Plant growth substances affect cells in their division and vacuole formation. They also affect flowering, fruiting and leaf fall.

Auxins are a group of plant growth substances made naturally by flowering plants. They can be made artificially in the chemical industry, for example **2,4 D** is an auxin type chemical used for rooting cuttings and as a weedkiller.

Response

Growth movements are seen to occur in flowering plants as:
- lengthening of stems and roots;
- bending or tropic movements of stems and roots;
- opening and closing movements of flowers and leaflets.

Tropisms are the bending growth movements of a plant stem and roots in response to an external stimulus from a specific direction. Bending occurs due to *unequal cell elongation* as a result of the effect of unequal amounts of plant growth substances in these areas. A *positive* response results in the plant bending *towards* the external stimulus. A *negative* response results in the plant bending *away* from the external stimulus.

Phototropism

This is the response of a plant to light. Shoots are *positively* phototropic in growing *towards* light. The majority of roots are not affected by this stimulus.

The theory of phototropic response

Experimental investigations into the mechanism of phototopism have been carried out on oat seedling coleoptiles or plumule sheaths. This involved collecting the auxin on agar blocks and hindering the diffusion of this material by means of mica discs or razorblades. The experimental work is summarised in Figure 8.1 and the theory, called the auxin theory, is summarised in Table 8.1.

One proposal made is that light produces an enzyme in the shoot which destroys the auxin. This prevents cell elongation on the light side but not on the dark side.

Figure 8.1 Summary of experimental work and the theory of phototropism

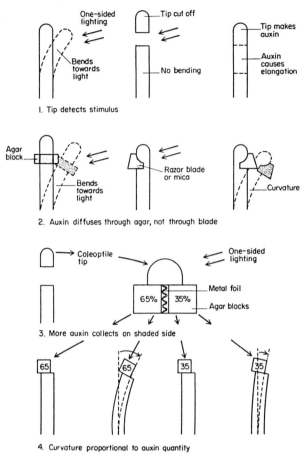

1. Tip detects stimulus

2. Auxin diffuses through agar, not through blade

3. More auxin collects on shaded side

4. Curvature proportional to auxin quantity

Table 8.1 Auxin theory of phototropic response

	Even illumination	Total darkness	Uneven illumination
1	Seedling shoot evenly illuminated on *all* sides	Seedling shoot grown in dark	Seedling shoot receives light on *one side*
2	Auxin distributed *evenly*	Auxin distributed *evenly*	Auxin distributed *unevenly*
3	Auxin produced in *smaller* concentration on *all* sides, or light destroys *same* amount of auxin	Auxin concentration same on *all* sides. *No* light to destroy auxin	Auxin concentration *greater* on *dark* side, or auxin is *destroyed* on light side
4	Auxin causes 'normal' *elongation* on *all* sides	Auxin causes *increased* elongation on *all* sides	Auxin causes *increased* elongation on *dark* side
5	Results in 'normal' upward growth. Shoot straight, sturdy and green	Results in *etiolation*, increased upward growth. Shoot spindly, weak and yellow	Results in *bending towards* light. Shoot bent, sturdy and green

Demonstrating phototropism

Phototropism can be demonstrated by germinating soaked oat grains in three small pots of compost. When the seedlings are 1 cm high:
(a) one pot is covered by a large box and all light is excluded;
(b) the second pot is covered by a box with a side opening, allowing one sided illumination;
(c) the third pot is not covered by any box and is the control.

Observations

After all the seedlings have been kept in these conditions and at 25°C for a few days, the boxes are removed and all the seedlings are examined for curvature, yellowing and tallness. Figure 8.2 shows the results.

Figure 8.2 To demonstrate phototropism in oat seedlings

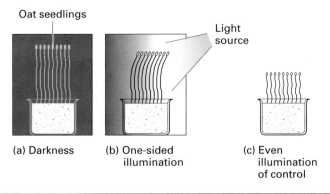

(a) Darkness   (b) One-sided   (c) Even
                      illumination   illumination
                                              of control

Geotropism

This is the response of a plant to gravity. Roots are usually *positively* geotropic and move *towards* the stimulus and shoots are *negatively* geotropic and move *away from* the stimulus.

The theory of geotropic response

Roots without a root tip or apex *do not* respond to gravity. Roots are more sensitive to auxin than shoots. Very small amounts of auxin *stimulate* root cell elongation whilst larger amounts *retard* elongation. This is the opposite effect to the action of auxin on shoots. Auxin collects on the lower side of the primary root in greater concentrations than on the upper side. This causes the upper side cells to elongate, causing curvature *towards* gravity.

Demonstrating geotropism

Geotropism in bean roots can be demonstrated by the following procedure.
1 Soaked beans are germinated by placing them between several thicknesses of rolled paper tissues that are kept continually moist. After 8 days the seedlings will have grown straight radicles.
2 The seedlings are pinned to the cork disc of a klinostat with their radicles projecting horizontally (see Figure 8.3).

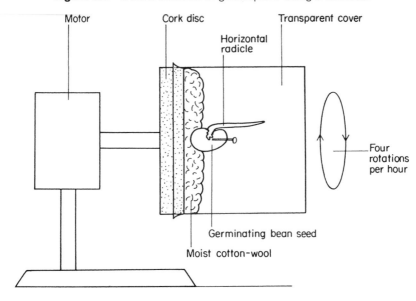

Figure 8.3 Demonstration of geotropism using a klinostat

Motor   Cork disc   Transparent cover

Horizontal radicle

Four rotations per hour

Germinating bean seed

Moist cotton-wool

3 The cork is covered with moist cotton wool and a transparent cover fitted over the cork.
4 The klinostat is allowed to rotate at four times an hour for 2 days then stopped, noting the shape and appearance of the bean radicles after this time.
5 The moisture in the cotton wool lining is renewed and the klinostat is left switched off for a further two days.

Observations

The rotating disc allows gravity to act equally on all sides of the growing bean radicle. The static klinostat allows gravity to act on one side of the bean radicle. The rotated seeds have straight radicles, whilst the static radicles show a curvature down towards gravity.

Questions

5 What factors responsible for the seasons, summer and winter, could act as stimuli to flowering plants?
6 Most people do not think that plants move. State three ways in which a shoot with flowers can show movement.
7 Why are experiments on tropisms performed on the shoots and roots of seedlings?
8 How are the mechanisms which produce phototropism and geotropism (a) similar, (b) different?

8.3 Sensitivity in mammals

Sensitivity is the detection of stimuli by receptors.

Stimuli

The stimuli affecting mammals include:
- **Physical stimuli:** changes in light intensity, temperature, pressure on the skin, tension in the muscles, gravity and sound.
- **Chemical stimuli:** changes in concentration of chemical substances – acids, bases, salts, pH, sugars, toxins and strong-smelling odours.
- **Biological stimuli** due to other mammals and organisms, predators and disease-causing pathogens, bacteria, viruses and fungi.

Receptors

Receptors are cells, or parts of cells, that are specialised to detect a stimulus and convert it into an impulse, or signal, for further transmission. Receptors are **energy transducers**, changing energy from one form into another form.

Mammals have **internal receptors** in muscles, tendons, joints, glands and arteries, and **external receptors** in the skin, eyes, nose, tongue and ears.

The external receptors are:
- **Light** receptors in the **eye**.
- **Chemical** receptors in the **nose** and **tongue**.
- **Sound** receptors and **gravity** detectors in the **ear**.
- **Temperature**, **pain**, **touch** and **pressure** receptors in the **skin** (see Figure 7.11).

8.4 Mammalian sense organs

The skin

The skin is an important sensory organ at the boundary between the mammal's internal and external environments. The receptors in the skin detect pressure changes, warmth and cold and pain.

The tongue

The tongue has sensory receptor cells grouped in **taste buds**. They detect the type and concentration of **sweet**, **sour**, **salt** and **bitter** chemical substances. The number of taste buds on the tongue depends on the species of mammal. A dog has 1700, a human has 9000 and a rabbit has 17 000. The taste buds are located in grooves in the tongue mucus membrane (see Figure 8.4a).

Figure 8.4 Receptor cells of (a) the tongue and (b) the nose

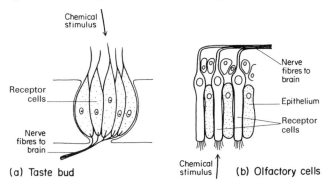

The nose

The nose has sensory receptor cells grouped as an **olfactory patch** in the nasal cavity (see Figures 8.4b). They detect changes in the type and concentration of volatile chemical substances. Most mammals have a keen sense of smell with the exception of humans, monkeys and seals.

Questions

9 Give an example of how each of the following stimuli may produce a response in a mammal – light, temperature, pressure on the skin, strong smelling substance, predator, virus.

10 Name three forms of energy which are changed in receptors to electro-chemical energy for transmission through the nervous system.

11 How may stimulation of the receptor cells that detect a bitter and sour taste prevent a mammal from becoming ill?

The human ear

The human ear, like the ears of other mammals, collects, concentrates and detects sound energy waves. The main functions of the human ear are in sensing sound and changes in the body position in maintaining balance.

Figure 8.5 Structure of the human ear

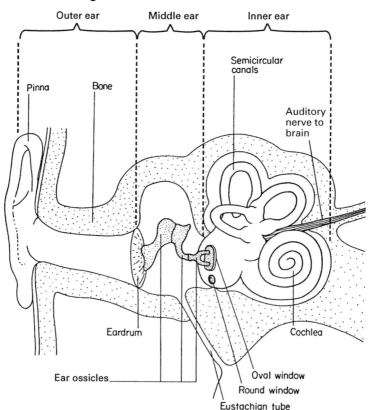

The structure of the ear

The ear structure (see Figure 8.5) may be considered in three parts: **outer, middle** and **inner** ear.

Outer ear

The **pinna**, found only in mammals, collects sound waves and directs them down the external **ear tube** towards the **eardrum** or tympanum. Many mammals, with the exception of humans, can turn the ear pinnae to locate sound sources. Glands similar to skin sebaceous glands produce wax.

Middle ear

The air-filled cavity communicates with the pharynx via the **Eustachian tube**, which allows air pressure on either side of the eardrum to be equalised. Yawning and swallowing help to open the Eustachian tube.

Sound waves cause *vibration* of the eardrum. This vibration is transmitted by three auditory **ossicles** (bones). They are the **hammer** (malleus), **anvil** (incus) and **stirrup** (stapes) bones. The stirrup bone fits into the **oval window**. On reaching the oval window sound energy from the eardrum has a pressure between 17 and 30 times that on the eardrum.

Inner ear

This is the fluid-filled part of the ear. It consists of a membraneous labyrinth which is composed of:
• three semicircular canals;
• three ampullae;
• the utricle and saccule.
• A coiled **cochlea** (Figure 8.6) containing the sensory **organ of Corti** which is concerned with hearing.

Figure 8.6 Structure of the cochlea

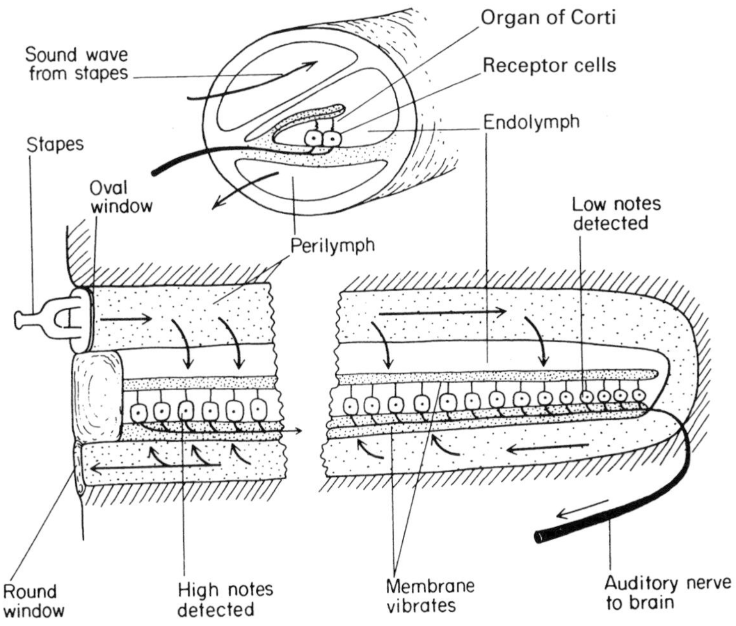

The fluid inside the inner tubes of the ear is called **endolymph**. The outer fluid surrounding the inner tubes is called **perilymph**.

Sound pressure enters the inner ear by the oval window and, after stimulating the different receptor cells, emerges from the inner ear by the **round window**.

Hearing

Sound is a form of energy. **Vocal cords** in the larynx are made to vibrate by transferring energy to them from the moving air and the resulting sounds form the human voice. Sound is conducted through air, water and solids. The speed of sound in iron is 15 times greater than in air, whilst in water it is five times greater than in air. Sound will therefore reach the human ear by air, bone or water conduction. We hear our own voices partly by air and bone conduction, and find that our recorded voices sound different. This is because the recording detects air-conducted sounds only.

Sound properties

The frequency of sound is measured in **Hertz** (Hz). Human ears detect sound frequencies from 20–20 000 Hz. Human conversation takes place over the 250–6000 Hz frequency range. Dogs are able to detect frequencies up to 40 000 Hz.

The **amplitude** of a certain sound at a particular frequency is an indication of its loudness or quietness. The **pitch** or quality of sound is determined by the brain region interpreting amplitude and frequency. The pitch of different sounds is comparable with the range of colours in a colour spectrum or rainbow.

Balance

In the ampullae of the semicircular canals and in the saccule and utricle are receptor cells which detect movements of the head and any changes in its position. The brain uses the information provided by these sense organs together with sense organs in muscles, tendons and joints to keep the body balanced.

Ear protection and care

- Loud intensity noise (e.g. at a disco or nightclub) can lead to hearing impairment through damage of the organ of Corti in the inner ear.
- Ears have few natural protective devices to control sound reaching the inner ear, apart from small muscles attached to the ear ossicles.
- Muffs and ear-plugs provide valuable protection in noisy workshops and when using noisy machines.

Questions

12 Using Figures 8.5 and 8.6, list the parts of the ear that the energy in a sound waves passes through from outside the ear to the receptor cells in the cochlea.
13 If the Eustachian tubes become blocked due to infection how may hearing be affected? Explain your answer.
14 How do ear muffs protect the ears?

The eye

The human eye collects, concentrates and detects light energy waves.

Structure of the eye

Figure 8.7 Vertical section through a human eye showing internal structure

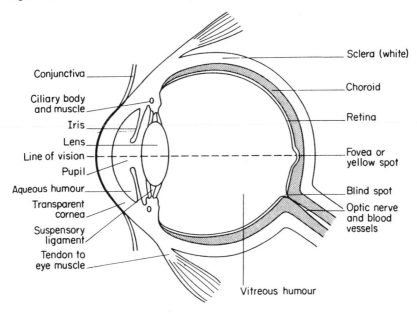

Surrounding structures

The eye **orbits** are cavities in the skull housing the eyeballs. Each eyeball is connected to the orbit by six muscles. These muscles allow different eye movements. The **eyelids** are skin folds which protect the front of the eye and distribute the tear fluid. The **conjunctiva**, the innermost lining of the eyelid, completely covers the front of the eyeball.

    Tear glands secrete the tear fluid which either evaporates from the eye surface or drains into the nasal cavity via a **tear duct** after cleaning and lubricating the eye surface.

The eyeball

The eyeball is almost spherical and has a wall composed of three coats:
- **Outer coat:** an almost opaque **sclera** and a transparent **cornea** connected in a continuous ring.
- **Middle coat:** a **choroid** layer which is well supplied with blood vessels providing nutrients and oxygen. An eye deprived of oxygen will be irreparably damaged. A black or dark-blue pigment makes the choroid layer totally opaque and light-proof. The circular **iris**, and **ciliary muscle** and **body** form the front portion of the middle coat.
- **Inner coat:** the **retina** consists of about 140 million receptor cells called **rods** and **cones** (see Figure 8.8). There are 13 times more rods (130 million) than cones (10 million) in the human retina.

The transparent and light-refracting parts

Light rays are bent or **refracted** as the light passes from the air through the solid and liquid parts of the eye.
- **Cornea:** This solid, transparent layer is in front of the liquid **aqueous humour**, a substance similar to tissue fluid, providing nutrients to the cornea which exchanges gases with the surrounding air by diffusion. The greatest amount of light bending or refraction occurs at the air–cornea surface due to the cornea's curved shape.

Figure 8.8 Rods and cones in the human retina

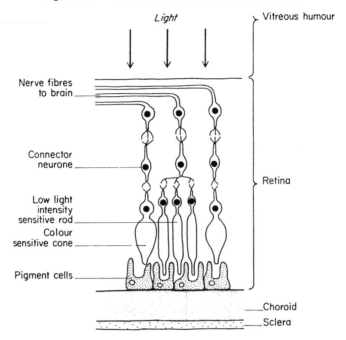

- **Lens:** This is a clear, biconvex disc held in position by **suspensory ligaments**, connected to the **ciliary body** and **muscles**. The eye lens, in contrast to a glass lens, is able to change its shape due to its natural elasticity.
- **Vitreous humour:** A clear, jelly-like material, the vitreous humour completely fills the eyeball behind the lens and ciliary muscles.

The spherical shape of the eye is maintained by constant pressure of its fluid components. Excess fluid drains from the eye interior via a small drainage canal.

The optic nerve

Nerves connect the retina receptor cells, which join together at one position at the back of the eye to form the **optic nerve** which connects with the region of the **brain** that interprets sight. The optic nerve leaves the eyeball at a point called the **blind spot**, where there are no sensory cells.

Focusing the image

Pathway of light rays

Light rays enter the cornea and are refracted at the air–cornea surface. This causes the light rays to **converge** and finally meet at a point on the retina. The amount of light that enters the eye through the pupil is controlled automatically by a **reflex** (see Section 8.5) which contracts and relaxes the iris sphincter muscles. In bright light the sphincter muscles *contract* and produce a *small* pupil. In dim light they *relax* and produce a *large* pupil.

After slight refraction in the aqueous humour, the light rays are refracted by the lens. The converging light rays pass through the vitreous humour to focus on the retina. The most sensitive part of the retina is called the **yellow spot** or **fovea**. It only contains cones.

Of the light entering the eye, 70% of the refraction occurs at the air–cornea surface and 30% in the lens.

Accommodation

Without a lens the refracted light rays from the cornea, reaching the eye from distant objects, would come to a focus point *behind* the retina. This image would be out of focus and be interpreted by the brain as a blurred picture. The human eye lens is able to focus the refracted light rays from the cornea and produce a clear, upside-down or **inverted** image, much smaller than the object on the retina. The ability of the eye to adjust its lens focusing power for distant and near objects is called **accommodation**.

Distant viewing

An eye at rest is focused on distant objects. The lens is thin and flatter or *less* convex and the ciliary muscles are relaxed, whilst the suspensory ligaments are stretched and taut (see Figure 8.9a).

Figure 8.9 Accommodation of the eye

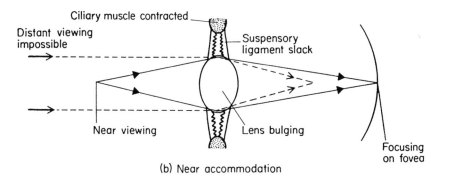

(a) Distant accommodation

(b) Near accommodation

Near viewing

Objects within 3 metres of the eye are focused by the lens becoming fatter and shorter or *more* convex, and by ciliary muscles contracting, causing the suspensory ligaments to slacken (see Figure 8.9b).

Eye defects

Long-sightedness – hypermetropia

This is caused by a *short* eyeball or loss of lens elasticity. Light rays from distant objects come to focus on a point *behind* the retina. The ciliary muscles are never at rest and are continually contracting in order to allow the person to read at close quarters. It is corrected by a **converging** or **convex** lens (see Figure 8.10a).

Figure 8.10   Eye defects and their correction

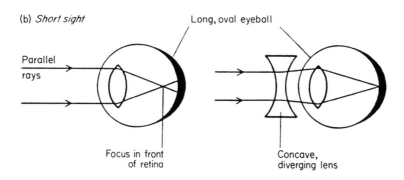

Short-sightedness – myopia

This is due to a *long* eyeball or increased corneal curvature. Light rays from distant objects come to focus on a point *in front of* the retina in the vitreous humour. It is corrected by wearing **diverging** or **concave** lenses in spectacles or contact lenses (see Figure 8.10b).

Retina functions

Light sense

Dim light of low intensity, or black and white light, is sensed by the **rod receptor cells**. The rods are present in greater numbers around the edges of the retina and absent from the fovea. At night-time, vision in dim light may be improved by looking at the object through the corner of the eye. Vitamin A, retinol, is an important nutrient which aids night vision. It is needed for the formation of visual purple by the rod cells. This substance is bleached by light, and increases in amount in darkness, making the rods more sensitive to dim light.

Colour sense

The component colours of white light are red, orange, yellow, green, blue, indigo and violet. Red, green and blue are **primary colours**, sensed by the **cone receptor cells**. The fovea consists entirely of cones. Colour-blindness is an inherited disorder of certain cone receptor cells.

Eye protection and care

The delicate structures of the eye are protected as follows:
- **Bony** orbit sockets and their lining of **lipid** padding protect the back and sides of the eyeball.
- **Eyebrows** and **eyelashes** keep out sweat and rain and shade the eyes in strong light.
- **Eyelids** protect from dust, and close together in strong light or danger by a blinking reflex (see Section 8.5).
- **Iris** circular muscles control the amount of light entering the eye by a reflex (see Section 8.5).
- **Tear glands** produce **tear fluid** by reflex following irritation by dust or on drying of the surface by evaporation. A bactericidal component, **lysozyme**, destroys airborne bacteria landing on the eye surface.
- Diets should contain sufficient **vitamin A** (retinol) for night vision, and **glucose** for energy provision.
- Ultraviolet and infra-red radiations damage the eye lens and retina. This is prevented by not looking into the sun, sunlamps, heat lamps or furnaces and by wearing protective goggles, dark sunglasses and tinted spectacles.

___ Questions _____

15 List all the structures that a ray of light passes through from outside the eye to a receptor cell in the retina.

16 Which parts of the eye change as a person walks from a sunlit street into a shady shop then out into the sunlit street again? How do the parts change and how do the changes help the person to see?

17 What changes take place in a person's eyes when they look down from a distant view to a map in front of them then back to the view to find their way? Why do the changes take place?

18 How is long-sightedness different from short-sightedness and why cannot the same kind of lens be used to treat both conditions?

8.5 Nervous coordination in mammals

Nervous coordination is concerned with the transmission of **nerve impulses** through nerve cells or **neurones**. The nerve impulse is generated from the stimulus in the form of a flow of electrochemical energy inside the neurone. The **neurone** or nerve cell is the functional unit of the nervous coordination system. Neurones are described as **excitable** because they are able to receive and transmit the nerve impulses but their excitability is slowed down by anaesthetics and also by drug, alcohol and solvent abuse.

Nerve impulse

Neurones are able to transmit impulses electrochemically. It is not a flow of electrons as in an electric current flowing through a copper wire. The impulse is generated by the alternate *inflow* of sodium ions into the nerve fibre and the *outflow* of potassium ions. The impulse is conducted very rapidly at a speed of 120 ms^{-1}.

The parts of a neurone

There are four main parts:
- the **cell body**, containing a nucleus and many energy-providing mitochondria and protein-making ribosomes;
- one **axon**, leading the nervous impulse *away* from the cell body;
- one or more **dendrons** leading the impulse *towards* the cell body from a number of **dendrites**;
- the axon **terminals** which relay the impulse to the next neurone.

A fatty **myelin sheath** usually surrounds the axon and dendrons in **myelinated** fibres but can be absent from other nerve fibres.

Neurone types

There are three main types of neurones:

Figure 8.11   Receptor and effector neurones

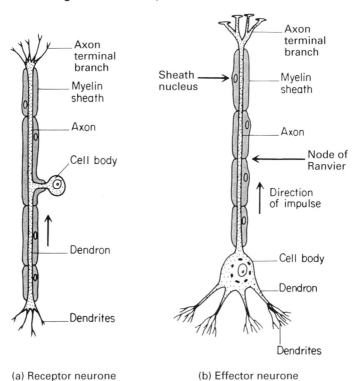

(a) Receptor neurone   (b) Effector neurone

Receptor neurones

These connect the **receptors** to the **spinal cord** or **brain**. A receptor neurone may have a single dendron either forming a free nerve ending or connecting with a receptor cell (see Figure 8.11a). The dendron and axon are both long fibres with a cell body midway between them.

Effector neurones

These connect the **brain** and **spinal cord** with the **effectors** which are normally muscles or glands (see Figure 8.11b). Impulses are collected by the dendrites and passed into short dendrons, grouped about the cell body, and from there they continue down a long axon. If the neurone is attached to a muscle the impulse passes to the end plate. If the neurone is attached to a gland the impulse passes into a fibre.

Intermediate or relay neurones

These are found in the **brain, sense organs** or **spinal cord**. They have many cell body branches. Neurones in the spinal cord are called relay neurones and brain neurones are called pyramidal neurones. An important feature is their great number of dendrites, allowing the neurones to make many interconnections with other spinal cord and brain neurones.

Receptor, relay and effector neurones link together to form a **reflex arc** (see Figure 8.12).

Figure 8.12   The arrangement of neurones in a reflex arc

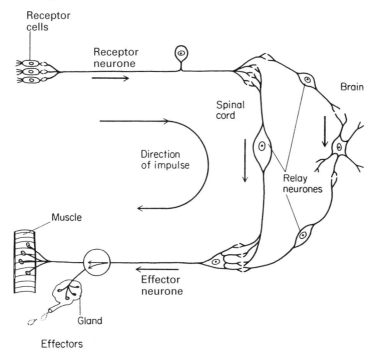

Synapses

Neurones conduct their impulses into neighbouring neurones across a gap between the terminal branch of a neurone and the dendrite of another neurone or effector membrane. This point is called a **synapse** (see Figure 8.13).

Figure 8.13 A synapse at the end of an effector neurone connecting with a muscle

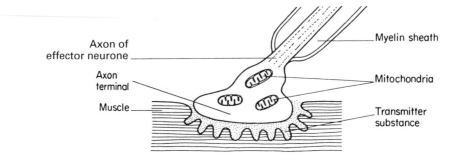

When the impulse reaches the synapse, it releases a **chemical transmitter** which crosses the **gap** by diffusion and enters the other neurone or effector. Certain drugs such as curare and organophosophorus insecticides (see Section 15.4) can block the gap and act as nerve poisons, causing muscle paralysis.

Nerve impulse pathways

The path along which a nervous impulse travels from receptor to effector is called the **reflex arc** (see Figure 8.14). **Reflex action** is a form of **protective** behaviour. It is a rapid, automatic, unlearned or inborn response to a stimulus. It is an instinctive action. Two examples are the blinking of the eye and the pupil reflex (see Section 8.4).

Figure 8.14 A simple reflex arc

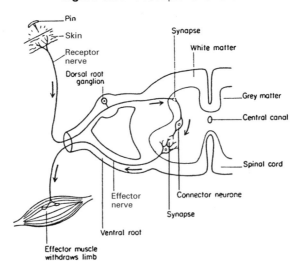

Spinal reflexes

These are reflex arcs involving connection between receptor and effector neurones either by a synapse in the spinal cord or through an intermediate neurone. This reflex action is seen in the 'knee jerk' or hand withdrawal from a hot surface:

Knee jerk reflex

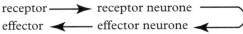

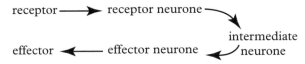

Hand from heat reflex

receptor ⟶ receptor neurone

intermediate
effector ⟵ effector neurone   neurone

Cerebral reflexes

These are reflex arcs involving a connection between receptor and effector neurones through intermediate neurones in the **brain**. This reflex does not involve association neurones. The cerebral reflex action is seen in various 'blink' reflexes, lens accomodation, iris muscle action and in the major reflexes of breathing and heartbeat.

The number of 'units' involved in a reflex action is at least four as in the 'knee jerk' reflex. The 'units' are receptor, receptor neurone, effector neurone and effector. In others there may be five 'units' with an additional intermediate neurone.

Conditioned or learned reflexes

When a nerve impulse travels in a reflex arc which involves brain **association neurones** to *decide* on what response is to be made to the original stimulus, the response is called a **conditioned reflex** (see Figure 8.15). Salivation or saliva secretion occurs by a simple reflex action caused by the touch of food inside the mouth affecting the receptors which stimulate the salivary gland effectors, via receptor and effector neurones:

food touch ⟶ receptors ⟶ receptor neurone

salivary glands ⟵ effectors ⟵ effector neurone
secrete saliva

The sight and smell of food can cause salivation. Similarly, a dinner gong sound can produce salivation.

Ivan Pavlov (1849–1936), a physiologist, carried out experiments by making dogs salivate at the sound of a ringing bell, which the dogs had learnt to associate with feeding times. The dogs were conditioned to associate the appearance of food with the bell ringing. They learnt, from information stored in the cerebrum, to salivate on hearing the bell even if no food appeared (see Figure 8.15).

Such conditioned reflexes are an important part of learning in vertebrate and some invertebrate animals. At least seven 'units' are involved in conditioned reflexes as illustrated below:

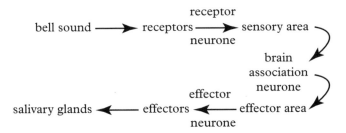

                              receptor
bell sound ⟶ receptors ⟶ sensory area
                      neurone

                          brain
                          association
                effector neurone
salivary glands ⟵ effectors ⟵ effector area
                  neurone

Figure 8.15 A conditioned reflex arc

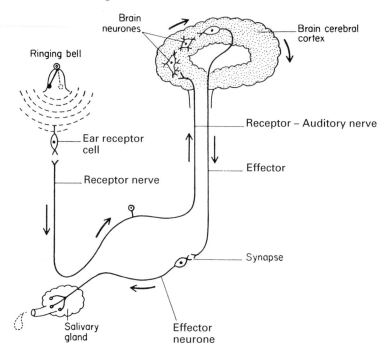

___ **Questions** _____

19 How is a neurone adapted to perform its task of collecting and relaying nervous impulses from one part of the body to another?

20 How can you tell the difference between a receptor neurone and an effector neurone?

21 How does curare cause paralysis?

22 Rearrange these components of a reflex action into their correct order: effector, stimulus, coordinator, response, receptor.

23 Describe two examples of reflex actions using the terms in question 22 but in the correct order.

24 How is a conditioned reflex action different from a normal reflex action?

25 When a goalkeeper saves a penalty, the commentator may say it is due to reflex action. Why is quick thinking a more accurate description?

The central nervous system

The central nervous system or **CNS** in vertebrates is composed of the **brain** and **spinal cord**.

The human brain

The human brain weighs about 1.5 kg or 2% of the total body weight. Lower vertebrates, fish, amphibia and reptiles show a division of the brain into forebrain, midbrain and hindbrain. These divisions are not easily seen in birds and mammals.

The following main parts are recognised in the human brain. They are the **cerebrum, cerebellum, medulla oblongata** and **thalamus** (see Figure 8.16).

Figure 8.16 Vertical section through the human brain

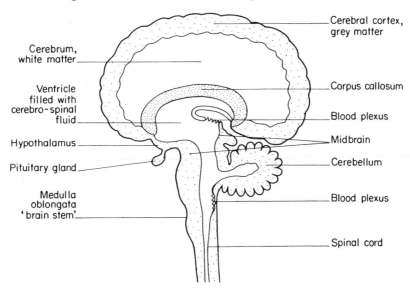

Cerebrum

Two **cerebral hemispheres** connected by a nerve tract form most of the human brain. The outer 3 cm of the cerebrum consists of neurone cell bodies forming the **grey matter** or **cortex**. The neurones are of the intermediate type with dendrites making thousands of interconnections with other neurones.

The surface of the cerebral hemispheres is greatly folded, thus increasing the surface area and allowing more neurones to fill the grey matter. Inside the cerebral hemispheres is the **white matter** composed of neurone axon fibres. There are spaces in the brain called **ventricles**. They are filled with a clear fluid called **cerebro-spinal fluid**. It has a similar composition to tissue fluid.

Functions

The surface of the cerebral hemisphere is insensitive to pain. The functions of different areas have been discovered by studying patients with accidental injury or disease and by experiments in which parts of the brain are stimulated with a current of electricity. From these studies areas in the brain have been mapped out as shown in Figure 8.17.

- **Sensory (receptor) areas** receive impulses from the receptors in the skin, eyes, nose and ears.
- **Motor (effector) areas** transmit impulses to the effectors in skeletal muscles or glands and to those effectors involved with speech.
- **Association areas** sort out, interpret and decide on action. It is the area where receptor impulses from the eye are made into a picture and where receptor impulses from the ears are understood. This area is concerned with learning, memory, personality, morals and ethics.

The outermost coverings of the brain are called the **meninges** and they provide protection and a continuous supply of glucose and oxygen. If deprived of these materials for 2 minutes, the brain can be irreparably damaged.

Continuous electrical activity, detectable by a recording instrument called an **electroencephalograph** (EEG), shows that the brain never rests even when the body is asleep or unconscious.

Figure 8.17 Sensory (receptor), motor (effector) and association areas of the brain

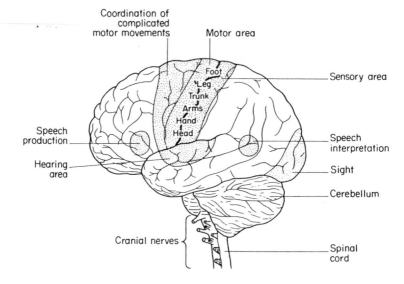

Cerebellum

This is similar to the cerebrum in having a folded surface to increase its grey matter.

Functions

It coordinates voluntary skeletal muscle activity in allowing a person to make accurate and smooth muscle action in walking, etc. It is linked to the cerebrum which starts muscle activity through its motor areas. Damage to the cerebellum causes a staggering, uncoordinated way of walking.

Medulla oblongata

This region is also called the 'brain stem' as it connects the spinal cord with the rest of the brain.

Functions

It is the automatic control centre for heartbeat, breathing, swallowing, coughing and sneezing. Sleep, loss of consciousness and cerebrum activity are also controlled by tissue in the medulla.

Thalamus

All receptor impulses entering the cerebrum must pass through the thalamus. Below the brain floor is the **hypothalamus**.

Functions

The thalamus is associated with pain and pleasure. The hypothalamus is the control centre for homeostasis described in Chapter 7. This region is also a control centre for the endocrine system by the **pituitary gland**.

Spinal cord

The spinal cord extends from the medulla or brain stem to a point near the 'tail'. It is surrounded and protected by **meninges** and **vertebrae**. It is a hollow tube; the hollow cavity is connected to the ventricles in the brain and is filled with cerebro-spinal fluid.

The grey matter of the spinal cord is located *inside*, whilst the white matter of axon fibres is *outside* the spinal cord, beneath the protective meninges. The spinal nerves are **mixed nerves** which have two roots. The **dorsal root** is connected with the receptor neurone whilst the **ventral root** is connected with the **effector neurone**. The position of the receptor neurone cell body is in a swelling on the dorsal root called the **dorsal root ganglion** (see Figure 8.14).

Functions

The spinal cord transmits receptor and effector impulses to and from the receptors and effectors, or relays receptor and effector impulses to and from the brain. Simple coordination can be performed for certain involuntary automatic actions called **spinal reflexes**.

Questions

26 How are the cerebrum and cerebellum (a) similar, (b) different?

27 (a) Why can a person still remain alive even though the cerebrum is badly damaged?

   (b) What quality of life would the person have if their cerebrum was badly damaged? Explain your answer.

Drugs and the central nervous system

Drugs used medicinally for their action on the central nervous system include:
• **painkillers**, such as aspirin, paracetamol, codeine and morphine;
• **sedatives**, which also act to induce sleep, including barbiturates and chloral.
• **anti-convulsants**, which are used to prevent epileptic fits and convulsions and include a range of sedative drugs and phenytoin.
• **stimulants**, such as caffeine in coffee, amphetamines and alcohol.
Incorrect use of these drugs is called **drug abuse**; continued use of drugs leads to drug addiction and degeneration of the brain and nervous system and the breakdown of normal behaviour.

8.6 Chemical or hormonal coordination

The ductless glands or **endocrine glands** of mammals (see Figure 8.18) secrete **hormones** or chemical messengers **directly** into the bloodstream. These glands are unlike the sweat glands and salivary glands which have tubes or ducts along which their secretions travel. Glands with ducts are called **exocrine glands**.

Hormones

Hormones are either **protein** (non-steroid) or **steroid** (non-protein) chemical substances. They are produced in response to either a nervous stimulus from the *external* environment or a change in the concentration of cell substances in the *internal* environment. When the hormones are released into the bloodstream they are distributed throughout the body wherever the blood and tissue fluid can penetrate. Each hormone has **target cells**. If these cells are in a gland they produce a **secretion**. The action of target cells may also produce growth, aid the reproductive process or take part in a homeostatic mechanism which regulates the concentration of the blood.

The human endocrine system

Figure 8.18 The human endocrine system

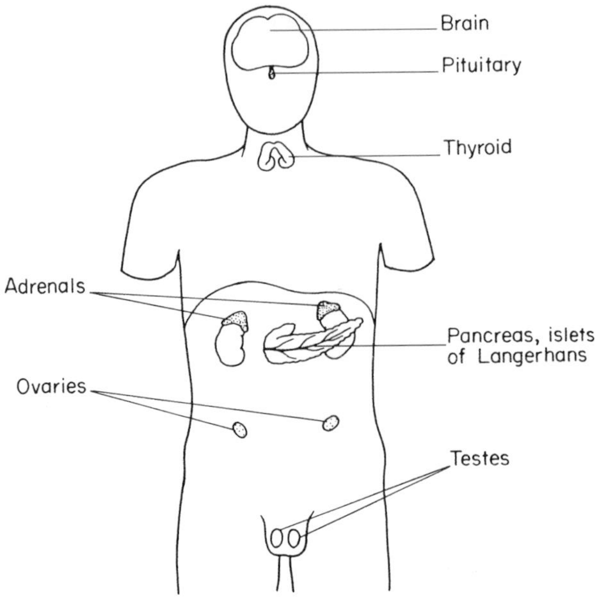

The pituitary

This gland produces hormones which stimulate the thyroid, adrenal, ovaries and testes to release their hormones. It produces a **growth hormone** which promotes the growth of the whole body. If *too much* growth hormone is produced, **giantism** results. If *too little* growth hormone is produced, **dwarfism** is the result. The pituitary also produces **anti-diuretic hormone** (ADH) (see page 175) and **oxytocin** and **prolactin** (see pages 271–272).

The thyroid gland

This gland is near the larynx. It controls **metabolism** and influences **growth**.

The adrenals

The outer part of the adrenal gland called the **medulla** secretes **adrenalin** (see page 167). The inner part called the **cortex** secretes **hydrocortisone** which controls the metabolism of lipids, proteins and carbohydrates.

The Pancreas

This organ contains groups of cells known as the **islets of Langerhans** which secrete mainly **insulin** and also a second hormone called **glucagon**. These hormones regulate the concentration of blood sugar (see Chapter 7).

Ovaries

These organs produce **oestrogen** which controls the development of the female secondary sexual characteristics such as growth of pubic hair, development of the breasts, widening of the pelvis and the development of the menstrual cycle (see page 267). The ovaries also produce **progesterone** which makes changes in the uterus after ovulation and during pregnancy (see page 273).

The testes

The testes produce **testosterone** which produces the male secondary sexual characteristics of growth of pubic and facial hair and a change in the voice.

The action of adrenalin

When the sense organs detect a frightening situation, nervous impulses pass to the adrenals and stimulate them to produce adrenalin. This hormone prepares the body to defend itself or to run away or for 'fight or flight'. Adrenalin increases the heart beat, causes vasodilation in the muscles and vasoconstriction in the digestive system and the skin, increases the respiration rate, makes the pupils dilate, and speeds up the conversion of glycogen to glucose.

Homeostasis in the endocrine and nervous systems

Homeostatic systems depend for their efficient operation on having a **feedback** system which allows information to pass by *negative* or *positive* feedback (see page 164). In the endocrine system, the feedback of information is slow, because the ductless glands must first release the hormones which travel via the bloodstream to reach the particular organ. It then takes time to restore a specific concentration to the norm and then excess of the hormone in the blood causes feedback to slow or stop further hormone production.

In the nervous system the feedback of information is very rapid. This is because the impulse is conducted electrochemically through neurones. Impulses from a muscle stretch receptor are rapidly transmitted to the brain, informing it of a change in tension. The adjustment does not 'waver' about a norm as in metabolism homeostasis.

Table 8.2 compares nervous and endocrine communication.

Table 8.2   Comparison of nervous and endocrine communication

	Nervous communication	Endocrine communication
1	Response to a stimulus, internal or external	Response to a stimulus, internal or external
2	Nervous response more rapid	Hormonal response slower
3	Impulse travels electrochemically long axons and chemically across synapses as transmitter substances, e.g. acetylcholine	Impulse travels chemically as hormones in blood
4	Response is specific in certain muscle or gland	Response is widespread over the body
5	Response is short-lived in reflex muscle contraction	Response is long-lasting in growth or metabolism
6	Does not influence specific chemical changes	Influences specific chemical changes and regulates metabolism, growth and reproduction

28 Which endocrine gland affects other endocrine glands directly?
29 How does each of the changes that adrenalin makes to the body help the body prepare for 'fight or flight'?

8.7 Irritability in plants and animals compared

Flowering plants respond to stimuli in a different manner from animals. The main coordination system in flowering plants is the action of **plant growth substances**. Table 8.3 compares irritability in flowering plants and animals.

Table 8.3 Differences in multicellular plant and animal irritability

	Multicellular plants	Multicellular animals
1	Respond to similar stimuli except sound	Respond to all stimuli including sound
2	No nervous system	Nervous system present
3	Stimuli received in definite regions – stem and root tip	Stimuli received by sensory cells and organs
4	Stimulus must be long-lasting	Stimulus need only be brief
5	Stimulus transmitted chemically – plant growth substances	Stimulus transmitted chemically – animal hormones – and electrochemically in nerve cells
6	Stimulus travels by translocation	Stimulus travels along nervous pathways – reflex arc
7	Response is slow in tropic response	Response rapid in reflex action
8	Response results in growth movement	Response results in locomotion or movement
9	Causes a definite long-lasting effect – elongation or bending	Causes short-duration effect

Question

30 Why do plants respond to stimuli in a different way to animals?

Summary

- Irritability is a characteristic of living organisms that is divided into three components – sensitivity, coordination and responsivity. (▶ 193)
- There is a range of stimuli to which living things are sensitive. (▶ 194)
- Plants coordinate their actions with plant growth substances and mammals do so with hormones. (▶ 194 and 215)

- Tropisms are bending growth movements of a plant stem and roots in response to a stimulus from a specific direction. (see page 175)
- Animals detect stimuli with their sense organs. (see page 199)
- Nerve impulses pass along three kinds of nerve cells or neurones. (▶ 208)
- A reflex arc is the path travelled by a nerve impulse from a receptor to an effector. (▶ 210)
- The central nervous system is composed of the brain and the spinal cord. (▶ 212)
- Drugs affect the working of the central nervous system. (▶ 215)
- Endocrine glands secrete hormones in mammals which coordinate some of the bodies activities. (▶ 215).
- Plants and animals show the characteristic irritability in different ways. (▶ 218)

9 Support and movement

___ Objectives ___

When you have finished this chapter you should be able to:
- describe a hydrostatic skeleton, an exoskeleton and an endoskeleton
- describe the structure and function of the chordate mammalian skeleton
- describe the structure of different joints
- describe the action of the skeletal muscles and the joint in the human arm
- understand how a bird is adapted for flight
- understand how a fish is adapted for movement through water
- compare movement in protoctists and animals
- compare movement in an earthworm and an insect
- compare support and movement in plants and animals

The body of every living organism needs support so that all of its parts may function properly. The means of support also provides a rigid material which in animals is used with muscle action to provide movement.

9.1 Hydrostatic and exoskeletons

Animals and protoctists have skeletons of three main types: **hydrostatic** skeletons, **exoskeletons** and **endoskeletons**.

Hydrostatic skeletons

Soft-bodied organisms such as *Amoeba*, *Hydra*, jellyfish and the earthworm are without hard tissues to support their bodies. Instead, their bodies have rigidity due to osmotic pressure exerted by liquid, mainly water, within the cell cytoplasm or body cavity, acting upon a stiff or elastic cell surface membrane or muscle.

Amoeba

Beneath the cell surface membrane is an outer layer of cytoplasm with a gel-like consistency – the **ectoplasm**. It can withstand the pressure of the fluid **endoplasm**. *Amoeba* can change its body shape by changing its cytoplasm from a gel into a liquid and flowing into a new position (see Figure 2.14).

Earthworm

Two layers of muscle tissue are arranged at right angles to each other in the earthworm's body wall: the **outer circular** muscle layer beneath the epidermis and the **inner longitudinal** muscle layer. **Coelomic fluid** fills the space between the gut and the body wall in each segment cavity. The fluid in each segment has a constant volume. Change in the body shape is by means of the body wall muscles. Simultaneous relaxation of the longitudinal muscle and contraction of the circular muscle makes the body long and thin. Longitudinal muscle contraction and circular muscle relaxation makes the body short and fat. This **antagonistic** action is shown in Figure 9.1.

Figure 9.1 Antagonistic muscle action in earthworm body movement

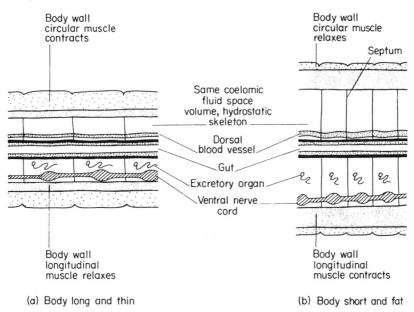

(a) Body long and thin   (b) Body short and fat

The coelomic fluid volume remains constant and fills the changing shape of the coelomic cavities. The coelomic fluid exerts **hydrostatic pressure** in response to the changes in surrounding pressure exerted by the gut and body wall.

Exoskeletons

Arthropods, such as the cockroach and the crab, produce a hard outer **cuticle** layer from the epidermis which forms the outer exoskeleton. Molluscs, such as oysters and snails, continually produce an exoskeleton in the form of a hard outer **shell**. Certain chordates (see Chapter 2) may have an exoskeleton, such as scales in fish or plates and shields as seen in reptiles and in particular in the tortoise.

9.2 Endoskeletons

Chordate animals have an internal **endoskeleton** mainly composed of **bone** and **cartilage**.

Cartilage

Cartilage forms the entire skeleton of cartilaginous fish such as the shark or dogfish. It also forms most of the skeleton in embryo vertebrates. Cartilage tissue is pliable. There are three types, each having specific physical properties. A glass-like cartilage called **hyaline cartilage** covers the ends of mammalian bones. The ear pinnae consist of **elastic cartilage**. The discs between spinal vertebrae are composed of **fibrous cartilage**.

Bone

Bone forms the internal skeleton of all chordates except cartilaginous fish. An intact whole bone consists of 45% water, 30% protein, 15% calcium salts and 10% lipids. The protein in bone is called **collagen**. The inorganic salts are mainly calcium phosphate with smaller amounts of magnesium chloride and calcium fluoride.

Bone has a low density due to its structure (see Figure 9.2). The outermost part of a bone is made from **compact bone**. Beneath the compact bone is an inner layer of **spongy bone**.

Figure 9.2 Longitudinal section through a mammalian femur, showing internal structure

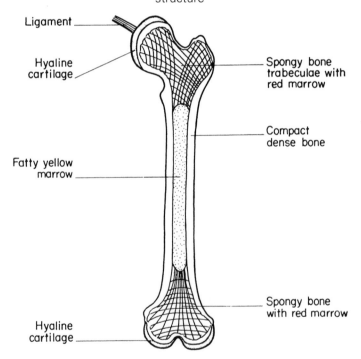

Marrow is of two kinds. **Yellow marrow** is a lipid substance inside the hollow long bones of the limbs. **Red marrow** is found mainly in spongy bone of the ribs, skull, sternum, vertebrae and pelvis. Red marrow makes the blood cells: red corpuscles, white corpuscles, platelets and lymphocytes.

The hollow long bones are stronger than a solid rod of equal length and weight.

Mammalian skeleton

A chordate skeleton (see Figure 9.3) is composed of:
- the **axial skeleton:** vertebral column, ribcage and skull;
- the **appendicular skeleton:** limbs and girdles.

Figure 9.3   The skeleton of a cat, *Felis domesticus* (Griffin Biological Laboratories)

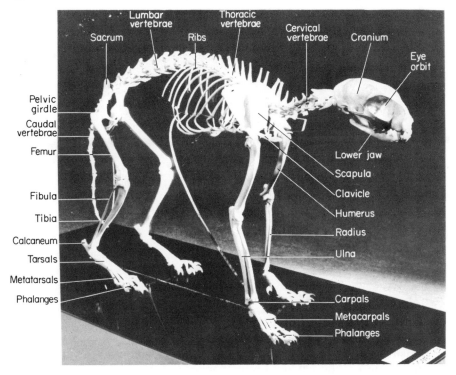

Axial skeleton

The **skull** consists of the **cranium,** surrounding and protecting the brain, with a fixed upper jaw or **maxilla** and movable lower jaw or **mandible**. Skull bones are spongy to reduce the skull density.

The **vertebral column** is made up of individual **vertebrae** separated by **intervertebral discs** or cushioning pads of fibrous cartilage. Vertebrae (see Figure 9.4) have the following general structure: a **centrum**, a **neural arch** containing the spinal cord, together with two **transverse processes** and a **neural spine** to anchor muscles. There are five different sets of vertebrae:

- **Cervical**, neck vertebrae (seven bones): the first, the **atlas**, supports the skull and allows a nodding movement; the second, the **axis**, allows skull rotation.
- **Thoracic**, chest vertebrae (12 or 13 bones): connect with the ribs and the sternum, forming a thoracic cage which protects the heart and lungs and allows respiratory movements (see Figure 9.4).
- **Lumbar**, waist vertebrae (seven bones): strongly built for muscle attachment.
- **Sacral**, hip vertebrae (five bones): these are joined together to strengthen the sacrum (see Figure 9.3).
- **Caudal**, tail vertebrae (variable number of bones): the number of bones depends on the species of mammal.

Vertebrae have an inverted 'T' or sectional structure, comparable to the T-girder in building construction. The complete vertebral column is arched in mammals in a manner resembling a bridge arch. The limbs can be compared to the bridge pillar supports.

Figure 9.4   Structure of a mammalian thoracic vertebra

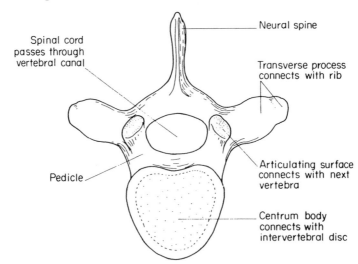

Appendicular skeleton

Pectoral girdle

This is made up of a **clavicle** or collar bone and a flat **scapula** or shoulder blade. It is attached by muscles and tendons to the thoracic vertebrae. The connection is a loose one, acting as a shock absorber to dissipate the upward force or thrust received by the forelimbs as they strike the ground.

Pelvic girdle

This is a strong structure, supporting the abdominal organs, made up of the **ilium**, **ischium** and **pubis** bones, and firmly connected to the sacrum by strong ligaments. In humans, extra strength is needed to absorb the thrust transmitted up the legs in walking and running.

Pentadactyl limbs

The skeletal structure of the forelimbs and hind limbs follows a similar general plan seen in the limbs of amphibia, reptiles, birds and mammals. The limbs have **five** (penta) fingers or toes, hence the name pentadactyl limb (see Figure 9.5).

Figure 9.5 The structure of the vertebrate pentadactyl limb (five digits)

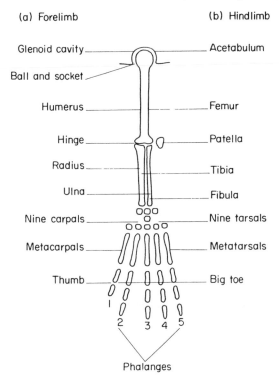

There is some variation in the number of digits and terminal bones in different vertebrates.

Joints

A joint is any part of the endoskeleton where two or more bones meet. Joints are of three main types (see Figure 9.6).
- **Immovable joints** are seen as sutures between the bones of the skull and the pelvic girdle, and between the pelvic girdle and the sacrum.
- **Movable cartilaginous joints** are between the vertebral centra, giving the spinal column flexibility by the intervertebral discs.
- **Movable synovial joints** allow considerable movement, as seen in the jaw, between the atlas and axis and between the bones in the fingers and toes. **Ball and socket** joints in the hip and shoulder allow limb movement in many directions. **Hinge joints** in the knee and the elbow allow limited movement in one direction.

Synovial joints have the following general structure. The bone ends are covered in smooth **hyaline cartilage** which is lubricated by the **synovial fluid**. This fluid overcomes any friction between solid surfaces in contact with each other. The lubricant is secreted by the **synovial membrane** which lines the joint capsule.

Figure 9.6 The structure of various moveable joints

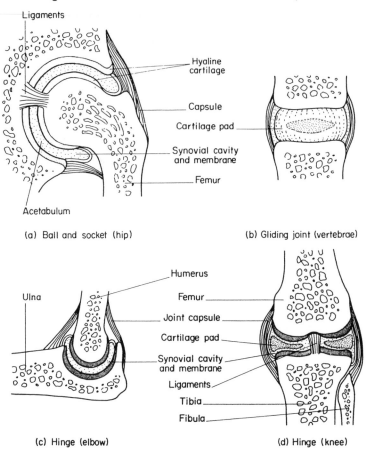

(a) Ball and socket (hip)

(b) Gliding joint (vertebrae)

(c) Hinge (elbow)

(d) Hinge (knee)

Tough fibrous **capsular tissue** forms the joint capsule. Joints are made stronger by **ligaments** which can be part of or separate from the joint capsule.

Ligaments are important connective tissues. **Fibrous ligaments** are made of collagen and connect bones together across most joints. **Elastic ligaments** are found only in a few parts of the body, e.g. the vocal cords.

One end of a synovial joint is concave, the other is convex. The concave part of the hip joint is called the **acetabulum**, whilst the concave part of a shoulder joint is called the **glenoid cavity**.

Questions

5 How do the skull and vertebral column provide protection for the brain and spinal cord?

6 How does the arrangement of the bones in the vertebral column help the body to move?

7 Why are the bones of the pectoral girdle much more loosely connected than the bones of the pelvic girdle?

8 How does each of these components of a synovial joint help the movement of a limb? (a) cartilage, (b) synovial fluid, (c) ligaments.

Bird skeleton

Figure 9.7 The skeleton of a domestic fowl, *Gallus* sp. (Griffin Biological Laboratories)

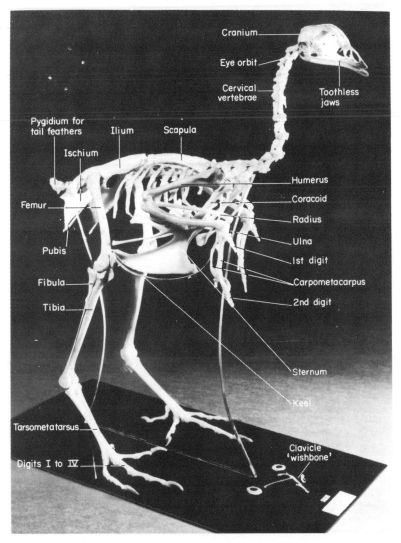

The following summarises the adaptations of a bird skeleton (Figure 9.7) for flight:

- The long bones are hollow and contain air sacs connected to the lungs, making the bird skeleton a low density structure.
- The sternum or breastbone has a deep keel to anchor the powerful pectoralis flight muscles.
- The pectoral girdle is strengthened by a strut-like coracoid bone.
- Tail or caudal vertebrae are reduced to overcome drag forces.
- Only three digits remain in the pentadactyl wing structure. The carpo-metacarpus bone is an elongated hand bone serving to support the primary flight feathers; the secondary and tertiary flight feathers are connected to the humerus and ulna bones.
- The hind limbs or legs have elongated foot or tarsometatarsus bones which keep the flapping wings clear of the ground and help in a fast 'take off'.

9 How do bones containing air cavities, the possession of a keel on the breastbone and few tail vertebra adapt a bird for flight?

Fish skeleton

The fish skeleton (Figure 9.8) is a weak structure compared to that of other vertebrates, much of its support being provided by the surrounding water.

Figure 9.8 Endoskeleton of a bony fish

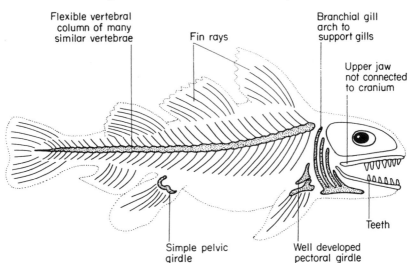

Flexible vertebral
column of many
similar vertebrae

Fin rays

Branchial gill
arch to
support gills

Upper jaw
not connected
to cranium

Teeth

Simple pelvic
girdle

Well developed
pectoral girdle

The vertebral column is composed of many similar vertebrae, bound together by
ligaments in a long flexible column, allowing sideways movement. Long transverse
processes connect with the muscle blocks.

The girdles are not connected to the vertebral column. Fin rays have web-like
membranes between them.

A gill arch supports the gills and gill covers. This gill arch is used in respiratory
movements.

10 Why is a fish skeleton a weaker structure than other chordate animal skeletons?
11 Why are fish vertebrae similar to each other when in the mammalian vertebral
    column there are different kinds of vertebrae?

9.3 Functions of a mammalian skeleton

- **Support:** The muscles, which are soft organs, need to be suspended and bones in
  the skeleton provide attachment points. The vertebral column is the primary means
  of support in vertebrates. Limbs act as props in raising the body off the ground. A

stranded whale is unable to raise its body off the ground and its internal organs are compressed by the upward force from the ground acting in response to the unsupported body mass.

- **Body shape:** The shape of a human hand or a bat's wing is a consequence of skeletal structure.
- **Protection:** Soft organs such as lungs, heart, spinal cord and brain are protected from injury by the thoracic cage, vertebrae and skull, respectively.
- **Blood formation:** Blood cells and lymphocytes are made in the red bone marrow.
- **Respiration:** Thorax and ribcage movements are important in external respiration.
- **Sensory:** The three auditory ossicles in the middle ear transmit sound.
- **Movement:** The bones provide levers for movement of part or the whole of an animal.

Question

12 How would a mammal's life be affected if it did not have a skeleton?

9.4   Movement in protoctists and animals

- **Cytoplasmic movement:** This is seen in *Paramecium* as the movement of food vacuoles from the gullet (see Figure 9.9).

Figure 9.9   Cytoplasmic and ciliary movement in *Paramecium* sp.

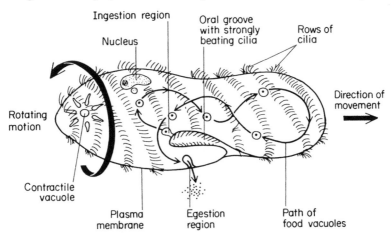

- **Amoeboid movement:** This is seen in *Amoeba* and white blood cells.
- **Ciliary movement:** This is the movement caused by the rhythmic beating action of **cilia** present in large numbers all over the *Paramecium* body surface (see Figure 9.9). Ciliated epithelium is found lining the lungs and oviducts of mammals. It serves to move fluid and solid particles. Ciliary movement can be responsible for moving the *whole* organism, as in the locomotion of *Paramecium* or in the movement of *small particles* as in the mammalian lung (see page 147).
- **Flagellate movement:** This is by means of one or two **flagella** which are 10 to 20 times the length of a cilium. They are seen in human sperm cells (see Figure 10.19) and some protoctists.

- **Muscular movements:** Muscle tissue consists of muscle cells; these are effector cells composed of contractile protein.

Types of movement caused by muscles

Involuntary movements

These are produced by **involuntary muscle,** also called **smooth muscle** (see Figures 2.7 and 9.10). This muscle tissue is found in the gut wall of animals, blood vessel walls and the urinary bladder wall. Involuntary or smooth muscle cells have a single nucleus, contract slowly and tire or fatigue slowly.

Figure 9.10 Smooth, involuntary muscle tissue of the gut and bladder wall

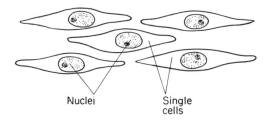

Nuclei   Single cells

Involuntary movements occur in the iris of the eye and peristalsis of the gut (see Figure 9.11), uterus and ureters.

Figure 9.11 Peristalsis in the oesophagus due to sequential contraction of involuntary circular muscle

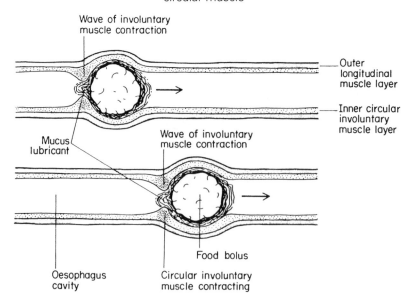

Heart movement

Heart or **cardiac** muscle contracts rhythmically by means of an in-built stimulus. The muscle does not tire or fatigue (see Figure 9.12).

Figure 9.12 Microscopic structure of heart or cardiac muscle tissue

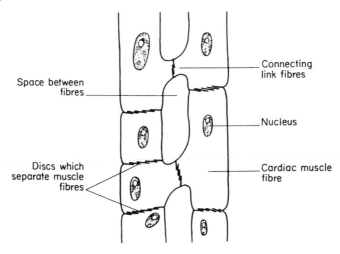

Space between fibres

Connecting link fibres

Nucleus

Discs which separate muscle fibres

Cardiac muscle fibre

Voluntary movements

These are produced by **skeletal muscle**, also called **voluntary muscle** (see Figure 9.13). This muscle is always connected to bones, forming a musculoskeletal system.

Figure 9.13 Skeletal or voluntary muscle

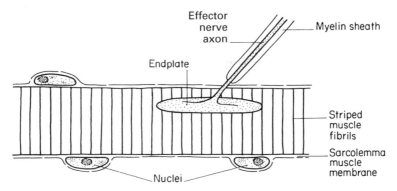

Effector nerve axon

Myelin sheath

Endplate

Striped muscle fibrils

Sarcolemma muscle membrane

Nuclei

Skeletal muscle is composed of many nuclei in a cytoplasm called **sarcoplasm**. It is not divided into cells. It shows cross-stripes or cross-striations and contains many mitochondria. Skeletal muscle contracts under the control of the central nervous system and the brain motor area. The muscle contraction is sudden and powerful and is subject to rapid tiring or fatigue.

Questions

13 What types of movement does *Paramecium* show?
14 How is the movement of cilia different from that of a flagellum like a sperm's tail?
15 In what ways are smooth (involuntary) muscle and skeletal (voluntary) muscle (a) similar, (b) different?

Skeletal voluntary muscle

The human body consists of 45% by weight of skeletal muscle and 15% bone; the remaining 40% is made up of organs (20%) such as the lungs, liver, skin and heart, and body fat (15–20%).

Skeletal muscles are connected to bones by **tendons**. The tendons join with the fibrous outer layer of bones called the **periosteum**. Tendons are similar to fibrous ligaments in being non-elastic and composed of collagen protein. The point where the skeletal muscle is connected to an immovable bone is called the **point of origin**. The movable bone connects by a tendon at the **point of insertion**.

Skeletal muscle contraction

Skeletal muscle is composed of tiny muscle fibrils. These are collected together and arranged lengthwise in a bundle of fibres. Several bundles of muscle fibres surrounded by a muscle sheath make up a whole muscle, such as the upper arm **biceps** muscle.

Effector neurones terminate as effector **endplates** in the muscle fibre. The nervous impulse releases a chemical transmitter which crosses the synapse (see Section 8.5). The chemical transmitter releases energy from ATP causing the whole muscle to contract.

--- Questions ---

16 What is the difference between the origin and the insertion of a skeletal muscle?
17 How is a tendon different from a ligament?

9.5 Support in mammals

Antagonistic muscles

Muscles, both smooth involuntary and skeletal involuntary, act in pairs, each one opposing the action of the other in the pair. As one contacts the other relaxes; such a pair of muscles is called **antagonistic muscles** (see Figures 9.1, 9.11 and 9.14).

Figure 9.14   Left forearm movement due to the antagonistic action of the biceps and triceps muscles

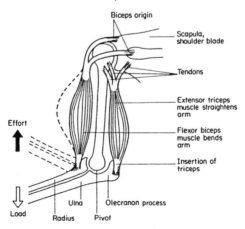

There are a great number of skeletal muscles in the human body. The following are names given to the main antagonistic skeletal muscles and their actions:
- **flexors** *bend* a limb at a joint, and **extensors** *straighten* a limb at a joint;
- **abductors** move a limb *away* from the body and **adductors** move a limb *towards* the body.

Skeletal muscles have two main functions:
- **Locomotor muscles** act in body locomotion, moving the whole of the body from one place to another.
- **Postural muscles** act in body support in maintaining the upright position of the trunk and in holding the head on the human body.

Receptors

There are receptors in skeletal muscles, tendons, ligaments and joints, detecting movement and tension changes. Muscles have **stretch receptors** that inform the central nervous system (CNS) of the state of contraction of the muscle, i.e. whether it is contracted or relaxed. The skin of the feet has **pressure receptors** which detect changes in the body position.

Muscle tone

When a human being is conscious, at rest and maintaining an erect posture, the skeletal muscles are in a slightly contracted condition, called **skeletal muscle tone**. If the person becomes unconscious or falls asleep, muscle tone is lost and the skeletal muscles relax completely.

Erect posture in human beings is achieved partly by the leg extensor postural skeletal muscles contracting, making the legs into straight supporting pillars. The muscle stretch receptors relay information by negative feedback to the brain, which supplies a stimulus to the muscle by an effector neurone sufficient to maintain the slight muscle tension or tone.

The force of gravity will tend to pull or bend the body. Consequently, the limb-straightening, quadriceps extensor muscles will contract. These are called the **anti-gravity muscles**. Complete body support is achieved by contraction of neck, back, thigh, hip and buttock extensor muscles.

Nerve impulses from the eyes and the balancing organs in the ears are also used along with those from the muscle receptors to help the body maintain an erect posture.

Questions

18 What changes take place in the arm when (a) the biceps contracts, (b) the triceps contracts?
19 How is information about movement and tension changes in one part of the body useful to the coordination activity of the central nervous system?

9.6 Locomotion in water – fish

Propulsion

Skeletal muscle blocks called **myotomes** are arranged on the right and left sides of a flexible vertebral column (see Figure 9.15). The pairs of muscles on each side of the column work antagonistically. As the right-side muscles *contract*, the left-side muscles *relax*, causing the body to bend towards the *right*. This continues as a wave of contraction down the length of the body to reach the tail. The **tail fin** provides a powerful pushing force or thrust against the surrounding water, the resultant action being a *forward* movement of the whole fish.

Figure 9.15   Muscle blocks or myotomes in a bony fish

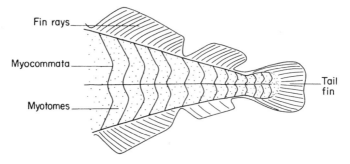

Body shape

A fish has a **streamlined** body shape with a rounded front at the head and a tapering body finishing as a pointed tail (see Figures 2.31 and 2.32). A smooth body surface helps to reduce **drag**. The fish scales provide the essential smooth surface, whilst the slimy mucus provides lubrication.

Stability

Fish have paired **pectoral** and **pelvic** fins, together with single **vertical** fins; these are located on the midline position of the back (dorsal) and ventral surfaces. Fins maintain stability and control the direction of motion (see Figure 9.16).

Figure 9.16   Fins in fish stability control

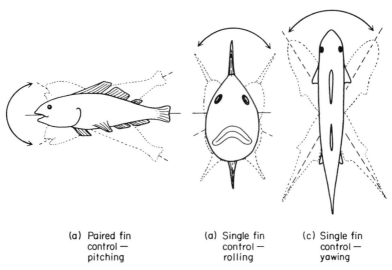

(a) Paired fin control — pitching

(a) Single fin control — rolling

(c) Single fin control — yawing

- **Pitching** is the alternate rising and falling of the head and tail and is controlled by the pectoral and pelvic fins.

- **Rolling** is the twisting and turning motion controlled by all fins.
- **Yawing** is the sideways swaying or the body in going *off course*. This is counteracted by the flat body sides and single vertical fins.

Support

The inclined paired fins act as hydroplanes and generate lift by the flow of water over the fins; this can help the fish to rise in the water. A bony fish is able to adjust its depth by changing its buoyancy, by inflating or deflating its **swim** or **air bladder**. This organ is not present in cartilaginous fish.

Question

20 How is the body of the fish adapted for movement through the water?

9.7 Locomotion on land

Earthworm

Body shape

The earthworm has a tapering body shape and its surface is lubricated by mucus secreted by the skin. The shape helps the earthworm to penetrate the soil and the moist surface helps overcome friction forces.

Propulsion

Propulsion is by means of alternate contraction and relaxation of the antagonistic muscles, the circular and longitudinal body wall muscles, which pass down the length of the body in waves (see Figure 9.1).

Chaetae (bristles) are inserted into the soil to increase the earthworm's grip on the soil as the waves of muscular contraction and relaxation move the body forwards.

Insects

Insects have six legs, each of which is a tube of chitinous exoskeleton material. The several parts of each leg are connected by flexible joints with a peg and socket pivot (see Figure 9.17).

Figure 9.17 The walking leg of an insect

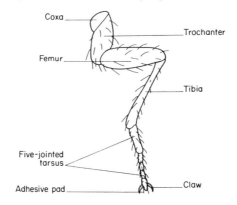

Inside each limb section are pairs of antagonistic muscles called **extensors** and **flexors.** Each muscle acts across the joint between leg sections (see Figure 9.18).

Figure 9.18   Antagonistic muscles in the leg of an insect

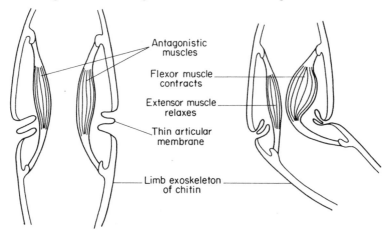

A walking insect will have three legs in contact with the ground at any one time, two on one side and one on the other, in a tripod. Other arthropods may leap, for example the flea and the grasshopper, or swim, for example the water beetle, or crawl, for example the centipede.

Mammals

Body shape

Mammals have most of the body mass concentrated in the trunk. The limbs have their mass concentrated in the thighs and are kept close to the trunk. The ends of the limbs are tapered towards the feet/hands.

The feet and hands at the limb extremities are worked by long tendons connected to muscles higher up the limb. This reduces the need for heavy muscles in the fingers and toes.

In order to reduce contact with the ground and also reduce friction, the feet are usually small and transmit considerable pressure or thrust, such as through the small hoof of a running animal.

Propulsion

Forelimbs or arms

In human beings, the forelimbs or arms are used for lifting weights. The biceps or flexor muscle bends the arm at the elbow, whilst the triceps or extensor muscle straightens the arm. In four-legged mammals, the triceps extensor muscle is used to straighten the forelimb and hold up the body. The forelimbs dissipate the upthrust on impact with the ground as in landing or falling forwards.

Hind limbs

In all mammals, the hind limbs provide the pushing force or thrust which moves the body forwards by foot pressure against the ground. The foot remains in contact with the ground by friction. Without friction the foot would make no contact, such as when feet slip on ice.

The hind limb produces a **power stroke** in straightening the limb by contraction of the extensor muscles and makes a **recovery stroke** by bending the limb with flexor muscles.

__ Questions __

21 How is the body of an earthworm adapted for movement through the soil?
22 How do the muscles and joints in an insect limb compare with those in the human arm?
23 How do the hind limbs provide the force to propel an animal forwards and how do the muscles work to keep it moving?

9.8 Locomotion in air – birds and insects

Birds

Body shape

A bird's body is **streamlined** and has a smooth surface provided by the feathers. Birds have three main kinds of feathers:
- **down:** small fluffy feathers for heat insulation;
- **covert feathers:** cover all the body and provide much of its streamlined shape;
- **flight feathers:** form the wing outer edge. Flight feathers have a short quill and a central shaft with a broad vane (see Figure 9.19).

Figure 9.19   The structure of a bird's wing showing bones and flight feathers

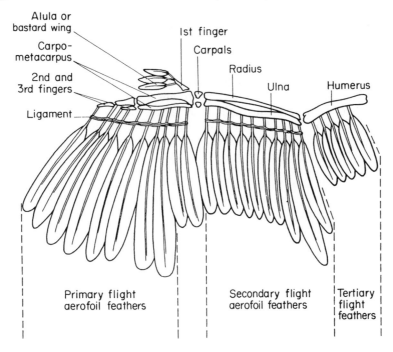

Propulsion

This is achieved by gliding and flapping flight.

Gliding

A wing seen in section at right angles to the wingspan has a sectional shape like an arched plate; this is called an **aerofoil**. In gliding, the wing is tilted with its front or leading edge slightly above the trailing or back edge. As air flows past the wing, it causes drag along the trailing edge.

The combined effect of lift and drag exactly balances the birds weight (see Figure 9.20). In time a gliding bird will lose height unless it experiences changes in wind speed or encounters warm air currents, when it will soar upwards on a thermal.

Figure 9.20   Forces acting on a wing aerofoil during gliding in still air

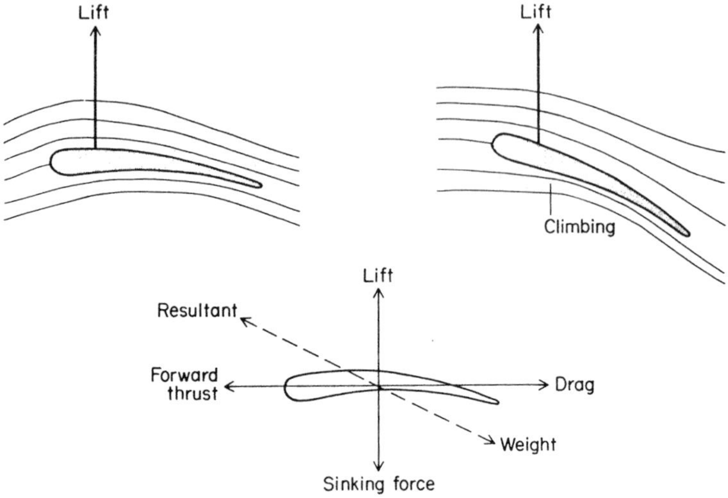

Flapping flight

Movement of the wings in flapping is by means of the antagonistic breast or **pectoralis muscles** (see Figure 9.21). The large pectoralis major muscle contracts and pulls the wing *downwards*. On this downstroke the wing presents a large surface area with closed vanes which generates lift and thrust, pushing the body forwards.

The contraction of the small pectoralis minor muscles produces the wing's *upstroke*, when the flight feathers part, allowing air to pass without resistance between the vanes and decreasing any thrust acting downwards.

Winds affect the **stability** of birds in the same way as water currents affect fish:

- **Yawing** or sideways swaying off course is controlled by the **tail feathers**.
- **Pitching**, nose diving or upward movement is controlled by **tail** and **wing feathers**.
- **Rolling** or twisting of the body is controlled by the **outspread wings**. Tail and wing feathers control braking in landing.

Insects

Insect wings are thin, flat plates of chitinous material stiffened by ribs or veins. They are seen to have an arched aerofoil structure in part of the downstroke. Some insects, butterflies, show gliding flight in addition to flapping flight.

Figure 9.21 Antagonistic muscles responsible for the up-and-down, flapping movement of a bird's wing

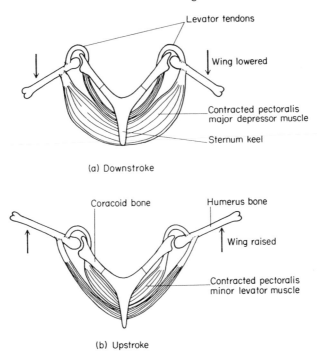

(a) Downstroke

(b) Upstroke

Insects generate lift and thrust propulsion force by flapping, using flight muscles which are of two kinds depending on the type of insect.

Direct flight muscles

These are a pair of antagonistic muscles called **depressor** and **elevator** muscles, joined to large slow-beating wings as in butterflies, locusts and dragonflies. The muscle origin is in the sternum of the thorax.

Indirect flight muscles

The **elevator** muscles are connected between the upper tergum and lower sternum of the thorax; their contraction *raises* the wing on the **upstroke**. The **depressor** muscles are connected longitudinally in the thorax and their contraction *lowers* the wing on the **down-stroke**. This type of flight muscle is seen in insects with small, fast-beating wings such as the housefly, wasp and bee (see Figure 9.22). On the downstroke the insect wing has an aerofoil profile for maximum lift and thrust. On the upstroke the wing is turned vertically with the leading edge directly above the trailing edge for the least resistance and lift.

Questions

24 How is the bird's body adapted to movement through the air?
25 A bird flaps its wings to get into the air, then glides before being blown by a gust of wind which makes it pitch and roll before it makes a safe landing. Describe the actions of the bird from the beginning to the end of its flight.
26 How is flapping flight of the bird and the bee (a) similar, (b) different?

Figure 9.22 Antagonistic muscles responsible for the up-and-down flapping movement of wings of insects with small, fast-beating wings

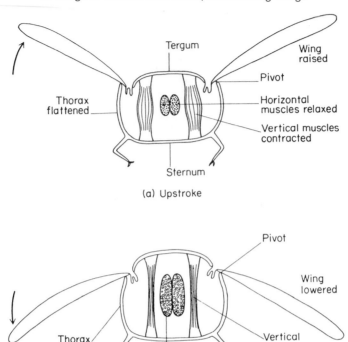

(a) Upstroke

(b) Downstroke

9.9 Support and movement in plants

Support in plants

Water **turgor pressure** in plant cell vacuoles acts against the rigid cellulose cell walls (see Section 5.1). Epidermis layers surround the multicellular plant, providing a firm layer which maintains the plant shape and withstands the outward-acting pressure generated by the inner tissues.

Collenchyma cells with cellulose thickening in the cell corners, provide mechanical support in leaf petioles and stems. **Lignin** thickening is present in the tubular, mechanical supporting tissues of xylem vessels and in sclerenchyma fibres protecting the phloem of vascular bundles.

The mechanical supporting tissues (sections 2.2 and 5.2) of **collenchyma, sclerenchyma** and **xylem vessels** are arranged in seed-forming plants:

- **cylinder-wise** in stems to resist bending forces by winds (see Figure 5.4);
- **centrally** in roots to resist pulling forces or tensions (see Figure 5.5);
- as a **network** in leaves to resist tearing forces by the wind (see Section 3.5).

Climbing plants obtain support from other plants by means of **twining stems** in the runner bean and honeysuckle, **tendrils** in peas and **hooks** in the blackberry.

Table 9.1 compares support methods in plants and animals.

Table 9.1 Comparison of support in plants and animals

	Plants	Animals
1	Firm cell cellulose wall – epidermis	Flexible cell surface membrane
2	Turgidity by high osmotic pressure (wilting = low osmotic pressure). Parenchyma cells of pith and cortex	Hydrostatic skeleton of fluid under pressure surrounded by muscles – *Hydra*, earthworm
3	Cellulose thickening in collenchyma cells – herbaceous plants	Exoskeleton as outer shell in molluscs, or as cuticle in insects
4	Lignin thickened cells in sclerenchyma fibres and xylem vessels of woody stems	Endoskeleton of bone and cartilage in vertebrate animals, together with skeletal muscle in posture
5	Climbing plants obtain support by twining stems, tendrils, thorns and prickles	Aquatic and flying animals obtain lift in various ways. Land animals have supporting stances
6	Stems resist bending force – support cylindrical. Roots resist pulling force – support central. Leaves resist tearing force – support a network of veins	Skeleton support is mainly central

Movement in plants

Plants are static and firmly rooted in the ground. They have no need of any methods of whole body locomotion which are essential to animals for them to obtain food and to find a mate for reproduction.

- **Cytoplasmic movement**: Cytoplasmic streaming of chloroplasts occurs in *Elodea* – Canadian pondweed.
- **Tropisms**: Tropic growth movements are described in Section 8.2.
- **Mechanical movements** occur in the opening of fern sporangia and certain ripe fruits such as legumes. These are **hygroscopic** movements due to water absorption.

Questions

27 Draw transverse sections of a stem and root and annotate your drawings to identify areas where the plant is receiving support from its cells.
28 'Plants don't move.' How true is this statement? Explain your answer.

Summary

- Animals and protoctists can support themselves by three kinds of skeleton – a hydrostatic skeleton, an exoskeleton or an endoskeleton. (▶ 220)
- The bones of the mammalian skeleton can be divided into the axial and appendicular skeleton. (▶ 223)

- There are several kinds of joint; the synovial joint provides movement in the limbs. (▶ 225)
- Birds have modifications to the vertebrate skeleton which enable them to fly. (▶ 227)
- The chordate animal skeleton provides support and protection, allows movement and makes red blood cells and lymphocytes. (▶ 228)
- There are three types of muscle – smooth (involuntary), cardiac (heart) and skeletal (voluntary) muscle. (▶ 230)
- The body of a fish is adapted for movement through the water with a flexible back-bone, a large number of antagonistic skeletal muscles and a streamlined shape. (▶ 228 and 233)
- Earthworms and insects use antagonistic muscles to move but have different types of skeleton. (▶ 235)
- Birds and insects have special flight muscles and use their wings to create lift and thrust. (▶ 238 and 239).
- Modifications to the basic cell structure provide support in plants. (▶ 240)

Life Continuation Processes

Reproduction

Objectives

After reading this chapter you should be able to:
- define a species
- describe asexual reproduction in certain living organisms
- distinguish between the different life cycles in flowering plants
- describe vegetative reproduction and artificial propagation
- describe sexual reproduction in flowering plants
- describe sexual reproduction in certain animals
- describe sexual reproduction in some chordates
- describe sexual reproduction in humans
- understand the different methods of contraception
- distinguish between the sexually transmitted diseases

10.1   Introduction

Genetic material appeared in the first living organism on Earth over 3000 million years ago and continues to be found in all living organisms today. Genetic material is passed on to new living organisms by different methods of **reproduction**.

Reproduction is the characteristic process of transmission of genetic material from parent to offspring. Organisms have a varied **life span** from a few hours to many years. Each organism undergoes a **life cycle** from the time of its formation from either one or two parents until the time of its death.

Reproduction is a characteristic of living organisms. It is essential to the continuation of genetic material and life, and in the continuation of the **species** of each particular organism. A species is a group of similar organisms able to reproduce amongst themselves, but not with organisms of other species. Individuals of the *same* species are composed of *similar* genetic material.

Asexual reproduction is a rapid method of increasing members of a species of plants and bacteria, fungi and protoctists from *one* parent. **Sexual reproduction** is a slower method and usually less prolific in increasing numbers in plant and animal species from *two* parents. In the tapeworm, sexual reproduction takes place through one individual which is an **hermaphrodite** (see page 55).

Reproduction ⟵ → species continuation
→ population increase
→ life continuation

┌─ **Questions** ──┐
│   │
│ 1 What is a species?   │
│ 2 (a) Which method of reproduction is better for colonising an area quickly? │
│   (b) State another advantage this method of reproduction has over the other │
│   method.   │
│   │
└──┘

10.2 Asexual reproduction

During asexual reproduction the genetic material of the nucleus divides by **mitosis** (see Section 11.4) producing one or more daughter cells, each having the *same* amount of genetic material as the nucleus of the parent cell. The main methods of cell division involving the *whole* body of certain organisms are described here. **Cloning** is a term used to describe the formation of a group of organisms of the same species by asexual methods involving mitosis and vegetative or artificial propagation. The clones are identical in the genetic material found in the nuclei or in **genetical composition** (see Section 12.4) or are said to be of the same 'strain'.

Binary fission

Binary fission is seen in protoctists, for example *Amoeba* (see Figure 10.1) and *Paramecium*, and in prokaryotes, e.g. bacteria, when the cytoplasm divides into *two* equal parts following the division of genetic material or **nuclear division**. The process occurs when the parent has reached a certain size (see Section 11.4).

Figure 10.1   Asexual reproduction in *Amoeba*

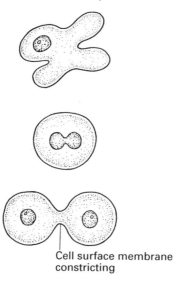

Cell surface membrane
constricting

Budding

Budding is seen in its simplest form in the fungus, yeast (see Figure 10.2) when the cell forms a small bulge into which cytoplasm and a nucleus pass. The bud is much smaller than the parent cell and eventually becomes detached.

Figure 10.2 Photomicrograph showing yeast cells, *Saccharomyces* sp., in the process of budding (The Distillers Co. Ltd)

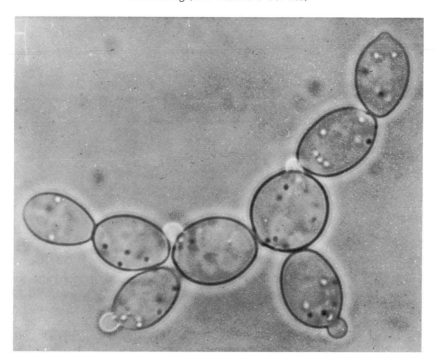

Hydra, a multicellular cnidarian organism, forms a a bud which later separates from the parent (see Figure 10.3). The flatworm, tapeworm, shows budding of proglottids from the neck of the scolex (see Figure 2.23).

Figure 10.3 Asexual reproduction by budding in *Hydra* sp.

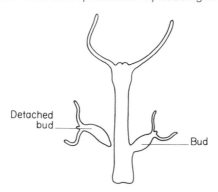

Detached bud

Bud

Spore formation

Fungi, liverworts, ferns and mosses produce enormous numbers of asexual **spores**.

Bacterial spores

Certain bacteria form a thick-walled spore inside the bacterial cell wall (see Figure 10.4). This, like most spores, can resist unfavourable conditions such as drought and extremes of heat. Many bacterial spores survive cooking temperatures and cause food poisoning (see Section 4.9). The light spores are carried by air currents to infect exposed food or wounds.

Figure 10.4 A bacterial spore, a means of survival in adverse conditions

Thick spore wall resistant to heat and chemicals

Nucleus

Bacteria cell wall

Spore

Spore-producing sporangia

The fungus, *Mucor*, or *Rhizobium* (pin mould) found on decaying bread and fruit, are able to produce a single hypha from their body or **mycelium** (see Figure 10.5). This erect hypha develops a swelling at its tip to form a **sporangium**, separated from the rest of the mycelium by a **columella**. A great number of spores are produced by mitosis and they are liberated on bursting of the sporangium and are carried over great distances by air currents.

Figure 10.5 (338B)

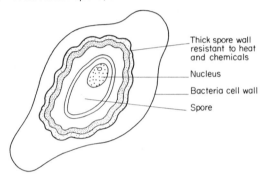

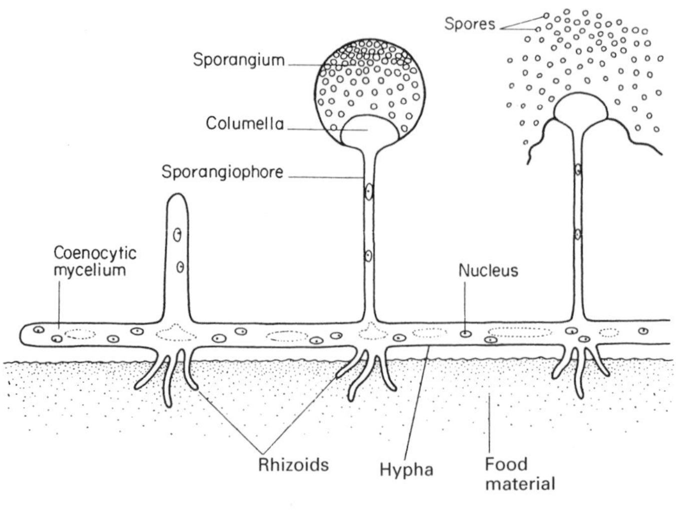

Spores

Sporangium

Columella

Sporangiophore

Coenocytic mycelium

Nucleus

Rhizoids   Hypha   Food material

If a spore lands on a suitable medium, it germinates to form a small hypha which eventually produces an extensive mycelium.

Questions

3 How is the budding of yeast different from binary fission in *Amoeba*?
4 Why have bacterial spores got thick walls but are also light in weight?
5 Why is the sporangium of *Mucor* raised above the mycelium?

10.3 Vegetative reproduction

Vegetative reproduction is a method of **asexual** reproduction in which **vegetative parts,** or organs (root, stem, leaf or bud) become detached from the parent plant body and then develop into another *complete* daughter plant. This method of asexual reproduction is seen in many flowering plants in addition to sexual reproduction (flowering and seed production), particularly the herbaceous flowering plants which are composed of soft tissues unlike trees and bushes which make harder woody tissues (see Artificial propagation, page 253).

Annuals

Annual herbaceous flowering plants complete their life cycle within one year from seed germination to seed dispersal, after which the parent plant dies. This is seen in the sweet pea, poppy and lettuce. **Ephemeral** flowering plants such as certain weeds like groundsel and shepherd's purse have several life cycles in one year.

Biennials

Biennials are herbaceous flowering plants requiring *two* years to complete their life cycle. An example is the carrot. In the first year the carrot produces vegetative growth of root, stem and leaves, and food is stored in the swollen tap root. In the second season the food reserve is used in early reproductive growth in producing flowers and seeds. This method is also seen in swedes, turnips, parsnips and beet.

Perennials

Many flowering plants survive from year to year by vegetative means other than dormant seeds (see Section 10.5). Herbaceous perennials include those flowering plants able to survive the winter as **bulbs, corms, tubers** or **rhizomes. Woody** perennials are those flowering plants which survive winter by losing their leaves, e.g. deciduous trees and shrubs. **Evergreen** perennials include grasses, rhododendron and holly, which lose their leaves throughout the year.

Questions

6 What is the difference between an ephemeral and an annual flowering plant?
7 (a) Why do some biennial flowering plants make good crop plants?
   (b) How is a biennial flowering plant different from a perennial flowering plant?

Winter twigs

Many deciduous flowering trees and shrubs **perennate** or survive the winter by shedding their leaves, to leave bare twigs. Trees have a main trunk whilst shrubs are much smaller and closely branched from ground level, as in privet, lavender and gorse.

A twig in winter has the following structure, as for example in the horse chestnut (see Figure 10.6). The **buds** are future shoots consisting of a short stem crowded with overlapping immature leaves. The buds are of two kinds: **terminal** or **apical buds** producing next season's stem growth, and **lateral buds** in the axil of a leaf scar, that will produce foliage leaves.

Figure 10.6 A horse chestnut twig, *Aesculus* sp., in winter

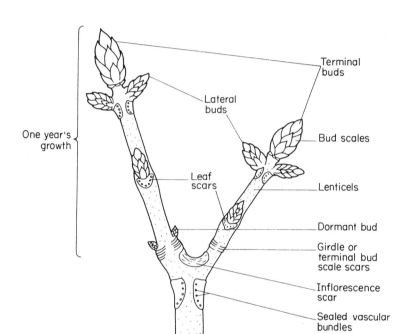

The buds are protected by sticky, waterproof **bud scales**; these leave a ring or girdle of scale **leaf scars** around the base. The distance between one set of scale leaf scars and the next is equal to one year's growth. A **saddle scar** or inflorescence scar indicates the position of the previous year's flowers.

The twig bark has numerous lenticels (see Figure 6.15). **Dormant buds** are reserve buds which develop if terminal and lateral buds die.

Leaf fall

The leaf scars on a twig indicate the point of attachment of the former leaf petiole to the stem. An **axillary bud** is always positioned in the former leaf axil. In autumn, certain excretory wastes, e.g. tannins, enter the leaves and an **abscission layer** develops close to the petiole stalk. The vascular tissue becomes blocked and the leaves, deprived of water, die; their chlorophyll pigments break up (see Figure 10.7). Loose cork cells form in the abscission layer closest to the stem, the petiole is severed and the leaves removed by the wind. The cork scar seals the leaf scar wound.

Figure 10.7 The process of leaf fall in a tree or shrub

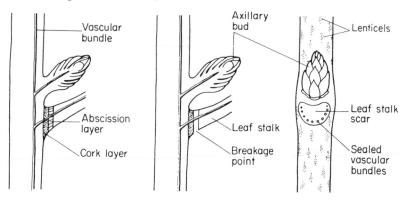

Questions

8 (a) How can you distinguish between a tree and a bush?
   (b) What is the difference between a deciduous tree and an evergreen tree?
9 How is a leaf scar made and how is it different from saddle and girdle scars?

Bulbs

Bulbs are perennating organs and a means of asexual reproduction in some flowering plants, e.g. tulip, hyacinth and snowdrop. Bulbs are swollen, underground buds in which the small disc-shaped stem bears a number of leaves with **fleshy** bases swollen with food reserves (see Figure 10.8). Between the fleshy foliage leaves are small **axillary buds**. A central **terminal bud** develops into the overground flower and shoot, using the food reserves for early spring growth.

Figure 10.8 The bulb of tulip as an example of an organ of vegetative reproduction

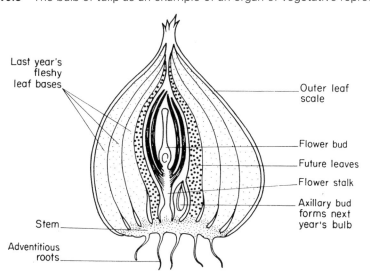

Food made by photosynthesis swells the axillary buds to form a number of daughter bulbs from one parent bulb. Reproduction therefore occurs by planting one bulb which at the end of its life-cycle produces a number of daughter bulbs separating from the parent by vegetative reproduction, each individual bulb perennating by surviving until the spring.

Stem tuber

Figure 10.9 The potato tuber as an example of an organ of vegetative reproduction

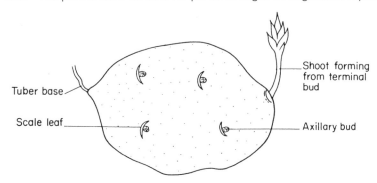

The potato is a stem tuber (see Figure 10.9) or swollen underground stem. It is identified by its numerous 'eyes' or axillary buds in scale leaf scars. One potato tuber, at the end of its life cycle, reproduces vegetatively in forming many daughter stem tubers, each of which separates from the parent plant, and perennates over the winter.

Table 10.1 compares the advantages and disadvantages of vegetative propagation.

Table 10.1 Comparison of advantages and disadvantages of vegetative propagation

	Advantages	Disadvantages
1	Rapid means of reproduction and 'spread'. Can keep down other species	Overcrowding and competition for space unless separated artificially
2	Offspring identical to parent and to each other. 'Varieties' or 'strains' can thus be preserved genetically in roses, apples, etc.	New 'varieties' cannot be produced by this method except by mutation (see Section 12.3)
3	Food storage organs allow perennation or survival in adverse conditions	Diseases typical of the species are rapidly transmitted and can decimate a crop

---

Questions

10 Look at Figure 10.8 and describe how the plant develops from the time it starts to use up the stored food in its leaf bases in the spring to its resting stage the following winter.

11 If one potato tuber was left in a garden plot and the plot was left untended, what would you expect to find after a few years? Explain your answer.

12 What features of vegetative reproduction are useful for the development of a new food crop plant?

13 How can the health of daughter plants be put at risk by vegetative propagation?

Artificial propagation

Vegetative reproduction or propagation can also be performed artificially by the methods of **budding**, **grafting**, by **cuttings** and by **culture** in different media.

Grafting

A graft in the form of a small vegetative part of a plant, for example a bud or a short stem apex (see Figure 10.10), is called the **scion**. This is grafted on to a different individual of the same species called the **stock**. In this way, a herbaceous perennial with certain advantageous features, such as prolific fruiting or a certain flower colour, is grafted onto a vigorously growing root stock, usually a wild variety of the species. For example, apple is grafted onto wild crab apple stock and roses onto wild briar stock. In grafting, the vascular bundle cambium of the stock and scion grow together, the two being bound together at a firmly made joint with waterproof waxed tape.

Figure 10.10   (a) budding and (b) grafting – two methods of artificial propagation

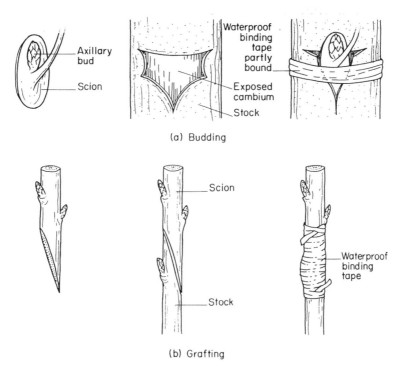

(a) Budding

(b) Grafting

Cuttings

This is a method of stimulating the formation of **adventitious roots** by means of root-forming auxins or plant growth substances. Leaves and stems of soft herbaceous or woody perennials can be propagated artificially by leaf or stem cuttings. The cut end of the cutting is dipped in water and then in a powder containing **plant growth substances**, before being set in sand or a special rooting compost which provides large air spaces for respiration as new root tissue grows. Dandelions propagate readily from small portions of root or stem cuttings.

Culture media

A sterile mixture of nutrients in water, either as a solution or gel, is kept at a steady temperature and a suitable pH and aerated with oxygen. It is used as a culture medium. The medium can be used to grow or reproduce bacteria, yeasts and unicellular algae, or to keep pieces of plant and animal tissue or whole organs alive.

Questions

14 In what ways is grafting different from taking a cutting?
15 How would you expect a cutting to develop if it was placed in a compost with small air spaces? Explain your answer.
16 Why do special culture media help in artificial propagation?

10.4 Sexual reproduction

Sexual reproduction occurs widely amongst living organisms. Sexual reproduction occurs between parent organisms each providing *half* the amount of genetic material which combines to form individual organisms of the same species. **Gametes** are sex cells, containing half the amount of genetic material of the parent organism. **Fertilisation** is the process of fusion of the sex cell or gamete nuclei, or a combination of genetic material to form a **zygote**, which later grows into a new organism having the *same* kind of genetic material as the parents.

Individuals may be either male or female, or hermaphrodite – having both male and female reproductive organs as in the tapeworm, earthworm and many flowering plants.

Gametogenesis is the process of forming gametes or sex cells, each having *half* the genetic material of the plant or animal parents.

Types of gametes

Isogamous gametes form from two different sexes but are *identical* to each other in form or size and shape. This type of sex gamete is seen in protoctists and fungi.

Anisogamous gametes form from two different sexes but are not alike. The **female** gamete is usually a large, non-motile cell called an **ovum** (plural ova), and it also has a reserve of food. The **male** gamete is small and motile. Fertilisation between anisogamous gametes is called **oogamy**.

Questions

17 What is as gamete?
18 What is an hermaphrodite?
19 What happens in (a) gametogenesis and (b) fertilisation?

10.5 Sexual reproduction in flowering plants

Flower structure

There are over 300 000 different species of flowering plants, each having a different flower structure (see Figures 10.11 and 10.12). A **complete flower** has *four* sets of

Figure 10.11 The structure of a flower of a typical dicotyledon

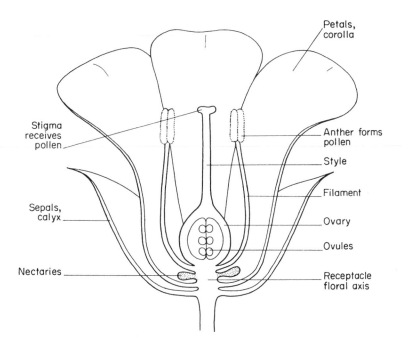

Figure 10.12 The structure of a flower of meadow grass, *Poa* sp., a monocotyledon

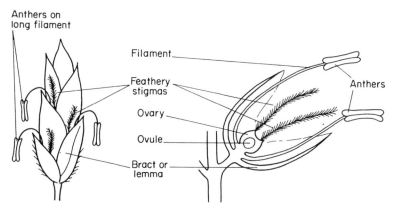

floral parts arranged in rings attached to a swollen stem tip called a **receptacle**, for example the buttercup flower. An **incomplete flower** may have a set of floral parts missing, as in the grasses.

The **calyx** is the outermost set composed of mainly green, leaf-like, protective **sepals**. The **corolla** is composed of brightly coloured, sometimes scented **petals**. They may be reduced or absent in certain incomplete flowers. The calyx and the corolla together are called the **perianth** and are indistinguishable in the bluebell, daffodil and tulip.

The **androecium** forms the male part of the flower. It consists of **stamens**. Each stamen is formed from a **filament** and an **anther.** The anthers split to shed their **pollen**. The **gynoecium** forms the central part, consisting of a **pistil** composed of **carpels** containing **ovules**, and a **stigma** connected by a **style**. Close to the base of the **ovary** are nectar-secreting **nectaries** in some flowers visited by insects.

Flowers are generally **bisexual** or hermaphrodite in having both male and female floral parts. Many flowers are **unisexual**, having either male or staminate flowers, or female carpellary flowers. Oak and hazel have male and female flowers on the same plant, whilst the holly and willow have male and female flowers on separate plants.

Flower variation

A plant may have a single, solitary flower on a stem, e.g. a daffodil. Others have clusters of flowers forming an **inflorescence**, as in the rhododendron. Petals are separate in one large group of dicotyledon plants including the rose, buttercup and wall flower, and joined together in tube form in other groups, as in primrose, foxglove and sage.

Questions

20 How are sepals and petals different?
21 Use Figure 10.11 and the text in this section to distinguish between a stamen, a stigma and a style.
22 The receptacle and the nectaries are found close together but have different functions. What are they?

Pollination

Pollination is the transfer of pollen from **anther** to **stigma**. Pollen grains are released from the ripe anthers and can be used in self- or cross-pollination. **Self-pollination** occurs when the flower stigma receives pollen from stamens of the *same* flower, or from stamens of other flowers on the *same* plant. The garden pea, used in heredity experiments, self-pollinates before the flower bud opens.

Cross-pollination occurs when the pollen from the stamens of a flower is transferred to the stigma of *another* flower of the *same* species. Cross-pollination will naturally occur between single-sex flowers, e.g. holly and willow.

Protandrous flowers have anthers which ripen *before* the pistil, for example the dead nettle, dandelion and foxglove. **Protogynous** flowers have pistils maturing *before* the anthers, for example the horse chestnut, plantain and bluebell.

Pollination agents

- **Artificial crossing:** This is carried out in cross-breeding experiments (see Section 12.7).
- **Wind pollination:** This is the means of transferring millions of pollen grains by air currents. Most trees and all grasses are wind pollinated by pollen of low density.
- **Insect pollination:** This occurs by insects being attracted to flowers by the flower's petal colour, scent and nectar. The sticky pollen is of a high density and clings to the insect's body bristles or hairs.

Table 10.2 compares insect-pollinated and wind-pollinated flowers.

Table 10.2   Comparison of wind- and insect-pollinated flowers

Flower part	Wind pollination	Insect pollination
Petals	Small, often green-coloured, unattractive, without nectar or scent	Large, brightly coloured, attractive, with nectar and often scented
Stamens	Anthers make large amounts of small, dry, low-density, smooth-surfaced pollen grains. Easily wind-borne. Filaments long, dangle the anthers loosely outside the flower	Anthers make small amount of large, sticky, high-density, rough-surfaced pollen grains. Clings well to insect body. Filaments are short, keeping anthers within the flower
Carpel	Stigma and style are large and feathery, hanging outside flower	Stigma is sticky and flat with short style inside the flower
Examples	Hazel catkins, grasses, willow	Bean, rose, apple

Questions

23 What is the difference between self-pollination and cross-pollination?
24 How have plants developed to prevent self-pollination?
25 Why does wind-pollinated pollen have a low density and insect-pollinated pollen have a sticky surface?
26 How is the flower of an insect-pollinated plant different from that of a wind-pollinated plant. Explain why the two kinds of flower are different.

Fertilisation

Gametes are male or female sex cells produced by the sex organs. They contain *half* (n) the amount of genetic material of the body cell nuclei and are called **haploid** (n), compared to twice the amount of genetic material present in body cell nuclei which are called **diploid** (2n). The process of gamete formation is called **gametogenesis**.

Fertilisation is the process of combination or fusion of male and female gamete genetic material to form a single diploid nucleus or a **zygote**, the first-formed body cell.

male gamete		female gamete		ZYGOTE
HAPLOID	+	HAPLOID	=	DIPLOID body cell
1/2 genetic material	+	1/2 genetic material	=	genetic material
n		n		2n

Pollen of the same plant species germinates on the sticky **stigma**, which provides a solution of sugars to feed the pollen cell. A **pollen tube** forms containing a **tube nucleus** and two **male gamete nuclei**. The pollen tube grows *downwards*, feeding on the tissues of the style, carrying the two male nuclei to the embryo sac.

The **embryo sac** has an opening called the **micropyle** and contains the **female gamete** and **endosperm cell nuclei**. Double fertlisation occurs in the embryo sac. The male nucleus first joins with the female nucleus to form the **embryo zygote**. The other male nucleus then joins with the endosperm nucleus to form the **endosperm zygote**. Endosperm is produced as a food reserve for the embryo by repeated division of the endosperm zygote by mitosis.

Figure 10.13 Life cycle of a dicotyledon flowering plant, showing pollination and fertilisation

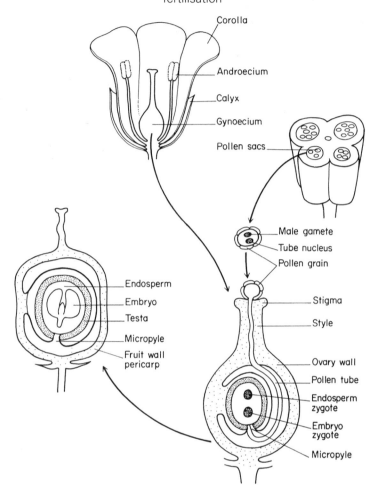

The embryo develops a **radicle, plumule** and one or two **cotyledons** and is surrounded by a **testa**. The placenta connects the seed with the fruit wall or **pericarp**.

Most flower parts wither and fall away, but some may remain attached to the fruit like the stigma and style in the apple or the sepals in the tomato. The receptacle may become fleshy as in the strawberry, or may completely surround the pericarp or 'core' as in the apple.

Questions

27 How does the male gamete nucleus reach the female gamete nucleus from the pollen grain?

28 What is produced by the double fertilisation in the embryo sac?

29 What happens to petals and stamens after fertilisation?

30 What does the ovary wall become after fertilisation?

The flowering plant life cycle is summarised as follows:

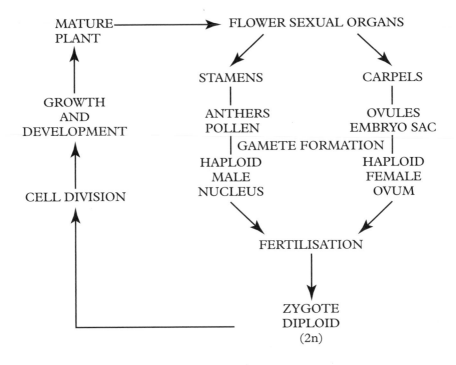

Fruits

The fruit wall or pericarp serves to protect the developing seed and to provide a food supply for non-endospermic seeds such as peas and beans. Fruits are also the means of **dispersing** the seeds. Dispersal is summarised in Table 10.3.

Table 10.3 Dispersal methods of fruits and seeds

Dispersal method	Features and example
Mechanical dispersion	Stresses develop within drying fruits, shrinkage causes fruit to split open and sides curl back flinging out seeds, e.g. balsam, lupin, gorse and other pods
Wind dispersal	1 Winged fruits and seeds with aerofoil structure, e.g. sycamore, lime, ash, elm and pine
	2 Censer mechanism for dry capsule fruits on long stalks, e.g. 'pepper pot', poppy, campion
	3 Parachute fruits, e.g. dandelion, thistle and willowherb
Animal dispersal	1 Hooked fruits cling to animal coats, e.g. goosegrass, burdock and agrimony
	2 Succulent fruits attract birds, seeds stick to feet, beaks, or pass through digestive tract and are distributed in faeces
Water dispersal	Water-resistant, air-filled fruit walls for buoyancy, e.g. coconut and water lily

The structure of fruits varies considerably. Some are succulent and juicy and others are dry. Some dry fruits split open when ripe (see Figure 10.14).

Figure 10.14 Examples of dry and succulent fruits: (a) sycamore, *Acer* sp., (b) rose hip, *Rosa* sp., (c) plum, *Prunus* sp., (d) oak acorn, *Quercus* sp.

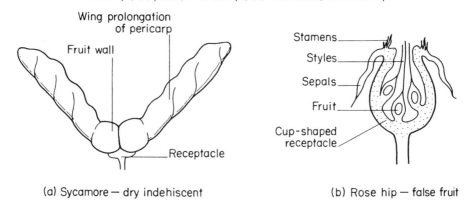

(a) Sycamore — dry indehiscent

(b) Rose hip — false fruit

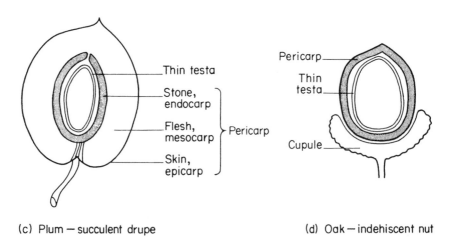

(c) Plum — succulent drupe

(d) Oak — indehiscent nut

Questions

31 How is mechanical dispersal different from wind dispersal?
32 What is the difference between wind pollination and wind dispersal?
33 How may animals disperse seeds?
34 What is the difference between a false fruit and a true fruit?
35 Distinguish between a berry and a nut.

Table 10.4 outlines the classification of fruits.

Table 10.4 Classification of fruits

Fruit type	Structure and example
False fruits	Develop from the receptacle which becomes fleshy: apple and pear, strawberry and rose-hip
True fruits	Wall of fruit, pericarp, forms from ovary wall
1 Succulent fruits	Brightly coloured, one-seeded, fleshy fruits
Drupe	Seed enclosed by woody stone, with flesh and skin, forming the pericarp: plum, cherry, peach
Berry	Many-seeded, without stone, with surrounding flesh and skin of pericarp: orange, tomato, gooseberry
2 Dry fruits	Either dehiscent, split open, or indehiscent, do not open, when ripe
(a) Dehiscent	
Legume	Pod: bean, peas and lupins
Capsule	Box-like fruit: poppy, chestnut
(b) Indehiscent	
Nut	With woody wall: hazel, acorn
Achene	One-seeded fruit: buttercup. True fruit: strawberry and rose
Caryopsis	Grass and grain fruit with seed coat and fruit wall forming 'husk'; oat, wheat, maize and grasses

10.6 Sexual reproduction in simple animals

Hydra

Hydra, a cnidarian, is hermaphrodite in having both kinds of gonads – **ovaries** and **testes** – which mature at different times on the same individual. This prevents self-fertilisation. Sperm are released into the surrounding water and are attracted to the ripe ovary of another *Hydra* where fertilisation occurs internally to form a zygote (see Figure 10.15).

Figure 10.15 Sexual reproduction in *Hydra* sp.

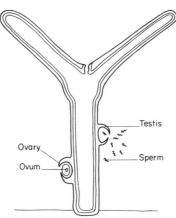

The embryo is then released. It is surrounded by a protective chitinous shell and in this way survives adverse conditions.

Earthworm

Annelids, e.g. earthworms, are hermaphrodite and reproduce entirely by sexual methods. Sperm is exchanged between two different individuals by **copulation** (see Figure 10.16); this prevents self-fertilisation.

Figure 10.16 Sperm exchange between copulating earthworms, *Lumbricus* sp.

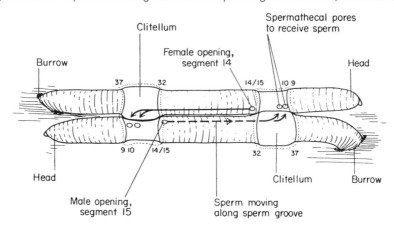

The exchanged sperm is used to fertilise the ova **internally** and these are laid as eggs in a cocoon. The young develop directly into worms without a larval stage.

Tapeworm

The tapeworm, a flatworm (see Figure 2.23), is hermaphrodite with the male gonads ripening before the female gonads. Both kinds of gonads are contained in the same ripe segment or **proglottid**. Self-fertilisation is therefore prevented. Cross-fertilisation occurs internally between proglottids of different worms, or different proglottids on the same individual. The fertilised eggs are surrounded by a shell and almost fill the ripe segments at the end of the tapeworm body.

Insects

Insects reproduce by copulation between a male and a female insect. The fertilised eggs undergo complete or incomplete metamorphosis (see Section 11.6). **Parthenogenesis** is the development of an ovum without it being fertilised by a male gamete. Aphids or greenflies produce eggs by mitosis which develop into wingless females. These can also reproduce normally by sexual methods.

Questions

36 How is self-fertilisation prevented in *Hydra*, the earthworm and the tapeworm?
37 How is parthenogenesis different from sexual reproduction?

10.7 Sexual reproduction in chordate animals

Fish

Fish, like most chordates, have separate sexes, male and female. Fertilisation in bony fish such as herring, salmon and trout is **external**. The eggs are small and are laid in great numbers into the surrounding water, where they are fertilised by sperm released over the eggs by the male. Fish 'soft roe' are male testes and 'hard roe' are the female's ovaries.

The young fish (fry) feed on plankton and no parental care is shown by either parent except in the stickleback.

Frog

The sexes are separate, female and male; the male is recognised by a pad on the first digit of the forelimb. In spring the female becomes swollen with eggs, each with a food reserve of yolk and coat of albumen. Mating occurs in the water. The males show sexual behaviour in croaking to attract females. The male mounts the female's back and sheds sperm in a steady stream over the ova as they are laid. The male and female frogs move about in this condition for several days until all the ova and sperm are shed. Fertilisation is therefore **external**.

Albumen swells in water and provides a protective coat, and spaces the eggs facilitating respiration. The eggs undergo metamorphosis as described in Section 11.6. During metamorphosis there is no parental care shown by either parent.

Bird

In birds the sexes are separate. The male is called the **cock** and the female is called the **hen**. The cock bird may usually be recognised by a brighter, more colourful plumage. In the spring the male bird claims a territory and the birds commence pairing. **Courtship** and **display** are elaborate forms of sexual behaviour, followed by **internal fertilisation** when the cock mounts the hen, passing sperm into the opening of the oviduct.

Nest-building involves the preparation of a structure made from many different materials, or a simple structure composed of a few stones or a hollow in the ground. A few eggs are laid, each egg (see Figure 10.17) composed of a central **yolk** food reserve with its **embryo** positioned in a germinal disc. Chalaza cords prevent rotation of the yolk sac. The egg white or **albumen** provides the developing embryo with water, and provides hydrostatic support and protection.

Figure 10.17 Internal structure of a fertilised bird's egg

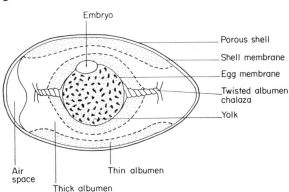

The shell is porous, allowing gas exchange into the air space which is essential for the chick's respiration. The young embryo becomes surrounded by an **amnion** or water-filled protective sac, and is connected to the yolk sac by an **umbilical cord**. A third sac called the **allantois** is well supplied with blood vessels. It functions in respiration and stores excretory waste.

After an **incubation period** when the eggs are maintained at a constant temperature by the hen bird, the young chick breaks out of the egg. A period of **parental care** follows after hatching, when the young offspring are protected and fed by both parent birds. Later they are taught to feed themselves.

--- Questions ---

38 What is the difference between external and internal fertilisation? Give examples of each kind.
39 What happens to the albumen in frogs' eggs? How does the change (a) increase the survival of the developing embryos, (b) speed up the development of the embryos?
40 How can birds maintain the size of their population by laying few eggs while fish and frogs have to lay large numbers of eggs to maintain their populations?
41 Why is the birds' egg shell porous?
42 (a) How would the development of the embryo be affected if the yolk could move freely inside the shell and rest against the shell wall?
   (b) What features of the egg keep the yolk in place?

10.8 Sexual reproduction in mammals

Mammals have separate sexes, with male and female individuals. Development of the embryo can be within a soft-shelled **egg**, as in the platypus, or immature, live offspring are produced which complete their development within a marsupium or **pouch**, as in the case of the marsupial mammals, the kangaroo and koala. Most mammalian embryos develop within the mother's **uterus** and are protected and nourished by embryonic membranes and a **placenta**. Such placental mammals are called **viviparous**. The rat, rabbit and human are examples of viviparous mammals. All young mammals feed on milk produced by the **mammary glands** of the mother. Mammals show considerable parental care of the young. This form of behaviour is seen in feeding, protection and teaching skills essential for their survival and life continuation.

10.9 The human reproductive system

The male

The male reproductive system is shown in Figure 10.18. The haploid **spermatozoa** are formed by **spermatogenesis** from cells lining the seminferous tubules, present as long coiled tubules within each **testis**, contained in a **scrotal sac** outside the body.

Each spermatozoon (see Figure 10.19a) consists of a **head** containing a nucleus with half the genetic material, and a short **neck** containing the energy-providing mitochondria needed to propel the **tail**.

Sperm enter the **epididymis** and are temporarily stored there and also in the **vas deferens**. **Ejaculation**, or expulsion of the sperm, occurs when the sperm is forced

Figure 10.18 The human male reproductive system

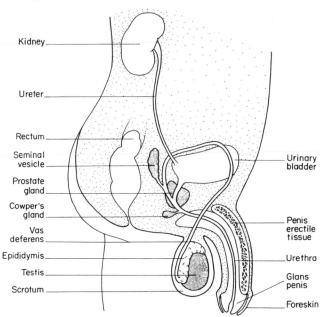

into the urethra by a reflex muscular action as a result of sexual excitement during copulation. Between 90 and 600 million sperm are produced in 3 cm³ of ejaculate or **semen**. This fluid contains alkali, sugars and proteins from the seminal vesicles, Cowpers (bulbourethral) gland and prostate gland.

The **penis** becomes engorged with blood and is rigidly extended by its hydrostatic skeleton.

Figure 10.19 (a) Spermatozoon and (b) ovum (not to scale)

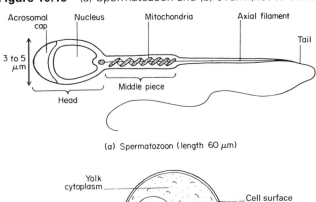

(a) Spermatozoon (length 60 μm)

(b) Ovum (diameter 100 μm)

Questions

43 What are seminiferous tubules and where are they found?
44 What are the structures through which the sperm pass as they leave the male?
45 What other substances are found in the semen and where are they added to it?

The female

Figure 10.20 The human female reproductive system

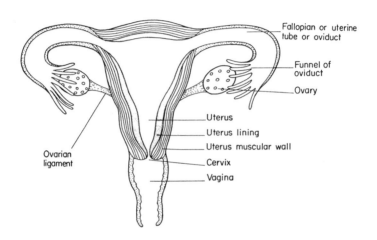

The female reproductive system is shown in Figure 10.20. Two **ovaries** lie within the abdominal cavity. Each ovary in a newborn female contains about 200 000 future haploid (n) ova, of which only about 400 will be released in a reproductive lifetime of about 30 years.

In a sexually mature ovary, one egg cell can develop each month within a fluid-filled **follicle**, which bursts, releasing the egg cell in the process of ovulation. Usually the two ovaries ovulate alternately each month. The tissue left by the follicle becomes a yellow body or **corpus luteum** (see Figure 10.21).

Figure 10.21 Ovum formation stages in the human female ovary

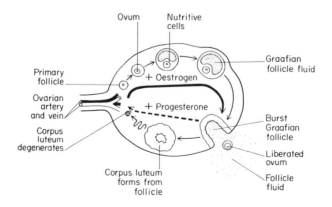

The egg cell falls into the body cavity and is collected by the **oviduct** funnel. In the oviduct, the egg cell becomes a mature ovum consisting of a nucleus surrounded by a yolk and cell surface membrane (see Figure 10.19b). The uterine tube or oviduct contains ciliated epithelium which moves the ovum into the uterus, where it may become **implanted** as an embryo if fertilised or leave by way of the vagina as an unfertilised ovum.

Questions

46 What is the path taken by the ovum after it has left the ovary?
47 Account for the differences seen in the egg and the sperm.

Sexual development

Puberty is the stage of the beginning of sexual maturity. In girls it occurs generally between 12 and 15 years of age and in boys between 13 and 17 years.

Secondary sexual characteristics

These appear in the growth and enlargement of the reproductive organs, namely penis, testes, vagina, uterus and ovaries. **Sperm production** and ejaculation commence in boys and **ovulation** occurs in girls.

The body changes in shape and appearance. Boys grow hair on the face, armpits and pubic region and the voice deepens, whilst limb and chest muscles develop. Girls develop hair on the armpits and pubic region, whilst the pelvic girdle widens and breasts develop. Skin becomes softer and fat deposits beneath the skin form and give shape to the body.

The sex hormones, **androgens** in males and **oestrogens** in females, are secreted by the developing gonads and are responsible for secondary sexual characteristics.

Adolescence

Puberty is that process of maturing of the reproductive organs which can last from 2 to 4 years. Apart from the sexual characteristics, it is seen as a growth spurt which continues in adolescence. This spurt stops in girls when the ovaries become functional and stops in boys at about 20 years of age. Puberty can be considered as the start of the adolescent period, between childhood and manhood (13–25) or womanhood (12–21).

Fertilisation

The menstrual cycle

The menstrual cycle (see Figure 10.22) occurs in the sexually mature female and is seen as a shedding of the uterus endometrium (lining) together with a quantity of blood in the act of **menstruation**. This happens at the beginning of the menstrual cycle.

- The uterus lining, the endometrium, breaks up and is discharged with blood and mucus as the menses by way of the vagina. This may last for 5 to 7 days.
- A follicle begins to develop and a new endometrium is formed. Ovulation with release of an egg cell occurs some 14 days after the start of menstruation.
- The ovum may or may not be fertilised and meanwhile the endometrium thickens. The remains of the follicle become a corpus luteum which secretes the hormone **progesterone**.
- Fourteen days after ovulation, unless fertilisation occurs, menstruation occurs again.

Figure 10.22 Summary of changes occuring in the female menstrual

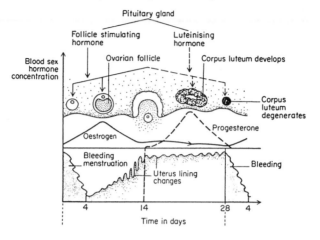

Questions

48 Look at Figure 10.22 and describe how the levels of oestrogen and progesterone vary during the course of the menstrual cycle. Relate these changes to developments in the ovary and the uterus wall.

Sexual intercourse

Sexual intercourse or copulation occurs when the erect **penis** is inserted into the **vagina** and the semen containing sperm is **ejaculated**. The sperm pass, partly by swimming and partly by the peristalsis of the female reproductive organs, through the uterus to reach the oviduct.

The **acrosome** or tip of the sperm's head digests the outer membranes of the ova and the *haploid* (n) sperm nucleus joins with the *haploid* (n) female nucleus to form a *diploid* (2n) zygote. No further sperm enter the fertilised egg cell due to the formation of an impenetrable fertilisation membrane. The human life cycle is summarised below.

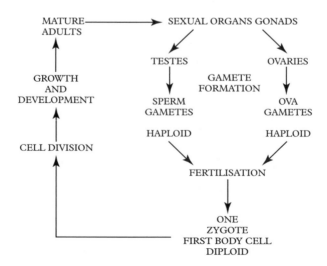

An egg cell or ovum is only present in the oviducts for a few days after ovulation. It is therefore necessary for sperm to be in the oviduct immediately before and during this period for fertilisation to take place. Sperm can remain alive in the oviduct for up to 3 days after intercourse.

Sexually transmitted diseases

Sexually transmitted diseases (STDs) are infections such as **gonorrhoea** (see Table 14.5), **syphilis** and **Herpes** sores. These infections can only be acquired through having sexual relations with an infected person, and usually appear as soreness, itching and pus discharge around the vagina and penis. **Scabies** or 'the itch' and **pubic lice** or 'crabs' are infestations passed from an infested person through close sexual or other contact.

AIDS

AIDS (Acquired Immune Deficiency Syndrome) can be acquired by sexual relations and by contact with infected blood entering cuts or scratches of an uninfected person. Symptoms of AIDS include swollen glands, tiredness, weight loss, breathlessness and dry cough.

Birth control

The following are methods of birth control or contraception in order of decreasing efficiency.
- The **oral contraceptive pill** is a mixture of hormones, oestrogen and progesterone. These two hormones are produced during the menstrual cycle. When taken as tablets over 21 days they prevent ovulation.
- An **intra-uterine device** (or IUD) is a metal or plastic strip or coil which is inserted inside the uterus. Its action prevents the implantation of the embryo.
- Barrier methods include the sheath or **condom**, the diaphragm or **cap** and the female condom. They are used with sperm-killing **spermicides**. All are made from thin sheets of rubber. The condom is fitted over an erect penis. The diaphragm fits over the cervix and the female condom lines the inside of the vagina.
- Unreliable methods include using spermicides alone or intercourse outside the ovulation period called the '**safe period**'.

Sterilisation methods

- **vasectomy** is an operation involving the severing, parting and tying of the vas deferens. Erection of the penis and ejaculation of sterile semen continue after the operation but there are no sperm in the semen.
- In **tubal ligation** the oviducts are cut, parted and tied, or the tubes are completely removed. It has no effect on the menstrual cycle.

Questions

49 How is fertilisation prevented by the (a) oral contraceptive pill and (b) the condom?
50 Which method of contraception does not prevent fertilisation?
51 The time of ovulation may be affected by the general health of the female. Why does this affect the reliability of the 'safe period' method of contraception?

Development and pregnancy

The fertilised egg or zygote becomes **implanted** in the endometrium of the uterus wall. The menstrual cycle stops, no further follicles develop and the corpus luteum becomes an important gland producing the hormone progesterone.

The zygote nucleus proceeds to divide by mitosis to form two, then four and ultimately a ball of cells called a **blastula**. If the two first-formed cells part they develop into **identical twins**. If *two* ova are fertilised together they form **fraternal twins**. The blastula develops into two main structures – the **embryo** and its **placenta**.

The placenta forms a close association between the embryo and the mother's tissues. Numerous villi with blood capillary vessels project into blood-filled spaces provided from the mother's blood. The great number of villi increase the surface area for exchange of substances by diffusion to and from the blood spaces.

Food nutrients, protective antibodies and water pass from the blood spaces into the villi, whilst the special form of haemoglobin in the embryo draws up oxygen rapidly from the blood spaces. Waste: carbamide (urea) and carbon dioxide, diffuse into the blood spaces for removal in the mother's circulation. The placenta acts as a gland in producing hormones called **gonadotrophins**, also oestrogen and progesterone, all of which control pregnancy.

Figure 10.23   The human embryo in the uterus

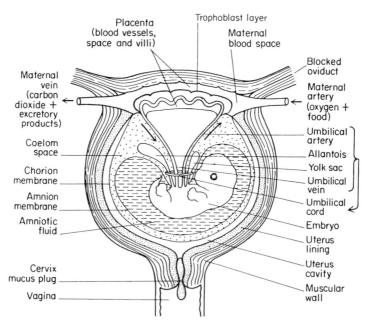

The **amnion** is a sac which is filled with **amniotic fluid**. After two months the human embryo (see Figure 10.23) is called a **fetus**, and is surrounded by shock-absorbing amniotic fluid. The fetus has a blood system and its heart pumps blood to the placenta by way of the umbilical artery in the **umbilical cord**.

Abortion is the termination of a pregnancy before the 28th week. Abortion is legal in certain countries when performed by qualified surgeons but is a criminal act if performed otherwise. **Miscarriage** is accidental abortion due to various natural causes.

52 What is the difference between identical and fraternal twins?
53 What is the purpose of the villi in the placenta?
54 Occasionally four or even five eggs may be fertilised at the same time.
   (a) How will this affect the growth of the placenta of each embryo and fetus?
   (b) How will the growth of the placenta affect the development of the embryo and fetus?

Prenatal care

Fetal prenatal care

The fetus requires the following as components of its constant homeostatically controlled environment:
- **Internal environment**:
  - nutrients for growth;
  - oxygen for respiration;
  - steady temperature or warmth.
- **External environment**:
  - protection from disease, mechanical injury and stress;
  - space to move about and exercise in.

The mother provides and maintains this steady environment by homeostatic control via the placental membranes and umbilical cord.

Mother's prenatal care

Pregnancy makes additional demands on the mother in extra growth and extra nutrition. This results in an overall increase in body weight of about 10 kg. Growth increase of the uterus, vagina and breasts occurs. Relaxation of the pelvic ligaments is necessary to accommodate the fetus and allow its exit at birth. Nutritional requirements increase in order that fetus, breasts, fetal membranes and reproductive tract can grow.

Health care is required with avoidance of alcohol, smoking and drugs. These can be transferred to the fetus across the placenta. **German measles** or rubella is a virus disease and causes deformities in the fetus if the mother is infected during the first three months of pregnancy. The main fetal defects are deafness, eye and heart disorders. Girls of 10 to 13 years are immunised with rubella vaccine in the UK.

Postnatal development

Postnatal development is the human development process *after* and *including* birth.

Birth

This is the process at which the fetus becomes separated from the mother. The act of birth is called **parturition** and occurs nine months after fertilisation; the period of time is called **gestation**.

Hormonal changes occur in the mother's blood system which involve an increase in oestrogen levels leading to contraction of the strong uterus wall muscles. Progesterone prevents this occurring during pregnancy. **Oxytocin**, a hormone secreted by the pituitary gland, causes uterus wall muscle contraction, which happens at regular intervals in 'labour'. The cervix dilates and the amnion bursts, releasing the amniotic fluid or 'waters'.

Figure 10.24 Birth

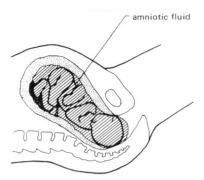

amniotic fluid

1 The amnion is about to rupture
('breaking of the waters')

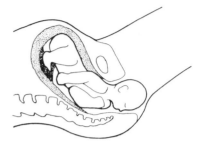

2 The uterine wall contracts, forcing out the head

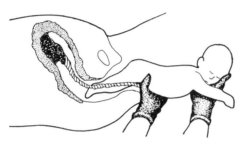

3 The baby is born

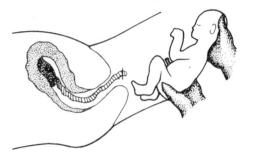

4 The umbilical cord is tied and cut

The baby is usually born head first and begins breathing unaided within a few seconds. Following the birth, the placenta is expelled as the '**afterbirth**'; the 'caul' is sometimes seen on the baby's head and is the remains of the amnion.

The **mammary glands** are stimulated to produce **milk** during the lactation period of breast-feeding by the hormone **prolactin** from the pituitary gland. The normal menstrual cycle begins again after the birth although it is suppressed by regular breast-feeding. Table 10.5 summarises the important sex hormones and their functions in women.

Questions

55 Describe the changes that take place to (a) the amnion and amniotic fluid, (b) the fetus and (c) the placenta during the birth process.

Lactation – breast-feeding

Breast milk composition

Breast milk differs considerably from the bottled milk feeds prepared from cows' milk. The following summarises the main components of breast milk compared to artificial bottle feeds made from cows' milk:

- Protein is mainly **lactalbumin** compared to cows' **caseinogen**. It is more easily digested than the tough casein curds. It also provides a different mixture of amino acids and more essential amino acids for human tissue formation.

Table 10.5 Sex hormones and their function in women

Gland or source	Hormones	Functions
Pituitary (brain)	FSH: follicle-stimulating hormone	1 Causes ovary follicles to develop
		2 Stimulates ovary to secrete oestrogen
	LH: luteinising hormone	1 Causes ovulation – egg release
		2 Causes corpus luteum to form from follicle
		3 Stimulates progesterone secretion
	Prolactin	In lactation, promotes milk formation and secretion
	Oxytocin	1 Causes uterus to contract and birth of fetus
		2 Causes milk flow
Ovary	Oestrogen	1 Causes uterus wall changes before ovulation
		2 Secondary sexual characteristics, breast development, pubic hair
	Progesterone (corpus luteum)	1 Causes uterus wall changes after ovulation
		2 Inhibits oxytocin
Placenta	Gonadotrophins, oestrogen, progesterone	1 Maintain the corpus luteum function
		2 Maintain pregnancy

- The carbohydrate present in both milks is **lactose**; breast milk contains 33% more than cows' milk.
- Vitamins A, D and C are present in *greater* amounts in breast milk.
- Minerals: **iron** content in both milks is *low*. Ample **calcium** is present in both milks. **Sodium** in cows' milk is four times that of human milk and could be harmful to the baby's kidneys which have to excrete the surplus sodium ions.

Colostrum

Colostrum is the first-formed milk produced in the first 5 days following birth. The protein in this milk is almost twice the amount of normal breast milk. This is mostly due to the presence of protective antibodies and extra amino acids for body tissue formation.

Advantages of breast-feeding

Breast-feeding has many advantages over bottle-feeding. They are summarised as follows:
- The composition is uniform and correct for human growth rate.
- It is more hygienic, being clean and sterile, and is prepared at the correct temperature.
- Protective antibodies are present in colostrum and to a lesser extent in later milk. It is non-allergic compared to proteins present in cows' milk.

• It is psychologically beneficial to mother and child in establishing close bonds.
(**NB**: Consumption of alcohol, certain drugs and tobacco smoking will all affect the milk composition and be passed onto the suckling child.

10.10 Parental care

The young baby is suckled on milk and gradually weaned onto a mixed, balanced diet. The home provides an environment where the parents can care for the child. Injuries due to accidents are reduced with parental supervision and the home offers some protection from disease.

Small babies have a high surface area/volume ratio and can lose heat rapidly. This is overcome by heat-producing brown fat layers in the skin. The homeostatic heat control is not fully developed in the newborn child.

The long period of parental care seen in human beings involves teaching by adults and learning by the child. Children learn through play and the examples set them by the adults in their family.

Summary

Table 10.6 compares sexual and a sexual reproduction.

Table 10.6 A comparison of sexual and asexual reproduction

Feature	Sexual	Asexual
Parents	Two parents	Single parent
Gametes	Male and female gametes produced by gametogenesis	No gamete production
Zygote	Diploid zygote formed by fertilisation	No zygote formation
Genetic material	Halved for haploid gamete formation	Same amount as parent cell nucleus
Offspring	Variation in offspring, with hybrid vigour	Offspring identical to parent, i.e. a clone
Rate	Not rapid	Rapid in favourable conditions
Population increase	Population numbers increase slowly	Population numbers increase rapidly
Organisms	Occurs amongst plants and animals	Mainly among protoctists, bacteria and fungi
Methods	By sexual organs, in flowers, testes and ovaries	By cell division, binary fission, fragmentation, budding, spores, vegetative reproduction and artificial propagation

- Examples of asexual reproduction are binary fission, budding, spore formation and cloning. (▶ 246)
- Flowering plants have life cycles which range in length from a few weeks to many years. Perennating organs allow some flowering plants to reproduce vegetatively. (▶ 249)
- Flowering plants can be propagated artificially by grafting and taking cuttings. (▶ 253)
- Flowers contain the sexual organs of flowering plants and are adapted for the transfer of the male gametes using insects and air currents. (▶ 256)
- After fertilisation in plants, seeds are produced enclosed in fruits which are dispersed mainly by the air currents an animals. (▶ 259)
- Some animals are hermaphrodites but do not fertilise themselves. (▶ 261)
- Chordates such as fish and frogs reproduce by external fertilisation and produce large numbers of eggs which receive no parental care. Birds reproduce by internal fertilisation and produce a small number of eggs. The eggs and chicks are cared for by the parent. (▶ 263)
- The human is a placental mammal in which internal fertilisation is followed by a nine-month gestation period. (▶ 264 to 274)
- The secondary sexual characteristics, menstruation and pregnancy are controlled by hormones. (▶ 267 and 273)
- There is a range of birth control methods. (▶ 269)
- Sexual intercourse can pass on infestations of lice and a range of sexually transmitted diseases including AIDS. (▶ 269)
- Young humans take many years to become mature and parental care is needed thoughout this time. (▶ 274)

Growth and development

Objectives

After reading this chapter you should be able to:
- describe the growth process and identify the types of growth in living organisms
- describe the structure and contents of the cell nucleus
- understand the process in which cells divide by mitosis
- explain the difference between mitosis in plant and animal cells
- understand the structure of genetic material, DNA, and the functions of m-RNA and t-RNA
- compare the seeds of monocotyledon and dicotyledon flowering plants
- understand metamorphosis in insects and the frog
- explain the advantages of dormancy
- describe the growth regions in plants and animals
- understand how to measure growth in different ways
- plot growth curves
- describe the factors that affect growth

During growth, genetic material must pass into new cells, and it must be shared equally. This occurs by a process of nuclear division celled **mitosis**.

11.1 Introduction

Genetic material is made during the **lifetime** of living organisms. This results in an increase in size of the organism's body or in other words, **growth**. Growth results in an increase in size of the organism and a change in its body form. Later in the life span there can be a decrease in size as parts of the organism wither away. When an organism **dies**, no further genetic material is made.

A spectacular change in form is seen during the early stages of certain organisms' life spans. For example, the change from a seed to a seedling then into a mature flowering plant. Similarly, many insects show profound changes in body form from egg into larva and finally into a winged adult. This change in body form is called **development**.

Growth is seen as body size increase.

11.2 Growth of an organism

Growth occurs in organisms when:
- the **input** of energy and raw materials *exceeds* the **output**;
- **anabolism** or synthesis *exceeds* the breakdown process of **catabolism**.

Atrophy, the reverse process of growth, resulting in decrease of body size occurs when:

- the **output** of energy and raw materials *exceeds* the **input**;
- **catabolism** or breakdown *exceeds* the synthesis process of **anabolism**.

Growth processes

- **Synthesis:** Manufacture of new genetic and structural material, plant cellulose and carbohydrates by **photosynthesis** and structural and functional proteins by **protein synthesis**, together with the **storage** of lipids and carbohydrates cause increase in body size.
- **Cell division:** This follows nuclear division. The multicellular human body results from the repeated **division** of one cell producing many millions of body cells.
- **Cell enlargement:** In flowering plants this occurs by the formation of large vacuoles.
- **Cell differentiation:** This occurs in multicellular organisms to form tissues, organs and systems of the complete body.

Types of growth

- **Determinate growth** is seen in organisms which stop growing when a certain body size is reached, or at certain ages; typical of annual plants, birds and mammals.
- **Indeterminate growth** is unceasing growth. It occurs in shrubs, trees, perennial plants, corals, fish and reptiles.
- **Intermittent growth** is growth occurring at certain intervals and is seen in arthropods such as insects and crustaceans which periodically shed their outer skeletons or moult in **ecdysis**.
- **Allometric growth** is the growth of different parts of an organism compared to the whole organism. Flowers grow faster than the vegetative parts such as the leaves and stems. The human brain halts its growth at about age 5 and the reproductive organs grow rapidly between 12 and 18 years of age.

The same growth pattern is seen in all the individuals of the same species and produces a similar external form or morphology of the species.

Questions

1 What is the difference between growth and atrophy?
2 How may an animal body increase in size?
3 How is growth in plants different from growth in animals?
4 How is cell differentiation different from cell division?
5 In an African swamp there is a greater variation in the sizes of the crocodiles than the hippopotamuses. Why is this?
6 A bird searches for an insect on the branch of a tree. How do the growth patterns of these three organisms differ?

11.3 The cell nucleus

The cell nucleus (see section 2.1) is spherical or egg-shaped and contains one or more **nucleoli**. It is the largest cell organelle. A nuclear **envelope** (nuclear membrane) surrounds the genetic material, separating it from the cytoplasm.

The genetic material of all organisms, except that of viruses and bacteria, is composed of **chromatin** which is chemically a protein, **histone**, and the nucleic acid, **DNA** or deoxyribonucleic acid. When the nucleus is about to divide, the chromatin becomes visible as long threads called **chromosomes**. They are invisible in a resting or non-dividing cell.

The nucleus is important in producing new genetic material and for controlling life in a cell. If an *Amoeba* (see page 50) is cut into three portions, one part with the nucleus and two without, the part with the nucleus lives and grows into a complete cell, whilst the two separate parts without a nucleus die.

The cell nucleus growth process – mitosis

Mitosis is the process of *equal division of genetic material* in a nucleus followed by the formation of two identical body or **somatic** cells. Two identical daughter cells form from the division of one parent cell. The time taken for mitosis of one nucleus varies with the species of organism and prevailing temperature. In *Drosophila*, the fruit fly, it takes 7 minutes. In humans it takes 100 minutes. Mitosis is a continuous process, but four main phases are recognised, each of which merges into the other (see Figure 11.1).

Figure 11.1 Mitosis or nuclear division in an animal cell

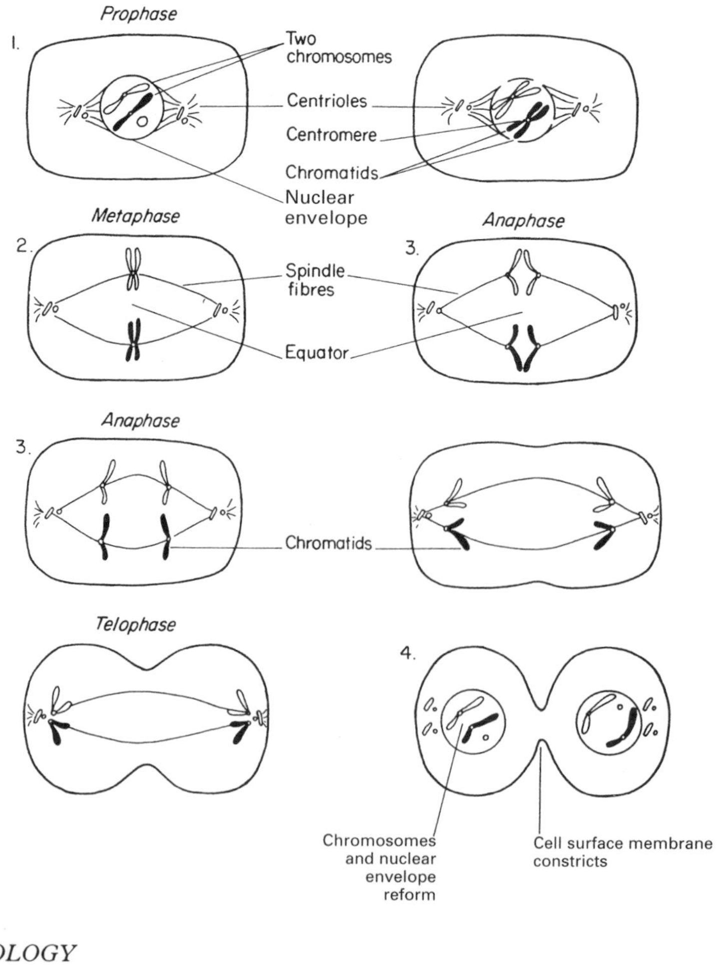

Prophase

- The nuclear envelope disappears.
- Chromosomes become visible as *two* lengthwise halves or **chromatids** which shorten and thicken.
- Nucleoli disappear.
- **Centrioles** in animal cells form poles which are connected to **spindle fibres**. Centrioles are absent in flowering plants.

Metaphase

- Chromatids arrange themselves on the cell equator position or **midline**.

Anaphase

- The chromatids are drawn apart and become chromosomes which move towards the *opposite* poles drawn by the spindle fibres.

Telophase

- Chromosomes reform by an uncoiling process, become thinner and finally disappear.
- Nucleoli reform.
- Nuclear envelope reappears.

__ Question _____

7　Produce a table showing the conditions, actions and positions of the parts of an animal cell during the stages of mitosis.

11.4　The cell division growth process

While the nucleus is dividing by mitosis, the cell cytoplasm and cell wall (in plants) are also dividing (see Figure 11.2). Animal cells and the protoctist *Amoeba* divide by a '**furrowing**' of the cell surface membrane and cytoplasm along the cell equator position (see Figure 10.1). Plant cells develop a **cell plate** along the cell equator position, which grows inwards to meet the opposite cell plate. Cellulose is then deposited on either side of the cell plate, the original cell having formed two daughter cells.

Figure 11.2　Cell wall formation following mitosis in a plant cell

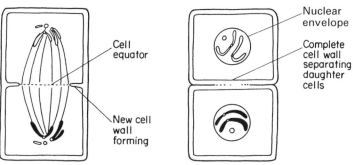

Cell equator

New cell wall forming

Nuclear envelope

Complete cell wall separating daughter cells

Cell growth

Following the process of nuclear and cell division, the new daughter cells enter a period of growth, occupying up to 80% of the cell's lifetime. During this period two important metabolic processes occur, namely **genetic material duplication** and **protein synthesis**.

The main life process of genetic material duplication

The **d**eoxyribonucleic acid (DNA) molecule (Figure 11.3) which constitutes a chromosome in all living things except viruses and bacteria is known to resemble a **double** coil or **helix** which can be compared to two intertwining springs or two sides of a twisted ladder.

Figure 11.3 A model of the DNA molecule showing its twisted form (Santoz Ltd)

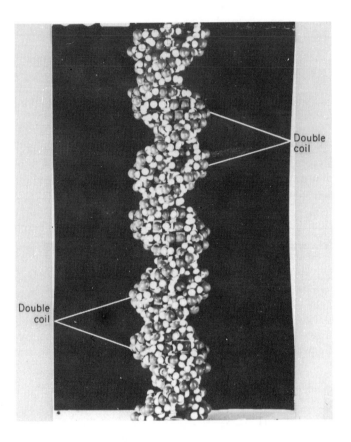

This knowledge is the result of James Watson and Francis Crick's researches. Each strand is composed of deoxyribose **sugar** and **phosphate** units, with thymine, adenine, cytosine and guanine **organic bases** forming rungs in between (see Figure 11.4).

Duplication or doubling of the DNA molecule takes place by untwisting and formation of single strands. Single strands pick up deoxyribose sugar, phosphate and organic base units from the nucleus and reform double DNA molecule coils (see Figure 11.5). The overall result is *doubling* in amount or duplication of DNA or genetic material in preparation for the next nuclear division by mitosis.

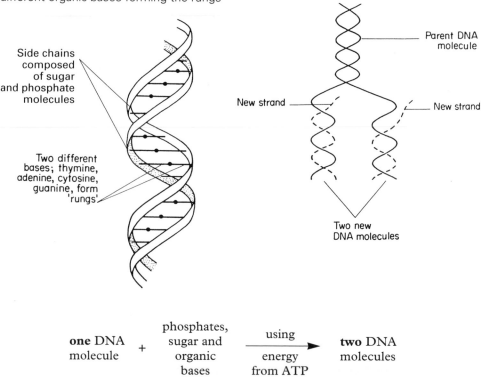

Figure 11.4 The DNA molecule resembling a twisted ladder, with pairs of different organic bases forming the rungs

Side chains composed of sugar and phosphate molecules

Two different bases; thymine, adenine, cytosine, guanine, form 'rungs'

Figure 11.5 The main life process of DNA duplication

Parent DNA molecule

New strand

New strand

Two new DNA molecules

$$\textbf{one}\text{ DNA} \atop \text{molecule} \quad + \quad {\text{phosphates,} \atop {\text{sugar and} \atop {\text{organic} \atop \text{bases}}}} \quad \xrightarrow[{\text{energy} \atop \text{from ATP}}]{\text{using}} \quad \textbf{two}\text{ DNA} \atop \text{molecules}$$

Nutrition provides the raw materials – phosphates, sugar and organic bases; **respiration** provides the energy.

Growth process of protein synthesis

The nucleolus increases in size and '**messenger**' **ribonucleic acid** (or **m-RNA**), resembling half a DNA molecule enters the cell cytoplasm and travels to the ribosome organelles. The 'messenger' RNA becomes fixed to the ribosome and 20 different cytoplasm amino acids are brought to it by '**transfer**' **RNA** (or **t-RNA**) to be made into new structural or functional proteins such as enzyme molecules or muscle protein.

The cell growth period is also a time when new chloroplasts and mitochondria are formed and a store of energy in ATP is created for use in the next mitosis of the nucleus. Table 11.1 gives a summary of work in the cell's lifetime or cell life cycle.

Importance of mitosis

• Genetic material or DNA duplication;
• Growth or increase in body size due to formation of new cells in multicellular organisms;
• Repair of damage and wounded tissue;
• Renewal of blood and epidermal cells in the skin;
• Asexual reproduction is essentially division or multiplication by mitosis.

Table 11.1 Summary of work in a cell's life

Period		Activities and duration
Growth	1	Normal cell functions: nutrition, excretion, respiration occur
	2	Mitochondria and chloroplasts increase in numbers
	3	Structural and functional proteins manufactured
	4	RNA and DNA manufactured
	5	This period occupies 80% of cell lifetime
Nucleus division	1	Nucleus divides during mitosis
	2	Contents of nucleus and cytoplasm mix together after nuclear envelope breaks down
	3	This period occupies 15% of cell lifetime
Cytoplasm division	1	The cell's lifetime ends in the formation of two new daughter cells
	2	Cells separated by fission in animals or a new cell wall in plants
	3	This period occupies 5% of cell lifetime

Questions

8 How is mitosis in a plant cell different from mitosis in an animal cell?

9 Why is the term 'rungs' used to describe the structures formed by the organic bases in a DNA molecule?

10 What happens to DNA before the nucleus divides mitotically and why does this change take place?

11 What is the function of (a) m-RNA and (b) t-RNA?

11.5 The growth process in seeds

Seeds form in the fruits of flowering plants as a result of fertilisation.

Seed structure

Surrounding the seed is a thick seed coat or **testa** providing protection against dehydration, mechanical damage and insect or microorganism attack (see Figure 11.6). The **hilum** is the scar of the seed stalk.

The embryo inside the seed consists of a **plumule** (future shoot), **radicle** (future root) and seed leaves or **cotyledons**. Monocotyledon flowering plants, such as grasses and wheat, have *one* cotyledon. Dicotyledon flowering plants such as the broad bean, have *two* cotyledons (see Figure 11.6).

Endosperm is a nutritive tissue which surrounds the embryo of maize and wheat, or it may be absorbed into the cotyledons of non-endospermic seeds such as the pea and bean.

Figure 11.6 Internal structure of (a) a dicotyledon seed and (b) a monocotyledon fruit grain

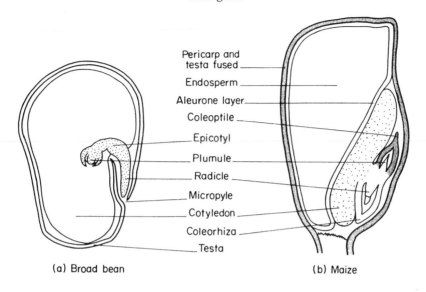

Pericarp and
testa fused

Endosperm

Aleurone layer

Coleoptile

Epicotyl

Plumule

Radicle

Micropyle

Cotyledon

Coleorhiza

Testa

(a) Broad bean   (b) Maize

Seed growth or germination

Seed growth or germination commences by water entering the seed through the **micropyle**, and the testa absorbing water. The seed's water content increases by 80% of the content of the dormant seed. Enzyme activity converts the seed food reserves into soluble simpler forms of glucose, amino acids, glycerol and fatty acids.

The testa splits and the radicle emerges. Following this the germination is of two types (see Figure 11.7).

- **Hypogeal** germination occurs in maize, pea and bean, when the cotyledons remain *below* ground and the plumule emerges.
- **Epigeal** germination occurs in the sunflower and cress when the plumule and cotyledons emerge *above* ground. The cotyledons can photosynthesize and function as leaves. Maize and cereal grain plumules are protected by a **coleoptile**.

Conditions for germination

Seeds need **water, oxygen** and **warmth** for germination. Red-coloured light encourages lettuce seed germination.

Questions

12 How are the seeds of a dicotyledon flowering plant and a monocotyledon flowering plant (a) similar, (b) different?
13 Describe the germination of the broad bean. How does germination differ in the sunflower?

Figure 11.7 Stages in germination of (a) broad bean, *Vicia faba* and (b) sunflower, *Helianthus* sp. seeds

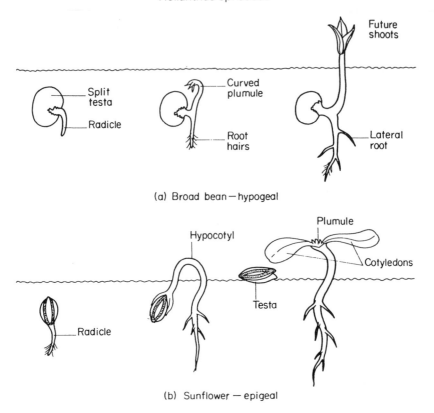

(a) Broad bean—hypogeal

(b) Sunflower — epigeal

11.6 The growth process of metamorphosis in animals

The fertilised eggs of most animals develop into an embryo, which following birth or hatching, increases in size or grows into an adult with little change in body form. Amphibians and certain insects show considerable changes in body form between the fertilised egg and adult.

Metamorphosis is the change in body form of a **larva** or immature organism into an **adult**. It is seen in the change from tadpole to frog, and from caterpillar to butterfly.

Insect metamorphosis

Insects show two different types of metamorphosis in their life cycles.

Incomplete metamorphosis

This is seen in the cockroach (Figure 11.8) locust and grasshopper (Figure 11.9) life cycles. The egg hatches into a **nymph**, a sexually immature miniature insect, lacking wings. The nymph shows intermittent growth by passing into an adult through several moults or **ecdysis** stages in which it sheds its complete outer skeleton. The stage

Figure 11.8 The life cycle of the cockroach, *Periplaneta* sp., showing incomplete meta-morphosis (Rentokil Laboratories Ltd)

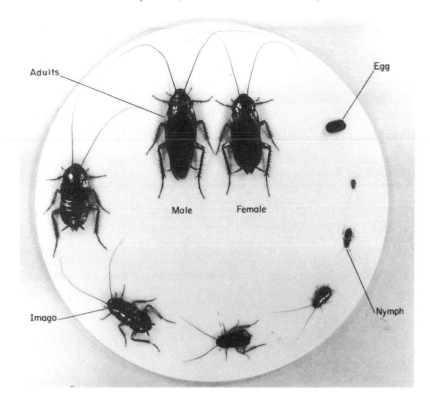

Figure 11.9 The life cycle of the grasshopper, *Acridiid* sp., showing incomplete metamorphosis

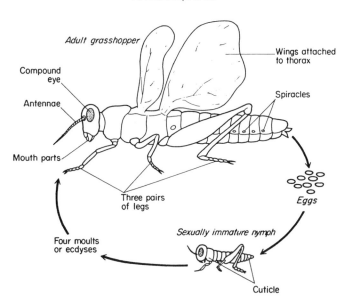

between successive moults is called an **instar**. It becomes a sexually mature adult after the last moult.

egg → nymph → adult
incomplete metamorphosis

Complete metamorphosis

This is seen in the housefly, moth, butterfly (Figure 11.10) and honey bee life cycles. The egg hatches into a very active larva different in form from the adult, being a maggot, grub or caterpillar. It feeds differently from the adult and passes through ecdysis stages to become an inactive, non-feeding pupa. Inside the pupa, tissue reorganisation occurs involving the destruction of certain tissues. The sexually mature adult insect emerges from the pupa.

egg → larva (maggot or caterpillar) → pupa → adult
complete metamorphosis

Figure 11.10 Life cycle of the large white butterfly, *Pieris brassicae*, showing complete metamorphosis

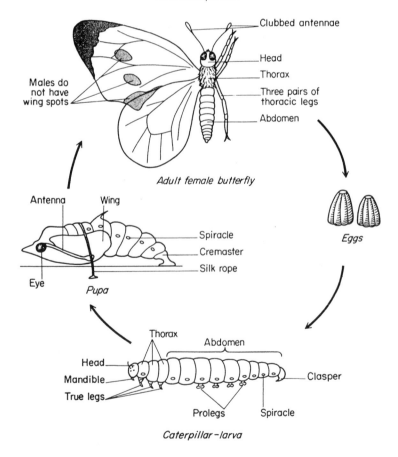

Frog metamorphosis

Stages in the metamorphosis of a frog are shown in Figure 11.11 and summarised in the following paragraphs.

Figure 11.11 Outline of the life cycle of the common frog, *Rana temporaria*

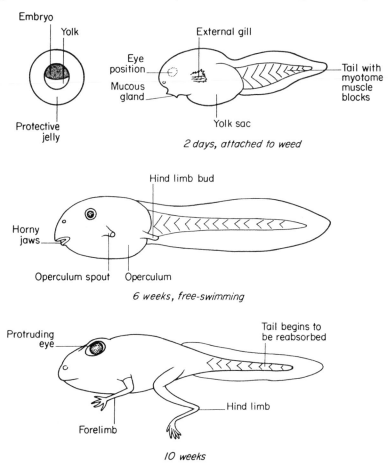

A **tadpole** (larva) emerges from the fertilised egg, attaching itself to pondweed. It feeds on its yolk food reserve and respires by **external gills**. A **mouth** develops, allowing the free-swimming tadpole to rasp microscopic algae off stone surfaces. **Internal gills** protected within an operculum cover replace the external gills.

Hind limbs appear followed by **forelimbs**, and the internal gills are replaced by **lungs**. Feeding is halted whilst the tail is absorbed, and a mouth and tongue develop, allowing a **carnivorous** diet to replace the **herbivorous** earlier diet.

Dormancy

Dormancy is a *resting* condition with a very low rate of metabolism in which growth ceases. It is seen in seeds, spores, buds, fruits and perennating organs such as bulbs, corms and tubers.

Dormancy is the means to survive adverse conditions of low temperature, lack of moisture in drought and wintery conditions. Seeds may have variable dormancy periods from 1 up to 100 years.

Some insects show a type of dormancy. The eggs of human fleas remain dormant in floorboard crevices. Butterfly pupae hang dormant under ledges and adult houseflies cluster together in wintertime in attic spaces. Amoebae **encyst** in times of drought and extremes of temperature.

Hibernation is a type of dormancy found in many different animals and is a means of survival over winter. **Aestivation** is a type of dormancy occurring in hot droughts in deserts in certain reptiles, fish and amphibia.

Questions

14 What is metamorphosis?

15 How is the metamorphosis of the grasshopper different from that of the butterfly?

16 How would you expect the body form and the behaviour of a 2-day-old tadpole to change over the following 10 weeks?

17 What is dormancy and why do some organisms have a dormant state?

11.7 Growth regions

Plants

There are three main growth regions (Figures 11.12 and 11.13) in flowering plants.
- cell **division** region;
- cell **elongation** region;
- cell **differentiation** region.

The first region is distinct from the last two and the last two merge into each other.

Cell division region

This is located in the **meristematic tissue** at the tip or apex of the root or shoot where cells divide after the mitosis of cell nuclei (Figure 11.13). The cells in this region are small and thin-walled with large nuclei and without a vacuole.

The shoot apex has a **terminal bud** and future leaf or **primordium**. The root apex has a protective **root cap** seen in Figure 11.13.

Cell elongation region

This is directly behind the cell division region, and is the part where cells elongate or expand (see Figure 11.13). Elongation occurs by the formation of a **vacuole** in the cytoplasm. Water is absorbed by osmosis causing the cytoplasm to increase in volume, resulting in the thin cell walls stretching. Elongation is controlled by plant growth substances.

Cell differentiation region

This is where the elongated cells develop the characteristic structure of the different tissues of the cortex, xylem, phloem and cambium. This results in the primary, unthickened structure of stems and roots described in Section 5.2. The result of **primary growth** is an increase in *length*.

Figure 11.12 Photomicrograph of elder, *Sambucus* sp., stem apex, seen in longitudinal section, showing where cell division and expansion occur (Griffin Biological Laboratories)

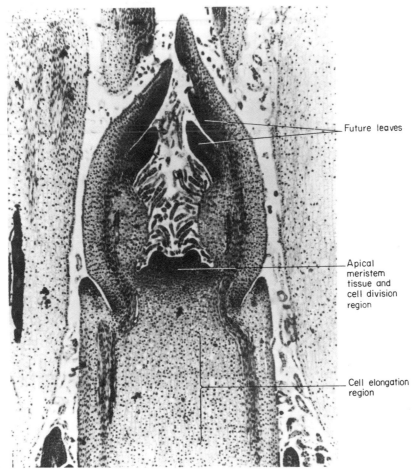

Future leaves

Apical meristem tissue and cell division region

Cell elongation region

Figure 11.13 Longitudinal section through a root showing cell division, elongation and differentiation regions

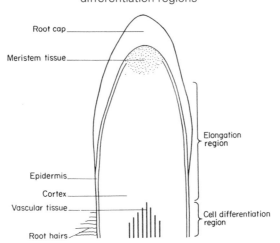

Root cap

Meristem tissue

Elongation region

Epidermis

Cortex

Vascular tissue

Cell differentiation region

Root hairs

Secondary growth (see Figure 11.14) occurs in woody perennials such as shrubs and trees, and some biennials. The main growing region is the continuous ring of **cambium**, a meristem producing rings of **secondary xylem** inwardly and **secondary phloem** outwardly. The xylem becomes **wood**, whilst the soft phloem becomes **bast** beneath the outer **bark**.

Figure 11.14 Transverse section through part of a woody stem showing secondary thickening

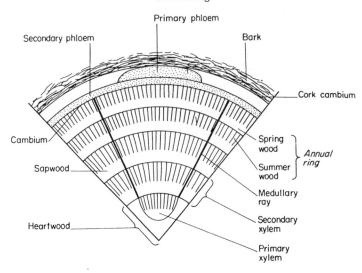

The overall result of secondary growth is an increase in *girth* or *diameter*. Splitting of the outer surface tissue is prevented by a **cork cambium** producing waterproofing, cork-filled cells in the bark (see Figure 6.15).

If the straight radicle of a germinating bean is marked with lines of ink then allowed to grow on, the elongation region will be seen as the region where the lines have become most widely separated (see Figure 11.15).

Figure 11.15 To show the elongation region in a broad bean seed radicle, *Vicia faba*

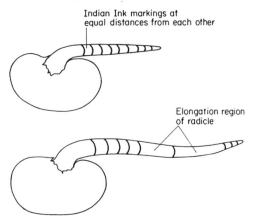

Animals

The fertilised egg nucleus undergoes repeated mitotic division to produce a mass of undifferentiated cells called a **blastula**. Differentiation follows by the cells moving into three main groups:
- **endoderm:** this forms the future gut;
- **mesoderm:** this forms the future muscle and blood;
- **ectoderm:** this forms the future skin and nervous system.

This process of differentiation is called **gastrulation**.

Mitosis occurs in almost all animal body cells, but is restricted to special meristem regions in flowering plants. Insects show intermittent growth in shedding the chitin exoskeleton by ecdysis. Mammals, including humans, show **allometric** growth when different body parts grow at different rates from the overall body growth rate. The human brain and eye grow more slowly than the arms and legs (see Table 11.2 and Figure 11.16).

Table 11.2 Comparison of height and mass amongst 2- to 18-year-olds

Age	2 to 10 years*	between 10 and 14 years (puberty)	after 14 years
Height (genetically controlled)	boys are taller than girls	girls are taller than boys	boys are taller than girls; girls complete their growth at 16 years; boys complete their growth at 18 years
Mass (mainly controlled by over- or under-nutrition)	boys are heavier than girls	girls are heavier than boys	boys are heavier than girls; obesity can occur in both due to over-nutrition

*During the first 6 months of life, the mass doubles. During the following 12 months the weight trebles.

Figure 11.16   Whole body growth compared to growth of a body part – allometric growth

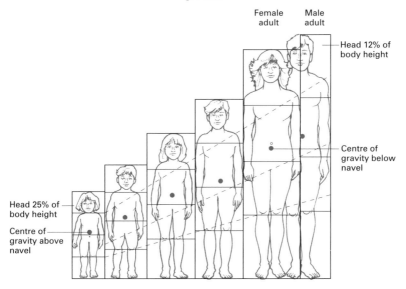

Human development can be summarised as shown below. Details of the development of the human are described in Chapter 10.

zygote → embryo → fetus → infancy/childhood → puberty

and

death ← senility ← ageing ← adulthood ← adolescence

Questions

18 Where do cells divide in the body of a plant?
19 How do plant cells elongate?
20 How is primary growth different from secondary growth?
21 What are the main tissues that are produced in differentiation in (a) plants,
    (b) animals?

11.8 Measurement of growth

Growth can be defined as *the permanent increase in biomass, cell numbers or body size as a
result of anabolic synthesis, cell division and cell expansion.* Growth of an individual organism can be measured by the following methods.

- **Total fresh weight**: This is measured when the organism is weighed at regular
  intervals, a method used for most large animals including humans. The method is
  influenced by the varying water content of the organism.
- **Dry weight**: This is determined by killing the organism and heating it to a constant weight at 110°C. This method is a more accurate one since it indicates the
  increase in weight due to synthesis of different materials, irrespective of the water
  content.
- **Volume**: This is determined by water displacement using an overflow can. The
  volume of water displaced by the organism is measured with measuring cylinders.
  Alternatively the volume can be calculated arithmetically from measurements on the
  surface of the organism.
- **Photographs**: Photographic methods of time lapse photography can be used to
  record graphically growth and movements in plants.

Growth curves

The results of growth measurements in individual organisms can be displayed as either
ordinary growth curves or rate-of-growth curves.

Ordinary growth curves

These are constructed in a graph using measurements of a particular character, such as
length or dry weight, plotted against time in days (see Figure 11.17). The graph shows
an **S** shape and is called a **sigmoid curve**.

Rate-of-growth curves

These are graphs constructed using growth increments or differences in a particular
character. For example, if growth is recorded in length as follows: 12, 14, 18, 26, 36,
52, 64 and 75 cm, the increments or changes will be 2, 4, 8, 10, 16, 12 and 11 cm.

Figure 11.17 'S'-shaped sigmoid growth curve of an annual flowering plant

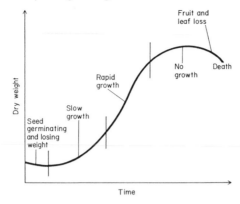

The first set of results is used to construct ordinary growth curves and the second set, using increments, is used to construct growth-rate curves. Growth-rate curves have a typical bell shape.

Questions

22 What are the advantages and disadvantages of using (a) the total fresh weight method, (b) the dry weight method to measure growth?
23 Why does the dry weight of a flowering plant fall before it produces leaves?
24 Draw a bell-shaped rate-of-growth graph and mark on it a period of (a) increasing, (b) decreasing, (c) maximum rate of growth.

11.9 Body growth factors

Genes

Genetic or inherited factors are controlled by **genes** located on the **chromosome**. Plants and animals inherit the trait or characteristic to grow tall or remain dwarfed as in garden peas (see Section 12.7).

Genes acting through the genetic code control the **structural** and **functional** proteins responsible for characteristics of the species.

Growth factors

Growth factors can be divided into two groups – **nutritional factors** and **growth substances,** as illustrated below.

GROWTH FACTORS

NUTRITION		GROWTH SUBSTANCES	
Plants: light, heat, water and nutrients; photosynthesis	Animals: proteins, minerals and vitamins, e.g. vitamin D	Plant growth substances: auxins – affect vacuolation and cell elongation	Animal hormones: pituitary – growth hormone; sex hormones – adolescent growth; thyroid hormone

over- and under-nutrition

Abnormal growth – chemicals and drugs

Carcinogenic chemicals such as tobacco tars cause cells to divide haphazardly to produce neoplasms or cancer tumours in mammals. **Anti-mitotic drugs** slow down mitosis of cell nuclei and are used in the treatment of cancer neoplasms. **Thalidomide** caused malformation of limbs in human embryos, hindering their normal growth due to the drug's teratogenic or malforming effect.

Light

Plants in shade or darkness grow tall and overcrowded plants tend to seek better positions facing the light. The amount of light also affects the rate of photosynthesis (see Section 3.4).

Mammals may be affected by not being able to form vitamin D in the skin. This results in abnormal growth and rickets in the young.

Temperature

Green plants only are affected and require an optimum temperature for photosynthesis (see Section 3.4).

Oxygen and water

Oxygen is essential for energy release in the growth of aerobic organisms. Water is an essential requirement for all living things to grow.

Question

25 What factors affect the growth of (a) animals and (b) plants?

Summary

- Living things grow by synthesis and duplication of new genetic material, followed by cell division and differentiation. (▶ 277)
- Some organisms stop growing when they reach a certain body size but others continue to grow throughout life. (▶ 277)
- The cell nucleus contains deoxyribonucleic acid (DNA) and a protein, histone. (▶277)
- Mitosis is a type of cell division in which two identical daughter cells are produced from one parent cell. (▶ 278)
- DNA is a molecule with a double helix structure which remains within the nucleus. m-RNA passes from the nucleus to the cytoplasm and t-RNA carries amino acids in the cytoplasm to the ribosomes. (▶ 280)
- There are two kinds of seed structure found in flowering plants but both take in large amounts of water when they germinate. (▶ 282)
- Insects and frogs change their body form as they grow into adults. This process is called metamorphosis. (▶ 284)
- There are three main growth regions in flowering plants. (▶ 288)
- Cells in animals differentiate into three main groups. (▶ 291)
- There is a variety of ways in which growth may be measured. (▶ 292)

- There are two kinds of growth curves – ordinary growth curves and rate-of-growth curves. (► 292)
- There is a range of factors that affect growth including the inheritance of genes. (► 293)

⬡ 12 Genetics

Objectives

After reading this chapter you should be able to:

- understand the terms phenotype and genotype
- understand what homologous chromosomes are
- distinguish between haploid and diploid nuclei
- understand chromosome variation
- explain the location and function of genes
- describe the factors which lead to gene mutation
- understand the essentials of genetic engineering
- explain the importance of meiosis
- understand the terms, symbols and Punnett squares used in the study of genetics
- describe a back-cross
- explain incomplete dominance
- explain how human blood groups are inherited and describe their compatibility
- explain how some disorders are sex-linked
- identify examples of dominant and recessive inheritance in humans
- explain how discontinuous and continuous variation differ and give examples of both

12.1 Introduction

Genetic material found within the nuclear envelope of plant and animal cell nuclei is inherited from either *one* parent **asexually** or from *two* parents **sexually**. It is then distributed *equally* by **mitosis** to all body cells (see Section 11.3).

Genetic material, DNA (deoxyribonucleic acid) provides 'instructions' for making **structural** and **functional** proteins (see Section 11.4) which give members of the same species (see Section 2.3) similar recognisable **external** (structural) and **internal** (functional) features which are called **characters** or **traits**. These are used to identify and classify species.

Sometimes the genetic material is changed (**mutated**) due to internal or external factors. This causes it to give different 'instructions' and leads to different kinds of proteins being formed and consequent small differences or **variations** in structural and functional characters or traits between members of the same species.

Human beings as a species share many similar main characters or traits which identify the species but they also have small differences seen in the individual, for example between brothers and sisters. This is **variation** within the species. **DNA typing** or **profiling** is the method of finding the DNA profile or composition for an individual person by examining samples of tissue containing cell nuclei. Every individual has a different DNA composition and DNA typing or profiling is used in crime detection through examination of samples of blood or other body materials from the scene of the crime.

Genetics is the study concerned with the methods of inheritance of genetic material and how it causes variation within the species.

Questions

1 How is genetic material inherited?
2 What kind of cell is a eukaryote cell? (see page 48)
3 What are structural and functional proteins? (see page 281)
4 How important are characters or traits? Explain your answer.
5 Why do members of the same species look different from each other?

12.2 Phenotype

The **phenotype** is a term describing the actual organism with respect to its *external* structural form and its *internal* physiology or function. The phenotype is the total of all the characters that the organism shows. The resemblances and differences in characters are the means of classifying organisms into species as described in Section 2.3. Members of a species will share certain characters resulting in their mutual resemblance and these will be inherited by their offspring.

There will be differences between members of the same species. This difference within a species is called variation and may or may not be passed on to the offspring (see Section 12.10).

Questions

6 If you observed a black rabbit with white paws and wrote down the features you could see, would you be describing the rabbit's phenotype? Explain your answer.
7 Describe the human phenotype.
8 (a) How can you tell if two organisms are members of the same species?
   (b) Where have their features come from?

12.3 Genetic material – the chromosomes

Thomas H. Morgan (1866–1945) was the first to investigate the structure and function of chromosomes by research on the fruit fly, *Drosophila melanogaster*. **Chromosomes** are composed of DNA and a protein, histone, described in Section 11.3. They are only visible when a cell nucleus is about to divide. Photographs of the dividing cell nucleus can be taken through a high-powered light microscope. These photographs are called **photomicrographs**. Figure 12.1 shows a photomicrograph of chromosomes from a specially treated human white blood cell nucleus which is being used to construct a

Figure 12.1 Left: a photomicrograph showing the full set of chromosomes in one human white blood cell nucleus. Right: the chromosomes being grouped in homologous pairs (World Health Organisation)

karyogram. This is a chromosome chart which is made by artificially arranging similar sized and shaped chromosomes together.

Chromosome number

Each chromosome is clearly visible in the karyogram and can be counted. Every body cell nucleus will have the same number of chromosomes. This is constant for each species of organism. Table 12.1 shows the number of chromosomes in some species of plants and animals. There are exceptions where certain individuals within a species may have a different chromosome number.

Homologous chromosomes

The karyotype or chart (see Figure 12.2) shows that in human body cell nuclei the chromosomes can be arranged in pairs, each similar in size and shape. These identical chromosome pairs are called **homologous** chromosomes.

Autosomes and sex chromosomes

The karyogram of a human body cell nucleus shows that the 44 homologous chromosomes can be paired together as **autosomes** whilst two chromosomes are different (see

Table 12.1 Chromosome numbers in various species

Species	Diploid chromosome number
Plants:	
Wheat	42
Peas	14
Broad bean	12
Tomato	24
Potato	48
Cabbage	18
Animals:	
Drosophila (fruit fly)	8
Housefly	12
Frog	26
Chicken	78
Cat	38
Dog	78
Human	46

Figure 12.2 Karyogram of chromosomes present in a cell body nucleus of a girl with Down's syndrome. The extra chromosome is shown with the homologous pair 21. Note the homologous paired XX chromosomes (Santoz Ltd)

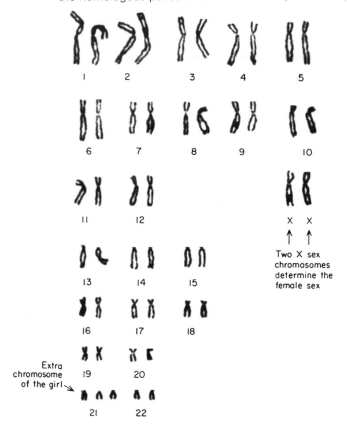

Figure 12.3). These **sex chromosomes** are called **X** and **Y** chromosomes. The **XX** homologous pair is usually found in females, whilst the **XY** chromosomes are usually found in males (see Section 12.10). The exceptions are seen in the nuclei of birds and butterflies where **XX** is in the male body cell nuclei and **XY** is in the female nuclei. The **X** chromosome of humans is not the same size and shape as the **Y** chromosome. It is much bigger that the **Y**.

Figure 12.3 The sex chromosomes

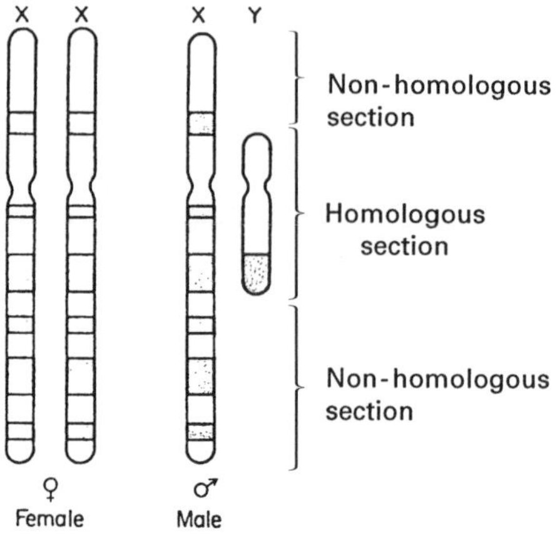

A human body cell nucleus will therefore have 22 pairs of autosomes and one pair of sex chromosomes, making a total chromosome number of 46. A woman has 23 matched pairs of chromosome whilst the 23rd pair in a man do not match.

Diploid and haploid nuclei

Diploid nuclei have the chromosomes in homologous pairs – seen in the body cell nuclei of animals and in vegetative cells of flowering plants. Haploid nuclei have only *one* set of unpaired chromosomes – seen in the gamete cells and in spores of algae, fungi, liverworts and mosses. Gamete nuclei in humans have 23 chromosomes – 22 autosomes and one sex chromosome. Haploid nuclei have *half* the chromosome number, or *one set* of chromosomes.

Diploid nuclei are given the symbol **2n** (2 sets) and haploid nuclei **n** (one set).

Questions
---

9 What is the difference between a photomicrograph and a karyogram?
10 What are homologous chromosomes?
11 Are the sex chromosomes in humans homologous? Explain your answer.
12 What is (a) a haploid nucleus and (b) a diploid nucleus?
13 What are the haploid and diploid numbers in humans?
14 Where would you find haploid and diploid nuclei?
15 Look at Table 12.1. Is there a relationship between the size of an organism and its diploid chromosome number? Explain your answer.

Chromosome mutations

Genetic material can be altered by natural or artificial means and so can cause sudden and permanent changes in a particular character. A change in chromosome structure or number is called **mutation** (see Figure 12.4).

Figure 12.4 One method of chromosome mutation

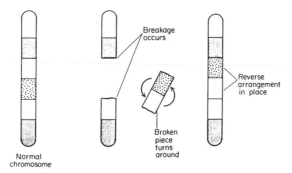

During the division of a nucleus in either mitosis or meiosis (see Section 12.5) pieces of the chromosomes may break off and are lost or become attached again to the wrong place in the same homologous chromosome or a different chromosome. This is an example of natural chromosome mutation. Artificial chromosome mutation can be achieved by means of radiation, for example X-rays, or by chemical substances such as colchicine, extracted from autumn crocuses. Agents causing mutations are called **mutagens**. Thomas Morgan (see page 297) was the first to perform artificial chromosome mutation by radiation. Since then gamma rays, ultraviolet rays and chemicals in tobacco have been found to be mutagens.

Down's syndrome

Down's syndrome may occur in about 1 in 1500 children and is seen as variable mental ability, slanted eyes, round, flat face and short stature. Karyotypes of a Down's syndrome child's white blood cell nuclei show that the number of chromosomes is 47 instead of 46. The extra chromosome is responsible for an increase of the total chromosome number to 47 (see Figure 12.2).

It is believed that 1 in every 100 human embryos has an abnormal chromosome structure. Many of these affected embryos are spontaneously aborted or miscarried.

Polyploids

Polyploids are cultivated plants such as tomatoes, marigolds, wallflowers, raspberries and wheat, which have extra sets of chromosomes. Polyploid nuclei can be **3n** (3 sets), **4n** (4 sets), etc. This gives the organism certain advantages in producing showy flowers or increased crop yields. Polyploid plants are produced by treating seedlings and cuttings with colchicine.

___ **Questions** _____

16 (a) What is a mutation?
    (b) When does a mutation occur?
    (c) How are mutations caused?

12.4 Genetic material – the genes

By his researches with the fruit fly, *Drosophila melanogaster*, Thomas Morgan showed that **genes** are the complex chemical material units of inheritance. They are arranged in a line along the chromosome and are capable of being reproduced and mutated.

Locus

Each gene occupies a short length of a chromosome at a position called its **locus**. Photomicrographs of chromosomes from salivary gland nuclei of *Drosophila* show transverse banding (see Figure 12.5). These bands indicate the position of certain genes, or their loci. The actual genes are not visible.

Every chromosome carries *many* genes and therefore organisms will have more genes than chromosomes. Over 800 different genes have been identified on the 46 human chromosomes. The human **Y** chromosome has at least one gene and the **X** chromosome about 200 genes.

Figure 12.5 Part of a large salivary gland chromosome of *Drosophila* sp., showing banding

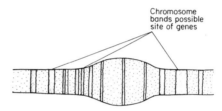

Alleles

When paired together, homologous chromosomes will have similar or different genes called **alleles** on the same relative position or locus. An allele is an *alternative form* of a gene. It controls the *same* characteristic but produces a *different* effect. Alleles for characteristics in the human, the guinea pig and the pea are listed below.

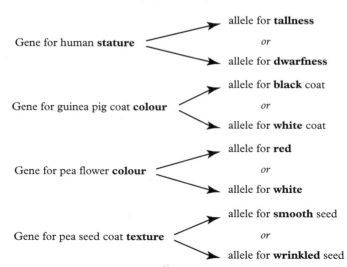

Composition

DNA is the main component of a gene, which can be duplicated or replicated by uncoiling of the double DNA coil. This occurs in chromosomes during cell growth (see Section 11.4) thus *doubling* the DNA content before the nucleus divides.

Gene function

The DNA of a gene controls protein synthesis (see Section 11.4). A gene can therefore control production of **structural** proteins forming muscles and bones or of **functional** proteins such as enzymes, hormones and antigens. The kind of protein to be manufactured, i.e. structural or functional, depends on the *'instructions'* provided by the messenger RNA which in turn come from the gene DNA in the nucleus. The 'instructions' provided by the DNA of a gene are called the **genetic code**. The differences between species such as dogs and cats are due to the different proteins made by the different genes.

Genes and characters

One gene pair may be responsible for determining a certain **character** or **trait**, e.g. *height* of garden pea plants, by controlling the reproduction of a certain protein. Often many gene pairs together are responsible for *one* character, e.g. the height of humans (see Section 12.10). The total gene composition in the nucleus of an individual is called its **genotype**.

Gene mutations

A gene can **mutate** and undergo a sudden change either in the amount or structure of the DNA material. Such changes do not happen often and the effect of the change will be seen in the phenotype. Most gene mutations show harmful characters as shown by the following examples and some may be lethal and kill the individual. Occasionally a gene mutation may provide beneficial characters.

Gene mutations may occur *naturally* in nuclear division, when the nucleotide bases interchange positions, or *artificially* by means of different mutagens such as X-rays, gamma rays or ultraviolet light. High temperatures and sudden changes in temperature are also known to cause gene mutation. Drugs and chemicals such as mustard gas, nitrous acid, alkali, LSD and even caffeine are believed to be mutagenic.

PKU (phenylketonuria) is caused by natural mutation of a gene which normally controls the production of an enzyme which changes the essential amino acid phenylalanine into tyrosine. When the gene is absent the enzyme is not produced and phenylalanine accumulates in body tissues of young children. This causes mental and physical disorders. All babies born in the UK have their blood tested to detect the condition. If PKU is found, diets are provided with foods low in phenylalanine which allows the child to develop normally.

Genetic engineering

Genetic engineering or gene manipulation involves the transfer of a gene or DNA segment from the chromosome of one species, the donor, into the chromosome of another different species or host. In this way the donor gives the host cell a new character or trait or the ability to code for making a special functional protein (see Figure 12.6).

The gene for insulin synthesis is transferred from a mammal chromosome and recombined in the chromosome of a harmless human gut bacteria, *Escherichia coli*,

Figure 12.6   Gene transfer from a human chromosome to a bacteria chromosome

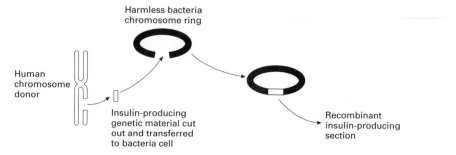

which is then grown in a sterile culture on a large scale by biotechnology (see Section 15.7). The transferred gene in the bacteria produces insulin in large amounts encouraged by favourable growth conditions, thus replacing the slower process of extracting insulin from pig and ox pancreas glands.

Figure 12.7   Genetic engineering used to produce insulin on a large scale

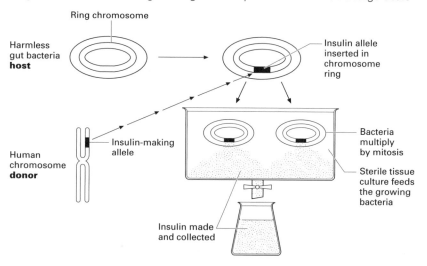

Other valuable functional proteins for humans are produced in this way. They include interferon, human growth hormone (see Section 8.6), rennin (Section 4.5) and certain antigens.

Questions

17 What is (a) a gene, (b) a locus (c) an allele?
18 How are DNA, a gene and a chromosome related?
19 Why is a cat different from a dog?
20 What is the difference between the genotype and the phenotype? (see also page 297)
21 How does a gene change when it mutates?
22 How are organisms affected by gene mutations?
23 What is genetic engineering and how is it useful?

Table 12.2 Summary of genetic material

Genetic material	Species	
	Human	Garden pea
Chromosomes: different number in different species	46	14
Genes: on specific chromosome positions; influence different general body characteristics	1 Stature – height 2 Eye colour 3 Hair colour 4 Hair texture	1 Stature – height 2 Flower petal colour 3 Seed colour 4 Seed coat texture
Alleles: alternative forms of a specific gene; influence a specific body characteristic	1 Tall/short height 2 Brown/blue eye colour 3 Red/dark hair colour 4 Straight/wavy hair texture	1 Tall/dwarf height 2 Red/white flower colour 3 Green/yellow seed colour 4 Round/wrinkled seed coat texture

12.5 Meiosis

Meiosis (Figure 12.8) is the nuclear division usually occurring in plant and animal reproductive organs, resulting in gamete formation. A **diploid** nucleus divides by meiosis to form four daughter nuclei, each of which is **haploid**. Meiosis causes the genetic material or chromosome number to be *halved*.

Figure 12.8 The behaviour of chromosomes during nuclear division by meiosis

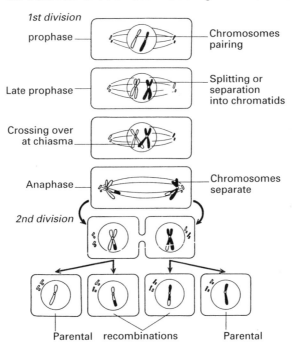

1st division
prophase — Chromosomes pairing

Late prophase — Splitting or separation into chromatids

Crossing over at chiasma

Anaphase — Chromosomes separate

2nd division

Parental   recombinations   Parental

Meiosis occurs in chordate animals during gametogenesis, before the sperm and ova are formed. Meiosis occurs in flowering plants during the formation of pollen and ovules. In humans, the chromosome number of diploid body cell nuclei is halved from 46 to 23 in the haploid gamete cell nuclei.

Stages in meiosis

Meiosis is a more complex process than mitosis. It involves *two* successive nuclear divisions after the process of duplicating genetic material (see Section 11.4). It is a continuous process described as taking place in up to 12 stages. The following is a summary of the main events.

First division

Pairing and separation: homologous chromosomes come together in pairs with similar genes opposite each other. The previously duplicated chromosomes with double DNA material form **chromatids**. During this pairing, some chromatids of homologous chromosomes break and rejoin at points called **chiasmata**. This causes **crossing over** with exchange of chromatid material between chromosomes from the parents. This will produce **heritable variation** in the offspring (see Section 12.10).

The nuclear envelope disappears, the chromosomes move apart and become separate and *two* new cells form.

Second division

The chromosomes of the two nuclei are completely halved or split lengthwise into two chromatids and move apart. They become surrounded by nuclear envelopes and form *four* haploid cell nuclei. During meiosis the chromosome number is halved and the alleles or genes separated. Each gamete has only one of a pair of alleles.

Table 12.3   Comparison of meiosis and mitosis

	Meiosis	Mitosis
Location	Reproductive organs: gamete formation in sexual reproduction	Body cells: somatic cell formation in growth, repair and asexual reproduction
Number of cells formed from one cell	four	two
Chromosomes in daughter cells	Haploid = half chromosome number = 23 in humans one set	Diploid = normal chromosome number = 46 in humans two sets – homologous chromosomes
Chromosome structure	Two are identical to parents; two are a new recombination with new composition	Identical to parent
Variation	Offspring show variation	No variation; offspring are identical – clones
Genetic material (DNA)	Halved	Remains the same

24 Look at Figures 11.1 and 12.8 and compare mitosis and meiosis. What are the main differences between them?
25 How does meiosis affect variation in a species?

12.6 Nuclear division and the life cycle

Gametogenesis

Chordate animals form gametes by **gametogenesis**. The diploid **2n** spermatocytes form the spermatozoa with haploid **n** nuclei by meiotic nuclear division. The female ovum with a haploid **n** nucleus forms by meiosis from a diploid **2n** oocyte cell. This process also forms three **polar bodies** which serve no further purpose and perish.

Fertilisation

This occurs between the male and female gamete during sexual reproduction (see Figure 12.9). Each haploid gamete provides one set of chromosomes to form a diploid **zygote** with a *double* set of paired homologous chromosomes. Fertilisation brings together the alleles or similar genes on opposite homologous chromosomes and *restores* the chromosome number of the species.

After a sperm has entered a vertebrate ovum, each nucleus duplicates its DNA content. Their nuclear envelopes break down and the chromosomes separate into chromatids by mitosis. The process continues with the chromatids forming new nuclei surrounded by nuclear envelopes and the zygote divides to form *two* body cells. Pairing of the chromosomes provided by either parent takes place at mitosis of the two body cells into four.

Figure 12.9 Schematic diagram showing fertilisation of vertebrate spermatozoon and ovum

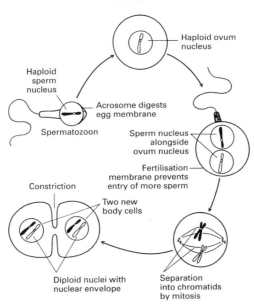

Differentiation

The zygote nucleus proceeds to divide many times, producing the different body tissues of organs and systems (see Chapter 11). Later the reproductive cells form which must divide by meiosis to form the haploid gametes in order that the chromosome number of the species is maintained.

Questions

26 When and where do meiosis and mitosis take place in the life cycle of a vertebrate animal?

27 How would the zygote be affected if meiosis did not take place?

12.7 Principles of inheritance

Gregor Mendel (1822–1884), a priest, carried out breeding experiments using garden peas in the 1850s. The result of his work provided an explanation of the basic mechanism of inheritance or **heredity**. Thomas Morgan founded the chromosome theory of heredity, which states that an individual's heredity is due to the genes received from its parents. Mendel called them **factors** and stated that these factors are transmitted to offspring causing a particular character or trait to appear in the phenotype.

Inheritance of contrasting characters

Mendel first studied the inheritance of one pair of contrasting characters or **monohybrid inheritance**, in garden pea plants.

Characters

Garden peas show several different **contrasting** characters also called traits, which are seen in the phenotype and are easy to recognise. These characteristics are flower **colour** – red or white; seed **colour** – green or yellow; seed **texture** – smooth or wrinkled; plant **stature** – tall or dwarf. These contrasting characters are due to the effects of alternative genes or alleles.

Pure lines

Garden pea plants are self-fertilising, and this occurs before the flower buds open. Surrounding the flower buds with plastic bags makes certain that cross-fertilisation does not occur. Self-fertilisation produces a **pure line** of the identical genotype. Any other variation will be due to the effects of the environment which are not heritable (see Section 12.10) or to mutation which is a rare event. Tall plants will produce successive generations of tall plants and dwarf plants will produce dwarf plants following self-fertilisation.

Pure line parents (symbol **P**) will carry pairs of identical genes on corresponding loci of homologous chromosomes and are called **homozygous**. When pairs of *different* genes or alleles are present on the corresponding loci of homologous chromosomes they are called **heterozygous**.

Dominance

Cross-fertilisation (symbol ×) is between pure-line parents with contrasting characters. For example, a tall parent crossed with a dwarf parent produces all tall fertile **hybrids** in the first generation of offspring, called the **first filial generation** (symbol **F1**).

Cross-fertilisation of garden pea plants does not occur naturally. It is achieved by artificial means by removing anthers from the flower while it is still in bud. This is called **emasculation** and is followed by surrounding the flower with a plastic bag. When the flower opens it is pollinated with pollen from *another* plant using a camel-hair brush and then enclosed in the bag. Plants of the **F1** generation show the dominant character; tall is dominant to dwarf, red is dominant to white flower colour, and yellow is dominant to green seed colour. This is due to the effect of one gene of the allele pair *suppressing* the effect of the other.

The **F1** plants are heterozygous with different gene pairs. This means they have a gene for tallness paired with a gene for dwarfness on homologous chromosomes. Note that characters are usually **dominant** or **recessive**. Less commonly an intermediate form is produced showing blending of the contrasting characters.

It is important to remember that a character can be the product of the effect of *one* gene or the interaction of *many* genes. It is therefore incorrect to describe a gene as dominant or recessive. It is the characters which show dominance. The dominant character will show in the heterozygote as a result of the genes or alleles controlling it.

Symbols are used as follows: the dominant character is given a capital letter, e.g. **T** for tallness, or **R** for red. The recessive contrasting character is shown as a lower case letter, e.g. **t** for dwarf and **r** for white. Table 12.4 summarises some dominant and recessive characters or traits in plants and animals.

Table 12.4   Some dominant and recessive characters or traits in plants and animals

	Dominant	Recessive
Plants		
1 Peas	Axial flowers	Terminal flowers
	Red flowers	White flowers
	Tall plants	Dwarf plants
	Smooth pod	Wrinkled pod
	Yellow seed	Green seed
	Smooth seed, also called round	Wrinkled seed, also called shrunken
2 Maize	Yellow kernel	White kernel
Animals		
1 Cattle	Black cattle	Red cattle
	Hornless cattle	Horned cattle
2 Guinea pig	Black	White
	Rough coat	Smooth coat
3 *Drosophila,* fruit fly	Red eye	White eye
	Normal wing	Vestigial wing
	Normal body colour	Ebony body colour

Questions

28 Why are contrasting characters easy to study?
29 What is the difference between the homozygous condition and the heterozygous condition?
30 Why should you refer to dominant and recessive characters rather than dominant or recessive genes?

The Punnett square

All the possible genotypes that form following random fusion of gametes at fertilisation can be displayed by using a **Punnett square** which is named after a geneticist. The following points must be remembered when constructing a Punnett square:
- A letter represents a contrasting character, e.g. **S**;
- A capital letter, e.g. **S**, represents a **dominant** character or allele;
- A small or lower case letter, e.g. **s**, represents the **recessive** character or allele;
- Male gametes are placed to the *left* of the female gametes. The gametes of both sexes are usually *encircled*, e.g. Ⓡ, Ⓣ;
- The genotype is represented by *paired* letters.

Gamete fusion and the Punnett square

A pure, tall parent plant provides gametes containing the gene **T**. A pure dwarf parent plant provides gametes containing the gene **t**. Using the Punnett square, the outcome of cross-fertilisation and all the possible fusions of gametes can be shown (see Figure 12.10).

Figure 12.10   Crossing pure tall and dwarf plants using a Punnett square

	P tall gametes	
Crossing	Ⓣ	Ⓣ
ⓣ	**Tt** heterozygous, tall	**Tt** heterozygous, tall
ⓣ	**Tt** heterozygous, tall	**Tt** heterozygous, tall

(P dwarf gametes on left side)

The haploid gametes carry one of the contrasting genes, **T** or **t**. Fertilisation produces diploid zygotes with the genes paired together as **Tt** in a heterozygous plant. All the offspring show the tall character in the phenotype.

Genotype and phenotype

Pure tall plants have the genotype or gene composition **TT**. F1 hybrids have the genotype **Tt**. The pure tall plants and the F1 hybrid tall plants have *different* genotypes but show the *same* external appearance in having the same phenotype.

Hybrid ratio

When F1 generation plants are allowed to self-fertilise, or animals to interbreed between brothers and sisters, a **second filial generation** (symbol **F2**) is produced. In the case of garden pea plants the flowers are enclosed in plastic bags to prevent cross-pollination.

The offspring seen in the **F2** generation are a mixture or assortment. The F1 hybrid tall pea plant produces tall and dwarf plants Out of a total F2 generation of 1064 plants, Mendel found 787 tall and 277 dwarf plants. This can be expressed as an arithmetical ratio 787/277 = 2.84 or 2.84 tall plants to one dwarf plant. This approximates to a ratio of 3:1. This 3:1 ratio shows the ratio of the phenotypes. The gene composition or genotype can be determined using the Punnett 2 × 2 square method.

The gametes of the F1 plants, **Tt**, are either **T** or **t**. This is displayed on the Punnett square to show all the possible fusions of F1 gametes (see Figure 12.11).

Figure 12.11   Selfing F1 Hybrids using a Punnett square

Selfing		F1 hybrid tall gametes	
		(T)	(t)
F1 hybrid tall gametes	(T)	TT homozygous, tall	Tt heterozygous, tall
	(t)	Tt heterozygous, tall	tt homozygous, dwarf

The phenotypes shown are 3 tall:1 dwarf; i.e. **TT, Tt, tT:tt**. This shows that there is a 3 in 4 chance of F2 plants being tall and a 1 in 4 chance of F2 plants being dwarf.

The genotypes are 1 homozygous pure tall (**TT**), 2 heterozygous hybrid tall (**Tt** or **tT**), 1 homozygous pure dwarf (**tt**). The ratio of genotypes is 1:2:1. Figure 12.12 shows the inheritance of seed coat texture in the garden pea.

Figure 12.12   The inheritance of a single character, seed-coat texture, in the garden pea, *Pisum sativum*

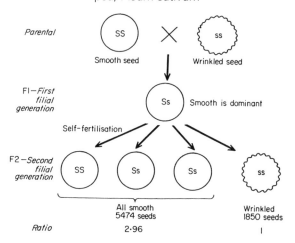

31 Explain using a Punnett square what happens when a pure line red flowering pea plant is crossed with a pure line white flowering pea plant.

32 Explain using a Punnett square what happens when F1 generation plants produced in the answer to question 31 are allowed to self-fertilise.

Law of segregation or Mendel's first law

The characters of an organism are controlled by genes occurring in pairs, which retain their individuality from generation to generation. Only one of a pair of genes goes into each gamete because they **segregate** or separate from others in meiosis.

The back or test cross

Two or more organisms may show the same phenotype but have different genotypes, for example red garden pea flowers may have different genotypes **RR** and **Rr**, being homozygous and heterozygous for flower colour, respectively.

In order to identify the genotype, it is necessary to perform a breeding experiment called a **back cross** or **test cross**. This involves crossing the F1 hybrid with the homozygous recessive parent. The parent will have one type of gamete, **a**. The genotypes to be identified are **Aa** and **AA**, and will have two and one type of gamete, respectively. The Punnett squares are constructed as in Figure 12.13 to show all possible gamete fusions.

It is therefore seen that a homozygous F1 hybrid, on test crossing with the homozygous recessive parent, produces all (100%) heterozygous offspring. A heterozygous F1 hybrid, on back crossing with a homozygous recessive parent, produces 50% heterozygous and 50% homozygous offspring.

Questions

33 Why is the homozygous recessive parent chosen for a test cross?

34 You are given a number of pea plants. How could you find out which pea plants were pure breeding for flower colour and which were hybrids?

Codominance

A cross between a red Marvel of Peru flower pollinated by a white one gives seeds which grow into pink flowers in the F1 generation (see Figure 12.14 on page 314). The self-fertilisation of these pink hybrids into an F2 generation produces a hybrid ratio of 1 red: 2 pink: 1 white flowers.

Codominance is seen in the phenotype of the heterozygote with characters blended and intermediate between those of either homozygote. Neither of the genes in the heterozygote exerts a dominance over the other.

The Punnet square (Figure 12.15a) shows all possible fusions of gametes from a homozygous red and white flower cross. All F1 are heterozygous pink phenotypes. The Punnet square (Figure 12.15b) shows all possible fusions of gametes by self-fertilisation of the F1 generation. The phenotypes are 1 red: 2 pink: 1 white. This shows there will be a 2 in 4 chance of flowers being red or white. The genotypes are 1 homozygous red (**RR**), 2 heterozygous pink (**RW**, **WR**), 1 homozygous white (**WW**). Table 12.5 shows some examples of codominance found in plants and animals.

Figure 12.13   (a) Homozygous hybrid **AA** back cross, (b) Heterozygous hybrid **Aa** back cross

(a)

P homozygous recessive gametes

	(a)	(a)
(A)	**Aa** heterozygous	**Aa** heterozygous
(A)	**Aa** heterozygous	**Aa** heterozygous

F1 homozygous hybrid gametes

(b)

P homozygous recessive gametes

	(a)	(a)
(A)	**Aa** heterozygous	**Aa** heterozygous
(a)	**aa** homozygous	**aa** homozygous

F1 heterozygous hybrid gametes

Figure 12.14 Illustration of codominance in flower colour of the Marvel of Peru plant, *Mirabilis jalapa*

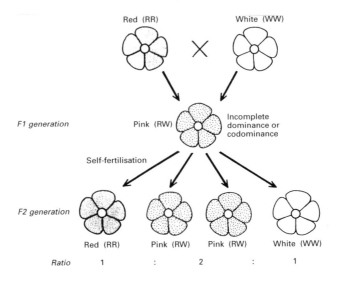

Table 12.5 Codominance in plants and animals

Parents (homozygous)		Offspring – F1 generation (heterozygous)
Plants		
Round spherical radish	× Long cylindrical radish	Oval radish
Red flower snapdragon	× White flower snapdragon	Pink flowered snapdragon
White flower Chinese primrose	× Red flower Chinese primrose	Pink flowered Chinese primrose
Red flower Marvel of Peru*	× White flower Marvel of Peru	Pink flowered Marvel of Peru
Animals		
Red Shorthorn cow	× White Shorthorn bull	Roan (red and white haired) Shorthorn calves
Black Andalusian cockerel	× White splashed Andalusian hen	Blue sheened Andalusian chickens
Normal haemoglobin	× Sickle cell anaemia	Sickle cell trait (page 335)

*Also called 4 o'clock flower as flower opens in late afternoon.

Questions

35 What are the possible fusions of gametes in crossing a red shorthorned cow with a white shorthorned bull?

36 What possible fusions of gametes would occur if you crossed (a) two roan cattle, (b) a roan cow with a white bull, (c) a roan cow with a red bull?

Figure 12.15 Punnett squares to illustrate codominance in flower colour. (a) F1 generation, (b) F2 generation

(a)

F1 generation

P red, homozygous gametes

	R	R
W	**RW** pink, heterozygous	**RW** pink, heterozygous
W	**RW** pink, heterozygous	**RW** pink, heterozygous

P white, homozygous gametes

(b)

F2 generation

F1 pink, heterozygous gametes

	R	W
R	**RR** red, homozygous	**RW** pink, heterozygous
W	**RW** pink, heterozygous	**WW** white, homozygous

F1 pink, heterozygous gametes

Karl Landsteiner (1868–1943) discovered that the four human blood groups result from the influence of three genes called **A**, **B** and **O**. Two of these can pair, producing the following combinations: **AA, AO, BB, BO, AB, OO**. Genes **A** and **B** show codominance with respect to each other, because both genes express their effects equally. Both **A** and **B** genes are dominant over the **O** gene which is recessive.

Four main blood groups are detectable in human beings:
- group A, due to **AA** or **AO** gene pairs;
- group B, due to **BB** or **BO** gene pairs;
- group AB, due to **AB** gene pairs;
- group O, due to **OO** gene pairs.

Children will receive one gene from each parent. Homozygous **AA, BB** and **OO** groups form if the genes from both parents are the same. Heterozygous **AB, AO** and **BO** groups form if the parents provide different genes. The *six* different gene combinations form the genotypes of the *four* phenotypes or the *four* blood groups.

Inheritance of blood groups can be shown by means of the Punnett square method (see Figure 12.16). If a male genotype **AO**, blood group A, has children by a female genotype **BO**, blood group B, the possible blood group genotypes of the children are as shown in Figure 12.16.

Figure 12.16 Punnett square illustrating blood group inheritance

	Male sperm	
	(A)	(O)
(B)	**AB** Group AB	**BO** Group B
(O)	**AO** Group A	**OO** Group O

Female ova

The male can only be the father of children with the blood group genotypes **AB**, **BO, AO** or **OO**. He will not be the father of any child with genotypes **AA** or **BB** from the female of genotype **BO**.

Questions

37 What are the possible phenotypes of children that could be produced by (a) a male of genotype **AB** and a female of genotype **OO**, (b) a male of genotype **AO** and a female of genotype **AB**?

Blood compatibility

When blood from two different people is mixed, one of two things may happen. No change may occur in which case the blood is **compatible**. If the blood is **incompatible**, clumping or **agglutination** of red cells, together with bursting of the cells, haemolysis, occurs. Agglutination is due to an **antigen** made by the red blood cell surface membrane reacting with an **antibody** in the other blood plasma. In blood transfusions only compatible blood can be used, since incompatible blood causes death.

The four blood groups differ by the presence or absence of the antigens **A** and **B** due to different gene influence in making the particular functional protein on the red cell surface membrane. The groups are named according to whether antigen **A** or **B** is present or not. People of group AB can receive all other blood types and are **universal recipients**. Group O blood can be given to all other blood types as a **universal donor** (see Table 5.2).

Table 12.6 gives a summary of the human blood groups.

Table 12.6 Human blood groups

Blood group (and percentage of UK population)	Antigen substance on red cell membrane	Antibody substance in plasma	Genotype		Phenotype	
A (UK, 40%)	A	Anti-B	AA Homozygous		A	Dominant
			AO Heterozygous			
B (UK, 10%)	B	Anti-A	BB Homozygous		B	Dominant
			BO Heterozygous			
AB (UK, 3%)	A and B	No antibody substance	AB Heterozygous		AB	Codominant
O (UK, 47%)	No antigen substance	Both anti-A and anti-B	OO Homozygous		O	Recessive

---

Questions

37 Can blood from a group A person be safely given to a person of group (a) A, (b) B, (c) AB (d) O?

38 Why can blood from a group A person be safely given to people of some blood groups but not to all?

12.9 The sex chromosomes in humans

Sex determination

Sex organs develop in the fetus under the direction of genes on the sex chromosomes. The **Y** chromosome has a gene which determines testes growth and development.

Female humans have the **XX** homologous pair of sex chromosomes. Since the **Y** chromosome is absent they will not be able to develop testes. Male human beings have the **XY** pair of sex chromosomes. Since the **Y** chromosome is present they *will* develop testes.

The Punnett square in sex determination

- Letters **X** and **Y** are used.
- Males have 50% **X** and 50% **Y** – heterozygous genotype.
- Females have 100% **X** chromosomes – homozygous genotype.

Figure 12.17 illustrates sex determination with a Punnett square.

Figure 12.17 Sex determination in humans

		Female chromosomes	
		(X)	(X)
Male chromosomes	(X)	**XX** Female	**XX** Female
	(Y)	**XY** Male	**XY** Male

The offspring following human sexual reproduction are 50% male (**XY** heterozygous) and 50% female (**XX** homozygous). Approximately equal numbers of boys and girls are born in a population. The ratio of males to females in a large population is 1:1 approximately.

Note: Sex changes are impossible in the human species because every body cell nucleus carries either the **XX** or **XY** chromosomes throughout the person's life. Surgical sex-change operations are cosmetic, involving the surgical removal of the male sex organs in a male to female sex change. The sex chromosomes remain the same in all body cells.

Sex linkage

This is the appearance of a certain character in one sex, namely the male. The **X** chromosome is longer than the **Y** chromosome in humans, and the **X** chromosome carries more genes than the **Y** chromosome. The sex chromosome genes are not all concerned with sexuality.

Haemophilia is a heritable disorder of the blood characterised by slowness or complete inability to clot during bleeding. It is caused by a recessive gene on the **X** chromosome (see Figure 12.18).

The haemophilia gene can be shown by **h** and is located at the end of the X chromosome shown as $\mathbf{X^h}$. In a $\mathbf{X^hY}$ male the recessive **h** gene exerts its effect since it is unpaired. In a female $\mathbf{X^HX^h}$ the gene is now paired with a homologous gene for normal blood clotting, represented by **H**, and its effect is masked. The normal gene **H** is dominant to the haemophilia gene **h.** The female is a carrier and a non-haemophiliac but her sons can inherit the disorder from her.

Figure 12.18   Haemophilia, showing the influence of the gene in sex linkage

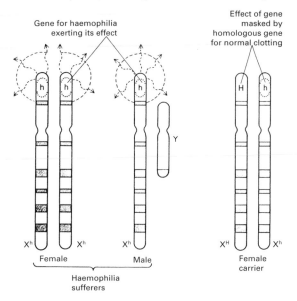

If the female has a haemophilia gene on each **X** chromosome and is **X^{h}X^h** she will be a haemophiliac. This is a very rare occurrence. Such females, if they survive birth, would be unlikely to survive their first menstruation.

Other sex-linked disorders include red/green colour blindness, Duchennes muscular dystrophy – a muscular wasting disorder, and inheritance of defective tooth enamel.

Punnett squares such as the one in Figure 12.19 can be used to forecast haemophilia in children through marriage of a female carrier **X^{H}X^h** with normal male **X^{H}Y**. The children of such parents can be normal males and females and carrier females with haemophiliac males.

Figure 12.19   Punnett square to illustrate inheritance of haemophilia

		Normal male gametes	
		X^H	Y
Carrier female gametes	X^h	X^{H}X^h Carrier female	X^{h}Y Haemophiliac male
	X^H	X^{H}X^H Normal female	X^{H}Y Normal male

Question

39 What children could be produced by a haemophiliac male and a carrier female?

12.10 Human species variation

Variation is the infinite number of resemblances and differences seen amongst living organisms of the same species. Variation is of two main kinds: **heritable variation** due to genetic characters inherited from the parents and **non-heritable variation** due mainly to the environment.

Inherited disorders

Inheritance of certain characteristics arising through the influence of one or more pairs of genes is seen in many heritable diseases or disorders in humans such as cystic fibrosis and Huntington's chorea. These heritable conditions may be dominant or recessive as shown in Table 12.7 and Figures 12.20 and 12.21.

Figure 12.20 Inheritance of cystic fibrosis – recessive inheritance

	Unaffected carrier mother Genotype = Cc gametes (ova): C	gametes (ova): c
Unaffected carrier father Genotype = Cc gametes (sperm): C	**CC** Normal, homozygote	**Cc** Unaffected, carrier heterozygote
gametes (sperm): c	**Cc** Unaffected, carrier heterozygote	**cc** Affected, homozygote

Figure 12.21 Inheritance of Huntington's chorea – dominant inheritance

	Unaffected mother Genotype = hh gametes (ova): h	gametes (ova): h
Affected father Genotype = Hh gametes (sperm): H	**Hh** Affected, heterozygote	**Hh** Affected, heterozygote
gametes (sperm): h	**hh** Unaffected, homozygote	**hh** Unaffected, homozygote

Table 12.7 Dominant and recessive inheritance in human beings

	Dominant	Recessive
Hair	White forelock Curly or woolly Dark and light colour	Normal hair colour Straight Red colour
Skin	Freckles Normal pigment	No freckles No pigment – albinism (1 in 20 000 affected)
Eyes	Brown iris Night blindness Normal colour vision	Blue iris Normal night vision Red/green colour blindness
Ears	Lobed ears	Lobeless ears
Mouth	Normal lip Ability to taste PTC (phenylthiocarbamide)	Hare lip Inability to taste PTC
Limbs	Extra fingers and toes – polydactyly Short fingers – brachydactyly Short limbs – in achondroplasia, dwarfism	Normal number of fingers and toes Normal finger length Normal limb length
Physiological	Resistance to tuberculosis (TB) Ability to convert phenylalanine Ability to decompose alkapton Normal mucus secretion in lung Huntington's chorea – progressive insanity and muscular spasm occurring in middle age (1 in 20 000 affected)	Lowered resistance to TB Phenylketonuria (Section 12.4) (1 in 25 000 affected) Alkaptonuria (1 in a million affected) Cystic fibrosis – abnormally thick mucus secretion in the lung, intestine and pancreas (1 in 2000 births) Normal condition

Amniocentesis is a technique for obtaining samples of amniotic fluid from around the fetus before its birth (see Section 10.9). This fluid contains cells from the fetus which can be examined microscopically by karyograms for signs of chromosome abnormalities that would cause Down's syndrome for example. Biochemical tests performed on the amniotic fluid can also show evidence of heritable disorders such as spina bifida. If tests of this kind are positive, a decision can then be made to terminate the pregnancy by legal abortion.

___ **Questions** ___

40 (a) What are the advantages of being able to make karyograms and perform biochemical tests on amniotic fluid?
   (b) What issues may the advantages raise?

Non-heritable variation

Non-inherited variation is caused by **environmental influences** other than mutagens, radiation and chemicals. The environment of an organism includes all the conditions in which it lives. These include the conditions in the internal environment and the external environment. The body cells have an internal environment due to tissue fluid at a constant composition, maintained in vertebrates by homeostasis (see Chapter 7). The external environment of an organism has two kinds of factors:

- **biotic factors** or external influences arising from the activities of other living plants and animals;
- **abiotic factors** or physical factors due mainly to climate, temperature, light, wind, moisture and also to mineral elements in sea water and soils (see Chapter 14).

The arrowhead plant

This shows two distinctly different phenotypes depending on whether it is grown in deep, running water when it develops long, narrow, grass-like leaves, or in wet soil when it develops broad, arrowhead-shaped leaves (see Figure 12.22). Grown together in the same constant environment, the offspring or daughter plants are indistinguishable from each other, showing that the phenotype is due to environmental effects.

Figure 12.22 Different phenotypes of the arrowhead plant, *Sagittaria sagittifolia*, showing aerial and submerged leaf forms

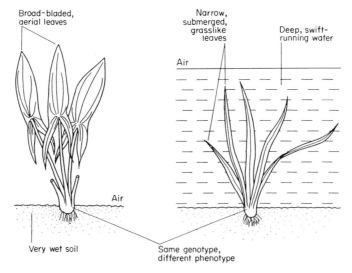

Clones

These are organisms reproduced asexually from a single-celled parent plant or animal, vegetatively from a single multicellular plant or by inserting the genetic material of a body cell into an enucleated egg cell and stimulating it to divide. The descendants will have the same genotype unless mutation occurs. A potato tuber can be cloned and reproduced by dividing one tuber into two parts. Each part will produce plants with different phenotypes if allowed to grow in different environments such as a soil complete with mineral nutrients and a soil deficient in nutrients. This shows that the variation is due entirely to the environment since the genotype is the same for both potato plants.

Twins

Identical or **monozygotic** twins are the result of the division of the zygote derived from one fertilised ovum into two (see Figure 12.23). Such twins are genetically identical in having the *same* chromosomes and genes in their cell nuclei. They are members of the same clone and have the same genotype.

Figure 12.23   Schematic diagram explaining the formation of identical and non-identical twins

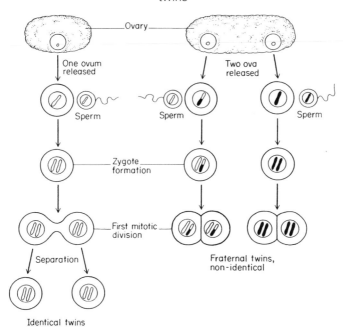

Fraternal or **dizygotic** twins result from the fertilisation of two separate ova. They differ genetically, having different chromosomes and genes like normal brothers and sisters do (see Figure 12.23). Table 12.8 compares identical and fraternal twins.

Questions

41 What are the factors at work on the growth of a tree on a cliff?
42 Why are there much greater weight differences in (a) identical twins reared apart compared to those reared together, (b) fraternal twins reared together compared to identical twins reared together?

Heritable variation

Dachshund puppies are bow-legged due to genetic causes. Other breeds of puppies can be bow-legged due to environmental causes such as a lack of vitamin D causing rickets. Dachshunds will always inherit bow legs unless a mutation occurs which alters this character, whilst rickets-affected dogs of other breeds will not pass on rickets to their pups provided that the pups are provided with a suitable diet. Table 12.9 compares some heritable and non-heritable characters.

Table 12.8 Comparison of identical and fraternal twins

	Identical twins	Fraternal twins
Origin	Monozygotic: develop from one fertilised egg or zygote, which undergoes fission into two embryos	Dizygotic: develop from the simultaneous fertilisation of two separate eggs
Sex	Always of same sex	Can be of same or different sex
Chromosomes	Same chromosome sets in cell nuclei of each twin	Different chromosome sets in cell nuclei of each twin
Genotype	Probably every cell has same set of genes in each twin or have identical genotype	Different sets of genes in cells, and will have different genotype
Environmental effect	Reared together Height: slightly different Weight: greater differences Reared apart Height: slightly different Weight: much greater differences	Reared together Height: greater differences Weight: much greater differences
Conclusion	Identical twins having same genotype, show differing phenotype with reference to weight which is mainly environmentally controlled. Height differences are slight and mainly genetically controlled.	

Table 12.9 Comparison of heritable and non-heritable characters

Heritable	Non-heritable
Genetic variations	Environmental variations
1 Due to genes	1 Due to environment, food, climate, soil, disease or other organisms
2 Reappear in offspring	2 Cannot appear in offspring
3 Mainly unchangeable in life-time. Exception: curing of stomach outlet constriction (pyloric stenosis) by surgery.	3 Sometimes changeable in lifetime
Examples	
(a) Achondroplasia – dwarfs with short limbs	(a) 'Thalidomide' limb malformation
(b) Hereditary blindness and deafness	(b) Blindness and deafness due to syphilis or German measles in mother
(c) Bow-legged Dachshund puppies	(c) Bow-legged puppies due to rickets – vitamin D shortage
(d) Dwarf pea plants	(d) Stunted pea plants from poor soil
(e) Tall pea plants	(e) Etiolated pea plants grown in poor light
(f) Dark and light skin colour	(f) Sun-tanned, light skin colour
(g) Hereditary night blindness	(g) Vitamin A deficiency – night blindness

Heritable or genetic variations are of two main kinds: **discontinuous** or qualitative and **continuous** or quantitative.

Discontinuous or qualitative variation

This type of variation shows *sharp differences* amongst individuals of a species. It does not show any intermediate forms; the characters contrast. Discontinuous or qualitative variation is seen in humans as the inheritance of one of the blood groups **A**, **B**, **AB** or **O** (see Section 12.8) and of eye colour. Animals show discontinuous variation in the inheritance of a smooth or rough coat as in certain breeds of dog, e.g. Jack Russell, and pea plants are either tall or dwarf. The environment cannot affect this type of variation, since eye colour, blood group, skin and flower colour are all unchangeable. The sex chromosomes are described in Section 12.3. In fertilisation they meet and pair randomly. This should result in half the zygotes being male and half being female. This is discontinuous variation of sex. In the United Kingdom 6% more boys are born compared to girls each year. The ratio of females to males in the United Kingdom population is 105 females to 100 males. Evidently there is a higher mortality of males at all ages.

Continuous or quantitative variation

This type of variation is seen as *very small differences* amongst individuals of the same species and many intermediate forms are seen. Continuous variation (see Figure 12.24a) is partly due to the effects of many genes of small effect. Characters controlled by the interaction of many gene pairs are called **polygenic**.

Figure 12.24 Graphical representation of (a) continuous and (b) discontinuous variations in populations

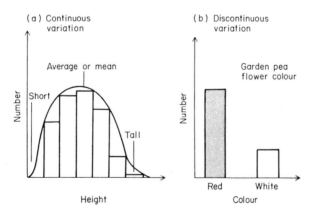

Continuous or quantitative variation is seen in the wide range of values for height and intelligence quotients in humans. This type of inheritance is caused by effects of polygenes and the environment and is called **multifactorial inheritance**. Discontinuous variations can be shown graphically and by frequency distribution bar charts (see Figure 12.24b).

___ **Question** ___

43 How is continuous variation different from discontinuous variation?

Summary

- A phenotype is a term describing the external structural form and internal function of an organism. (▶ 297)
- Chromosomes are arranged in pairs in body cells. (▶ 298)
- Mutations can occur in chromosomes. (▶ 301)
- Genes are located on chromosomes and contain DNA. (▶ 302)
- Mutations can occur in genes. (▶ 303)
- Meiosis involves a rearrangement of material in the chromosomes and the production of haploid nuclei in gametogenesis. (▶ 305)
- Fertilisation produces a diploid nucleus from the fusion of the nuclei of the male and female gametes. (▶ 307)
- The inheritance of characteristics can be explained using symbols and Punnett squares. (▶ 308)
- Some disorders are linked to the sex chromosomes. (▶ 317)
- There is a range of inherited characteristics in humans. (▶ 321)
- Some variation in a species is non-heritable. (▶ 324)
- There are two kinds of heritable variation – discontinuous and continuous variation. (▶ 325)

13 Evolution

Objectives

After reading this chapter you should be able to:
- describe environmental conditions on the early Earth and explain how we think that they have changed
- describe how fossils form and explain their importance in the study of evolution
- understand how the studies of comparative anatomy and classification are useful in the study of evolution
- compare the theories of Lamarck and Darwin
- explain how genetics shows how evolution can take place
- describe an example of natural selection in action
- explain how a new species may form

13.1 Introduction

The Earth is one of nine major planets revolving round the sun and is estimated to have come into being about 4750 to 5000 million years ago. The first living organism emerged over 3000 million years ago and today the Earth is inhabited by more than two million known species. In this chapter the origin of the species and how the present day plants and animals came into being are considered.

The process of **changing** the **genetic material** has occurred gradually and continuously since life started on Earth. Heritable variations have been a means of survival when external conditions changed on Earth. These variations became characters or traits and are seen in the different species alive today.

13.2 Chemical evolution

Evolution is a process of **gradual** development from simpler forms through a series of changes. Chemical evolution is one means of explaining the origin of life on Earth through the formation of complex organic compounds like the DNA molecule (see page 280) which is capable of reproducing itself. The other alternative explanations include a supernatural origin of life, and the landing of 'life seeds' on Earth from some other planet.

Chemical evolution explains the formation of the complex organic compounds from simpler chemical elements or compounds. Alexander Ivanovich Oparin, a biochemist, first produced an account of the chemical origin of life in 1936. He suggested that the early Earth had an atmosphere of methane and ammonia and that by using energy from the sun, life evolved from this.

Stanley Lloyd Miller in 1953 passed electric discharges or electrical energy into a mixture of methane, CH_4, ammonia, NH_3, water, H_2O and Hydrogen, H_2. The product was an organic mixture consisting of sugars, organic bases, fatty (alkanoic) acids and amino acids, all of which are important biological components of living organisms.

Biochemistry shows that living organisms are composed of similar organic compounds, in particular proteins and nucleic acids. These could have been synthesised in conditions prevailing over 4000 million years ago on Earth and formed by a process of **chemical evolution** into a nucleic acid substance capable of replicating itself (see Sections 11.4 and 12.4).

A complex organic mixture would provide the essential nutrients and amino acids in a **primitive soup**. In this soup the first living cells may have been evolved as virus-like organisms able to function in an atmosphere without oxygen. They could survive **anaerobically**. Once a living organism had been formed from non-living organic compounds, more living things would arise by **biogenesis**. Living organisms may have had a common ancestry in the soup of organic materials.

13.3 Evidence of organic evolution

Organic evolution is concerned with the evolution of populations of living organisms. Organic evolution is a **continuous** and **irreversible** process of change in heritable characters. It occurs by the interbreeding of organisms and is seen in the changes shown by successive generations. The first bacteria-like organisms fed on a ready-made soup. Later it was necessary to photosynthesise food and a form of algae was evolved. These primitive algae produced food for themselves and others by photosynthesis. They also produced **oxygen** which accumulated in the air for the first time in the Earth's history.

Evidence for the existence of primitive algae, related to existing blue-green bacteria, has been found as microfossils in South African rocks over 3000 million years old.

Fossil evidence

Fossils are either the actual remains or traces of organisms that lived in ancient geological times. The organisms may be embedded in sand, resin or ice, or an impression or cast is made of the body parts, the tissues being replaced or petrified by silica or calcium carbonate minerals.

Most fossils are found in sedimentary rock – sandstone and limestone – which forms as a settling and compacting or waterborne sediment of sand, silt, mud and animals.

Identical fossils are found in rock beds or strata formations laid down at the same geological time. These **index fossils** indicate the geological age of the rock formation. Rocks and fossils can also be dated by radioactivity measurements.

Fossil fuels include coal, oil and natural gas and are formed respectively from the remains of plants and minute shellfish or molluscs.

A study of fossils can provide a visual record in a complete series showing the evolution of an organism. The horse provides an example of such a study (see Figure 13.1).

Figure 13.1 A fossil record of hind leg bone evolution of the horse, *Equus* sp.

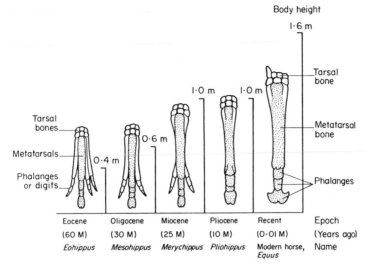

Certain **link fossils**, for example Archeopteryx, which was a bird-like animal with teeth, provide a link with the reptile ancestry of birds.

Microfossils are the oldest fossils of early microbe life which were the main kind of life until 500 million years ago. Fossil study or **palaeontology** shows that organic evolution was a gradual process spanning the last 3000 million years of Earth's history (see Figure 13.2). Evolutionary changes were not linear (in a straight line) but a branching process of descent from a common ancestor seen in an **evolutionary tree**.

Figure 13.2 Fossil record of plants and animals shown in a geological time chart relating to the 'great ages' of different groups of organisms

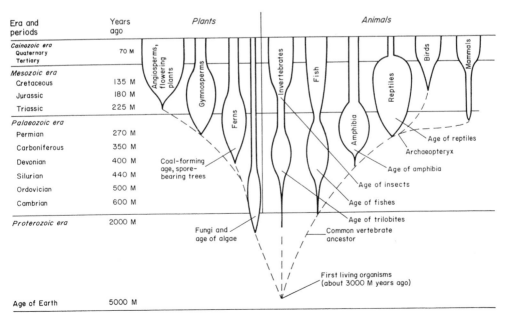

Questions

1 What is a fossil and where did most fossils form?
2 Why are identical fossils found in the same bed of rock?
3 (a) How has the footprint of the horse changed over the last 60 million years?
  (b) How has the structure of the hind leg bone changed over the last 60 million years?
4 Which type of organisms could be found on the Earth (a) 600 million years ago, (b) 350 million years ago, (c) 225 million years ago?
5 Which living organisms were present in a greater number of species 180 million years ago than they are today?

Evidence from comparative anatomy

The **pentadactyl limb** (see Section 9.2) evolved as an adaptation to living on land and became the means to walk, run and fly. Pentadactyl limbs are **homologous** organs, having the *same* origin in a common ancestor, but have developed *different* functions (see Figure 13.3). An insect wing is an **analogous** organ, having the *same* function as the wing of a bird and bat but with a *different* origin.

Figure 13.3   Adaptation of the vertebrate pentadactyl forelimb and examples of homologous limbs

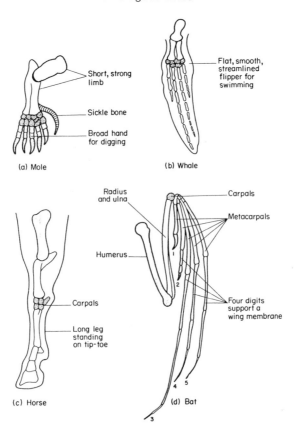

In the early development stages, vertebrate embryos are almost indistinguishable. The embryos of fish, amphibians, birds and mammals look very similar. They have homologous organs seen as **gill clefts** or branchial arches (see Figure 13.4).

Figure 13.4 Early embryos of (a) a bird and (b) a human showing gill clefts, a feature of all early vertebrate embryos

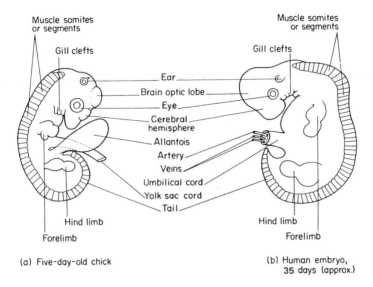

(a) Five-day-old chick

(b) Human embryo, 35 days (approx.)

The flower parts of a flowering plant are homologous. They are considered to have evolved from leaves, to form sepals, petals, stamens and carpels.

Questions

6 What features of the animal limbs shown in Figure 13.3 suggest that they may be related?

7 Look at Figure 9.5 and the limb of the mole and the bird in Figure 13.3. If the pentadactyl limb in Figure 9.5 represents the limb of these animal's common ancestor, how do you think evolution took place in (a) the leg of the mole and (b) the wing of the bird?

Evidence from classification

The science of classification of organisms is called **taxonomy** and is a means of classifying over 2 million different kinds of organisms. The organisms are classified on the basis of possessing certain similar characters. If differences appear slight, biochemical tests are used to determine close relationship. Blood agluttination tests can be used to indicate relationships between primates, e.g. chimpanzees and baboons (see Figure 13.5). Studies on the genetic material (DNA or deoxyribonucleic acid) of different species can show how closely or how distantly they are related.

Extinct fossils and present-day living organisms, when classified together, show kinship and ancestry which indicate the closeness of their evolutionary descent. It also shows the divergent or branching nature of the process of evolution.

Figure 13.5 Classification of primates showing relationship by descent

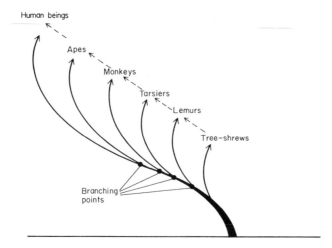

— Question —————————————————————————————————————

8 Some people believe that a 'missing link' between humans and apes may be alive today. If a new primate were found today with ape-like and human-like characters, could it really be described as a 'missing link'. Explain your answer.

13.4 Theories of evolution

Special creation

The theory of special creation states that all species had a separate origin without any relationship between the different species and the species have remained unchanged. The account in the Book of Genesis was written over 4000 years ago.

Lamarckism

Jean Baptiste Lamarck (1744–1829) put forward the theory of **acquired characters** as an attempt to explain the mechanism of organic evolution. The theory stated that all living organisms were related. Variation in the phenotype was developed due to 'use and disuse' of body parts. The changes that occurred as a result of use or disuse were considered to be heritable.

Fossil evidence shows that the ancestor of the giraffe had a short body and neck. According to this theory the long neck evolved from the habit of the giraffe of reaching for leaves higher up in the trees. This was disproved by cutting off the tails of breeding mice and finding the offspring that were born from them all had normal tails.

In Section 12.10 it is shown that variations due to the environment are non-heritable.

Neo-Lamarckism

Supporters of this theory believe that environmental influences affect self-reproducing substances in the cell cytoplasm called **plasmagenes**. This gene-like substance is passed on to affect daughter cell characteristics by **cytoplasmic inheritance**.

Darwinism

The theory of organic evolution by natural selection was proposed by Charles Darwin (1809–1882) and Alfred Wallace (1823–1913). In its simplest form, this theory stated two facts:

1 A species produces more offspring than can possibly survive.
2 There is variation within a species population.

Darwin proposed a mechanism for organic evolution along the following lines:

- A large majority of individuals do not survive and consequently the numbers of a species population remain fairly constant.
- The survivors in any population survive because of being better adapted to reach resources in the environment such as food.
- The survivors in the struggle for existence have certain favourable variations.
- The favourable variations are heritable, now known to be due to changes in the genotype or genetic structure.
- The heritable variations will be more pronounced in future generations until eventually a new species is formed.
- Nature or the environment selects those individuals best fitted for survival by natural selection.

The ancestral giraffes which showed the variations of being taller and in having longer necks than the others survived in times of food shortage. They also had an advantage in being the first to see predators and flee and thus to survive and reproduce.

Question

9 Select a limb from Figure 13.3 and suggest how it may have come to evolve using Darwin's reasoning.

13.5 Genetics and evolution

In Chapter 12 the origin of variations in individuals was explained as being due to the effect of gene and chromosome mutations and also due to recombination following sexual reproduction. Heritable variations are of two kinds: **discontinuous variation** (qualitative inheritance) and **continuous variation** (quantitative inheritance). These are the variations which take part in organic evolution.

In the same way as the gene is the unit of inheritance, the breeding population, instead of the individual, is the unit of evolution. This aspect of genetics is called **population genetics**.

A population is a group of individuals of the same species of all ages interbreeding freely in a certain area. For example, a rabbit population in a rabbit warren, or a bacteria population in a Petri dish culture. The modern evolutionary theory, or **neo-Darwinism**, is concerned with combining the theory of natural selection with modern genetics and relating it to populations rather than individuals.

The Gene pool

Every individual in a population has its own set of genes. These, together with the genes from other individuals in the population, will form the **gene pool**. This pool or reservoir provides genes which pair with others in a random manner in fertilisation.

The genes that are most numerous in the gene pool will tend to occur most frequently in a particular population.

When a population is reproducing it will draw genes from the gene pool. Those genes present in greatest numbers in the gene pool will occur most frequently in the next generation of the population.

Factors affecting the gene pool

Mutation

Mutation (Sections 12.3 and 12.4) is a sudden change in the gene or chromosome structure. If genes or chromosomes had never changed, new species would not have been formed and evolution would not have occurred. Mutation is a source of variation in a population.

Migration

Migration is caused by individuals of the same species entering and leaving the population by **emigration** or **immigration**. In this way sets of genes are removed or added to the gene pool.

Artificial selection

This can be brought about by plant and animal breeders, who remove or add individuals to their breeding populations. This is done by selecting a certain phenotype which affects the gene pool through its genotype. Humans exercise selection in the breeding of plants and animals in modern horticulture and agriculture and increase the rate of formation of new varieties.

Inbreeding

Inbreeding of plants is achieved by **self-fertilisation**, whilst in animal husbandry, carefully selected stock with strong phenotypes are interbred with closely related animals and may be back-crossed (see Section 12.7).

Outbreeding

Outbreeding is performed between individuals who are less closely related. Cross-fertilisation in plants produces an F1 generation with hybrid vigour which improves or increases a certain character by producing e.g. large fruits or increased resistance to disease. Outbreeding of animals is done to increase the strength of animal stocks in the same species.

Natural selection

The fossil record shows that certain species of plants and animals which evolved in the past do not exist today. Natural selection operates through changes in the environment, allowing certain individual phenotypes which are best adapted to live, whilst those in the population least adapted or handicapped perish. The survivors of natural selection go on to reproduce.

Natural selection can be regarded as a test for all members of a population with the conditions of the test being set by the environment. Natural selection is operating on the *phenotype* which indirectly selects certain *genotypes*. These genotypes in turn contribute to the gene pool. Those that perish and fail to reproduce have their genes removed from the gene pool. It is therefore evident that populations will evolve provided that they possess heritable variations. These variations arise from mutations and recombination of genes in meiosis and fertilisation.

Sickle cell anaemia

Sickle cell anaemia is a heritable disease due to the mutation of a recessive gene on an autosome. It changes normal red blood cell haemoglobin into **sickle cell** (SC) haemoglogin. The genotype for normal haemglobin is shown by **HH** and the two recessive genes for sickle cell anaemia are shown as **hh**. These are both homozygous, whilst **Hh** will represent the heterozygous condition which is called the **sickle cell trait**. This is an example of codominance (see page 312).

HH = normal haemoglobin; **hh** = sickle cell anaemia; **Hh** = sickle cell trait. If blood samples are placed on a microscope slide and oxygen excluded from it by adding a drop of liquid paraffin oil, the red blood cells use up their available oxygen. The **Hh** and **hh** red blood cells develop a sickle shape, whilst the normal **HH** red blood cells do not.

People with **Hh**, the heterozygous condition, have the sickle cell trait and experience above-average breathlessness during exercise or at high altitudes. It is not an illness however and has no symptoms.

People with **hh**, the homozygous recessive condition, suffer sickle cell anaemia. They are severely anaemic, feel tired and ill and suffer bouts of severe body pain. They are also prone to colds, jaundice, leg ulcers and rheumatism.

Intermarriage of sickle cell trait **Hh** people may result in their children being 25% **hh** sickle cell anaemia sufferers who die young from malaria and do not reproduce; 50% **Hh** with sickle cell trait who can transmit the disease and 25% **HH** normal individuals. The sickle cell anaemia **hh** children, lacking protection from malaria, do not live to reproduce. They do not contribute to the gene pool and their gene frequency in the gene pool is reduced. In time the gene will be eliminated. In some East African populations the sickle cell gene has a high frequency with 4% of all children born being of **hh**, i.e. with sickle cell anaemia. There is also a higher frequency of sickle cell trait **Hh** in the population. The reason for this is that the sickle cell trait **Hh** people have a selective advantage in being able to survive in greater numbers because of a resistance to malaria. Normal **HH** people have a lower resistance than **Hh** people. In this way the **Hh** sickle cell trait people are naturally selected and the frequency of the **h** sickle cell gene increases in the gene pool, allowing the **Hh** people to survive and reproduce.

Note that it is only the sickle cell trait that gives some protection against malaria. Sickle cell anaemia and normal haemoglobin do not.

Questions

14 (a) How could the gene that produces sickle cell anaemia be eliminated from the gene pool?
   (b) What environmental factor prevents the gene from being eliminated from the gene pool in some East African populations?

Polymorphism is a term used to describe distinctly different **heritable** forms within one freely interbreeding species. Examples are seen in red/green colour-blind people and normal colour visioned people and the sickle cell anaemia/trait and normal haemoglobin. The four human blood groups (see Section 12.8) are further examples of polymorphism.

Peppered moth

The peppered moth *Biston Betularia* (see Figure 13.6) is an example of polymorphism, being found as two distinct forms, the black or **melanic** form and the light or **peppered** natural form.

Figure 13.6   Light and dark forms of the peppered moth, *Biston betularia* (National Museum of Wales)

In 1848, 99% of the moth population of Manchester was of the light form whilst only 1% was of the black form. Almost 50 years later 99% of the Manchester moth population was of the black form and only 1% was light. The environmental change during the 50 year period was mainly due to air pollution and darkening of the tree trunks with sooty grime. The light form moth was easily visible to predatory birds which ate them. The black form was camouflaged, making it difficult for the birds to see them. Natural selection promoted the black form of the peppered moth whose gene frequency in the gene pool of the particular population increased. The gene frequency of the light form decreased since it was unable to survive the predation and reproduce.

The gene responsible for the black form gave it a **selective advantage** on soot-covered tree trunks in the polluted air of industrial towns. In the clean air of the countryside on lichen-covered tree trunks the light form of peppered moth has a **selective advantage** over the black form. Predatory birds such as thrushes are able to see the black form more easily against the unpolluted tree trunk.

Resistance

Bacteria, insects and rats develop a resistance to antibiotics, insecticides such as DDT and rodenticides such as 'Warfarin', respectively. Heritable resistant forms develop which have an increased fitness, being able to survive increased amounts of chemical substances.

Question

15 Many towns have lost the industry that used to pollute the air. How would you expect this lack of air pollution to affect the populations of the peppered moths in the towns? Explain your answer.

New species formation

Selection acting on heritable variations in a changing environment will encourage new forms of the same species. This is seen in polymorphism in the peppered moth.

A species consists of a population of organisms able to breed amongst themselves. **Speciation** is a process of forming a new species. This may arise by preventing new forms from interbreeding. It can be achieved by various means of **reproductive isolation**. Populations can be kept apart by **geographical isolation**, as in the case of the Galapagos finches, whilst different breeding seasons prevent populations of the same species in the northern and southern hemispheres breeding together.

The herring gull and lesser black-backed gull are considered to be extreme forms of the same species which interbreed in some parts of the northern hemisphere and are unable to do so elsewhere. Interbreeding is possible between varieties and is less possible between subspecies which are well on the way to forming a new species by maintaining reproductive isolation. Speciation shows the **divergent** or branching nature of evolution with new species evolving from a common ancestor and is summarised in Figure 13.7.

Figure 13.7   Summary of events leading to speciation

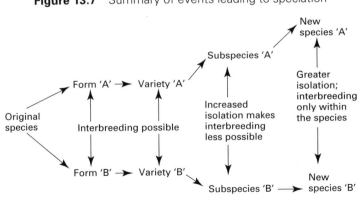

Summary

- It is thought that living organisms evolved from the chemical conditions on the early Earth. (▶ 327)
- Fossils, comparative anatomy and the study of classification provide evidence for evolution. (▶ 328)
- The theories of evolution include special creation, Lamarkism and Darwinism. (▶ 332)
- When genetics is applied to the study of evolution the concept of the gene pool helps to show how evolution could take place. (▶ 333)
- Studies on sickle cell anaemia and the peppered moth show natural selection at work. (▶ 335)
- The isolation of varieties of a species can lead to the evolution of new species. (▶ 337)

THEME V

Interrelationships

14 Natural ecosystems

After you have read this chapter you should be able to:
- describe the main components of an ecosystem and the main processes that take place in it
- identify the abiotic factors in an ecosystem
- compare aquatic and terrestrial habitats
- identify the biotic components of an ecosystem
- describe the different kinds of ecological interactions in an ecosystem
- describe a food chain and a food web
- explain the structure of a biotic or trophic (food) pyramid
- explain how energy flows through an ecosystem
- describe the carbon and nitrogen cycles
- describe the water cycle
- understand the balance between photosynthesis and respiration
- describe the growth of a population and the effects of limiting factors
- understand that the microorganisms in an ecosystem affect other living organisms in a variety of ways

In all previous chapters our attention has been given to the **individual** or single organism and how it maintains life through life support, life protection and life continuation processes. In this chapter we learn how millions of *different* organisms live and interact together, firstly as groups or populations of the same species, and secondly as groups or communities of different species.

Apart from organisms being affected by each other, we shall see that the organisms are also affected by their non-living surrounding environment. In this study of inter-relationships or interactions, it is important to look upon human beings as *very primitive* people living very close to nature in a primitive or uncivilised way of life.

14.1 Principles of ecology

Ecology is the study of the interrelationships between living organisms and their external environment. The **external environment** is composed of the surroundings in which an organism lives. There are two components to the external environment: the living environment and the non-living environment.

The **living** (**biotic**) environment includes the diversity of over 4 million species of living organisms. Their classification is described in Chapter 2. The **non-living** (**abiotic**) environment includes the components of soil, water, air and the sun's radiation.

Environmental factors are all the components of the external environment which can change. They include the temperature, the mineral content of the soil and the living organisms in the environment. The factors can be classified into **chemical** and **physical factors**. Chemical factors include water, minerals and air. Physical factors include light and temperature.

The **habitat** is the place where each individual organism is found or lives. For example the habitat of water-weed is a pond and the habitat of the earthworm is the soil. Water, soil or the land are the main habitats. They can be divided up into **aquatic** habitats and **terrestrial** habitats.

14.2 The ecosystem

An **ecosystem** is a self-supporting unit. It is composed of:
- many different species of **living** organisms;
- **non-living** components, e.g. water or land.

These two components interact together and exchange energy and materials. This can be summarised as:

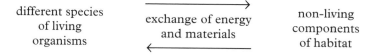

| different species of living organisms | exchange of energy and materials | non-living components of habitat |

The **biosphere** is that part or zone of the Earth's surface and atmosphere where living things are found. It is also the **global ecosystem**. It has two main components:
- **aquatic ecosystems**, composed of living organisms interacting with a watery environment, in oceans, lakes, rivers, ponds and puddles;
- **terrestrial ecosystems**, composed of living organisms interacting with the components of Earth materials, films of water and air; in woodlands, grasslands, deserts, hedges and rubbish dumps or in the soil.

Ecosystem structure and function

The ecosystem is an **open system** with inputs, throughputs and outputs. The global ecosystem has:
- **inputs** of energy from the **sun**;
- **throughputs** which involve interchange or exchange of energy between the Earth's living and non-living materials. This occurs during feeding, mineral cycling and in growth processes;
- **outputs** of heat energy into outer space from the **Earth**.

This can be summarised as:

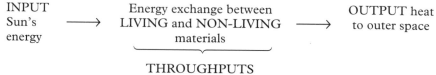

INPUT Sun's energy ⟶ Energy exchange between LIVING and NON-LIVING materials ⟶ OUTPUT heat to outer space

THROUGHPUTS

The composition of an ecosystem and the processes occurring in it are represented in Figure 14.1.

Figure 14.1 Summary of ecosystem composition and processes

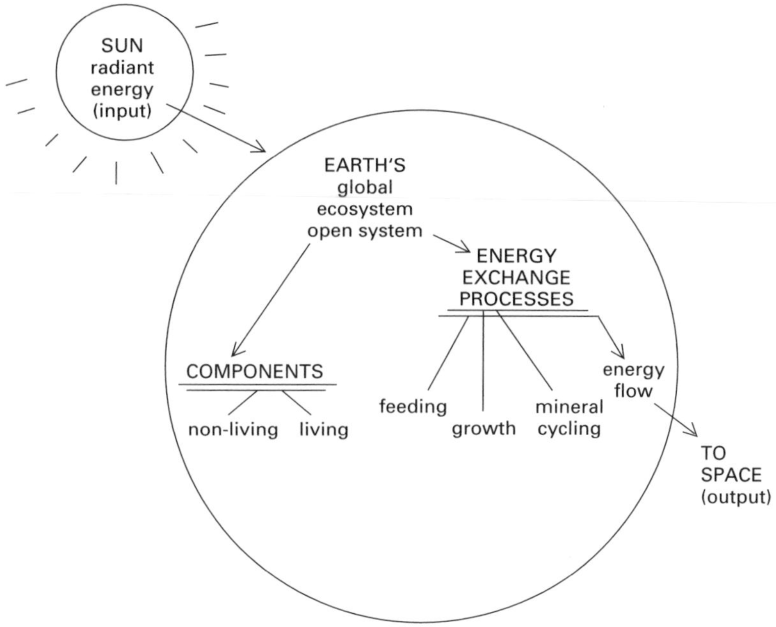

Questions

1 What are the biotic and abiotic components of a woodland?
2 What are the environmental factors that may affect a living organism in a pond?
3 What is the difference between a woodland ecosystem and the biosphere?

The abiotic ecosystem component

The **non-living** environment is composed of the following abiotic factors: solar energy which provides heat and light, water, air and soil.

Climate

The climate is due to changes in the abiotic factors of solar radiation, rainfall and wind. Table 14.1 shows how the abiotic factors associated with climate can be measured and their effects on living organisms.

Soil

Soil can be part of the abiotic environment for some organisms and a habitat for others. The life of many organisms depends directly or indirectly on soil.

The following are some of the main constituents of a soil.

Mineral components

Mineral components of soil are sand, mineral salts, silt, clay and gravel. They are all produced by the **weathering** of rocks. The rocks weather through the action of wind

Table 14.1 Summary of abiotic factors and their effects

Abiotic factor	Measurement and general effects on organisms
Light Sun is main source	Camera light meter Essential for **photosynthesis**
Temperature Sun is main source	Thermometers – maximum and minimum Life can only exist within range 0°C to 50°C
Water Lakes, oceans and rainfall are main sources	Rain gauge – funnel and measuring cylinder Essential for: • **body structure** 40–60% is water • **body support** – turgidity
Wind and *air* Wind due to changing air pressure	Wind gauges measure wind direction and speed Effects: • **Transpiration** rates increase with speed of wind; evaporation and sweat increased • **Oxygen** is present in air 20% by volume, in fresh water 1%, and about 0.85% in seawater • **Carbon dioxide** is present in air about 0.03%
Humidity or air moisture	Hygrometers Affects rate of transpiration, sweating and evaporation

and water erosion, temperature changes and particularly the freezing of water in rock crevices which splits the rock into fragments.

Soil water

Soil water is a dilute solution containing various soluble **inorganic mineral salts**. It forms a film around the mineral soil particles such as grains of sand and clay. The film of water moves upwards through the soil air spaces by **capillarity** and **evaporation** into the air.

Soil air

Soil air is found in the spaces between the soil particles. It consists of less oxygen and more carbon dioxide than atmospheric air. Water-logged soils have no soil air.

Soil air is needed for aerobic respiration of plant roots, soil organisms and certain aerobic microorganisms.

Soil acidity and alkalinity

A soil will have a certain pH value. **Alkaline** soils are associated with chalky soils. **Acid** soils are found in poorly drained, peaty and water-logged soils.

Soil temperature

A suitable temperature is essential for seed germination and plant growth.

Organic content

Organic content of a soil is due to **humus** which is the end product of decay of plant and animal material. This decay is brought about by soil microorganisms, fungi and bacteria.

Humus is an important source of nutrients, providing nitrates and ammonia for protein synthesis in green plant nutrition. It is also a food for soil organisms such as detritivores, earthworms and insect larvae.

Humus is a colloidal, organic compound which has water-absorbing properties and retains water by **hygroscopic** action. The adhesive properties of humus improve sandy soils and also cause clay particles to cling to its fibres.

Soil drainage and permeability

The water-holding capacity and drainage of a sandy, clay or peaty loam soil differs depending on the mineral content, air spaces and organic content.

Questions

4 How do the abiotic factors affect the lives of (a) plants, (b) animals?
5 When a soil becomes water-logged (a) how are the plant roots and the lives of soil organisms affected? (b) how is the pH of the soil affected?
6 How does the humus content affect the soil and the organisms in it?

Aquatic and terrestrial habitats

An environment which experiences very slight changes in its environmental factors is called a **stable environment**. An environment which experiences considerable changes in its environmental factors is called an **unstable environment**.
- **Aquatic environments** of rivers, ponds, lakes and oceans are **stable**, showing only a small variation in composition of the abiotic environment. For example, the temperature and the mineral content of an aquatic environment changes only slightly.
- **Terrestrial environments** on land are **unstable**. For example, the temperature may vary greatly over a 24 hour period and throughout the year.

Table 14.2 shows how the abiotic factors of aquatic and terrestrial environments compare.

Table 14.2 Aquatic and terrestrial habitats compared

Abiotic factor	Seawater	Fresh water	Terrestrial
Light	Varies with depth	Varies with depth	Varies greatly with day or season
Temperature	Slight variation	Slight seasonal variation	Varies greatly with day or season
Oxygen	0.83% – constant	1.0% – constant	20% – constant
Water	Constant	Constant	Very variable with season
Minerals	Constant	Constant	Variable

Question

7 What is meant by (a) a stable environment and (b) an unstable environment? Give examples of each kind.

Biotic components of the ecosystem

The following are important biotic environment terms:

- **Species:** a group of organisms sharing many similar characters and having the same general phenotype. They have the same gene and chromosome composition and can interbreed with each other. The species is the **biological unit** of ecological study. There are over two million known species of organisms. They contribute to the diversity of living organisms which comprise the biotic environment of an ecosystem (see Chapter 3).
- An **individual** such as a human being, a dog or a grass plant is one member of a certain species.
- A **population** is a group of many individuals of the *same* species. The individuals in a population can be of different sexes and ages as in the human world population.
- **Community:** the collection of *different* species of organisms found interacting in a particular habitat.

  Community + Abiotic environment = Ecosystem

Question

8 How can you distinguish between (a) a species and an individual, (b) a population and an individual, (c) a population and a community, (d) a community and an ecosystem?

14.3   Ecological interactions

Ecology is concerned with interactions between organisms. The main ecological interactions are:

- **Competition:** the interaction between two organisms which need the same environmental factor or resource which is in short supply. For example light, water, warmth, shelter, mates for breeding, territory and food.
- **Predation:** a prey–predator interaction. It is found between different species – the **predator** – which catches, kills and eats another different species – the **prey**.
- **Symbiosis:** a very close relationship between two different species of organism. It includes the **host–parasite** relationship in which one species (the parasite) benefits at the expense of the other (the host) and **mutualistic** symbiosis in which both species benefit from the relationship.

(See also the summary of feeding methods on page 368.)

Competition

This ecological interaction between organisms competing for an environmental factor in **short supply** is of two kinds: population competition and community competition.

Population competition

This is the interaction between individuals of the *same* species in a population. It is also called **intraspecific competition**. For example, individual meadow grass plants will compete with each other for light, water and mineral nutrients. Rabbits in a warren will compete with each other for food, shelter, territory and mates for breeding.

The consequences of competition within populations are:

- numbers will *rise* to a peak and then *fall*. This can be displayed as a histogram or graph (see Figure 14.12).

- migration will occur with organisms being *dispersed*. This will prevent an increase in the death rate.

Community competition

This is the interaction between individuals of *different* species in a community. It is also called **interspecific competition**. For example, dandelion and daisy plants growing on a grass lawn or sheep and rabbits grazing together.

The main result of competition within a community is that one species will *flourish* or become dominant in numbers whilst the other species will *decline* or be displaced. For example, rhododendrons are not native UK plants but their introduction into the UK has caused a displacement of native species. A comparison of competition within populations and communities is given in Table 14.3.

Question

9 How is intraspecific competition different from interspecific competition?

Table 14.3 Comparison of competition within a population and a community

	Intraspecific competition	Interspecific competition
Occurrence	In populations	In communities
Number of species involved	One e.g. rabbits or meadow grass plants	More than two e.g. rabbits and sheep or meadow grass plants and daisies
Result	Numbers increase to a peak and then decrease Death rate increases Individuals move out or emigrate from habitat	One species increases in numbers and becomes the dominant species The other species declines in numbers and becomes extinct
Data display	Data is continuous, shown in a histogram or graph	Date is discontinuous, shown in a bar chart
Consequence	Survival of fittest individual and its alleles contributes to natural selection (see Chapter 12)	Change of the species contributes to change in a community leading to succession

Predation

The predator–prey interaction can be displayed as population cycle graphs where the numbers of each species rise and fall in cycles over time (see Figure 14.2). There are many examples of the predation interaction. These include owls and voles, buzzards and frogs, ladybirds and greenfly, sparrowhawks and hedge sparrows and foxes and rabbits. The results of predation in an ecosystem include:

- **Selective removal** of the old, weak or diseased prey leaving the fittest prey to survive.
- **Reduction** of the prey population, preventing overcrowding and providing more space for the organisms that survive. For example reducing the size of a rabbit population gives the individuals that survive more space for grazing.
- **Birth rates** of predators are *lower* than birth rates of prey. This ensures that there are fewer predators than prey in the community.
- Predation is an important **regulator** of population size.

Figure 14.2 Predator–prey population cycles

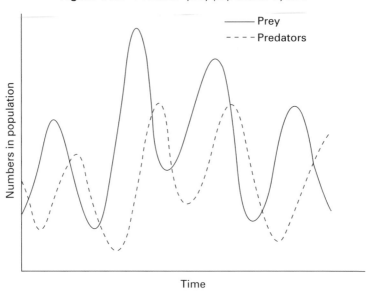

Questions

10 If a predator was removed from an ecosystem, how would the prey species be affected?

11 If a predator's birth rate increased how would this affect the prey species?

Succession

- **Succession** is a series of changes in the community in a particular habitat over a period of time. In succession, one group of organisms surpasses or outstrips another group and may displace them completely from the community.

- **Colonisation** is the formation of a community on previously uninhabited areas, such as bare rocks, sand dunes, quarry refuse, newly made ponds and reservoirs. For example a gravestone is colonised by a community of blue-green bacteria, then algae and lichens. This can be represented as:

$$\text{bare rock} \rightarrow \text{blue green-bacteria} \rightarrow \text{algae} \rightarrow \text{lichens}$$

Cleared woodlands or forests cleared by fire, show succession in the establishment of the following series of communities.

$$\begin{array}{ccccccc} \text{annual} & & & & & & \\ \text{herbaceous} & \rightarrow & \text{grasses} & \rightarrow & \text{shrubs} & \rightarrow & \text{trees in} \\ \text{weeds in a field} & & \text{in grassland} & & \text{in shrubland} & & \text{woodland} \end{array}$$

- A **climax community** is the final or stable community which is not replaced by another community. For example, the trees in a woodland. The dominant group of trees has surpassed, or outstripped, the other groups of shrubs and grasses.

12 How do colonisation and succession lead to a climax community?
13 An area of woodland was cleared of its trees and bushes and turned into grass-land for grazing livestock. After a while the land was no longer used by humans to graze their livestock or for any other human activity. What do you think happened to the grassland? Explain your answer.

14.4 Feeding processes in communities

Continuous interchange of energy and chemical materials takes place between the living biotic and non-living abiotic environments in a community or ecosystem. The four main processes occurring in the community or ecosystem are feeding, energy flow, nutrient element cycling and population growth.

Feeding

Biotic components

Feeding is the process of obtaining organic food with a high energy content.
- **Producers** (see Chapter 3) are green plants able to make their own organic food from solar energy and simple raw materials. Since green plants are self-feeding they are also called **autotrophic** organisms.
- **Consumers** are mainly animals which obtain their high energy content organic food by ingestion or **feeding** on producers or other animals. They are also called **hetero-trophic** organisms (see Table 14.4).
- **Decomposers** are organisms that cause decay, biodegradation or decomposition of dead plant and animal remains.
(See also the summary of feeding methods on page 368.)

Table 14.4 Consumers compared

Organism	Organic food source
Herbivores: zooplankton, caterpillars, locusts, rabbits, goats	Autotrophic plants
Carnivores: beetles, sharks, cats, lions*	Other heterotrophic animals
Omnivores: humans	Autotrophic plants and heterotrophic animals
Detritivores: earthworms, tubeworms, woodlice, millipedes	Dead plant and animal material
Scavengers: rats, gulls, carrion crows	Dead animal remains
Parasites: protozoa, viruses, bacteria, roundworms, arthropoda, e.g. human fleas	Mainly previously digested food in the body fluids of living organisms

*Certain green plants, e.g. sundew, feed on insects. This supplements normal photosynthesis.

Feeding interelationships

The interdependence of living organisms on each other is shown in the following feeding interrelationship between organisms:

```
                    SUN'S ENERGY
WATER and                ↓
CARBON      →       PRODUCER      → CONSUMER → DECOMPOSER
DIOXIDE                  ↑
                    MINERAL IONS
```

Food chains

A food chain is a series of stages of different feeding methods in which energy and food nutrients are passed from producers to consumers. Each different feeding stage is called a **trophic level**. The *number* of organism, or their *mass*, or their *energy content* are not shown in a food chain. The length of the food chain is variable.

A grassland ecosystem could have the following food chain:

$$\text{grass} \rightarrow \text{sheep} \rightarrow \text{human being}$$

A shorter food chain could be:

$$\text{wheat} \rightarrow \text{human being}$$

The trophic level is the method of feeding at each stage in the series or chain. It is described as the **first, second** or **third** trophic level along the food chain or as T_1, T_2 or T_3. Some long food chains (see Figure 14.3) have fourth and fifth trophic levels which can be represented as T_4 and T_5.

Food webs

Simple food chains are rarely found in an ecosystem since more than one species of organism can feed together at the same trophic level (see Figure 14.4). At the first trophic level are the **producers**. These are all the green plants and unicellular algae. At the second trophic level are all the **primary consumers**. These are the herbivores and include insect larvae, aphids, greenfly, snails and slugs, rabbits and field mice. At the third trophic level are the **secondary consumers**. These are the small birds and the insectivorous beetles. At the fourth trophic level are the **tertiary consumers** – the hawks, owls and foxes.

Questions

14 Where does the energy come from to produce food?

15 What are the basic components of the food in the chain?

16 How would the feeding behaviour of the small birds in Figure 14.4 be affected if the population of insect larvae declined?

17 (a) How would the spraying of insecticide alter the food web in Figure 14.4?
   (b) Suggest how the spraying of insecticide would affect the populations of the organisms that survived. Explain your suggestions.

Biotic or trophic pyramids

Trophic pyramids diagrammatically indicate the number, mass of organisms or amount of energy at each trophic level in a food chain. Each section of the pyramid is equivalent in size to the trophic level. Figure 14.5 shows how the food chain:

green plants $\rightarrow$ aphids $\rightarrow$ ladybird $\rightarrow$ frog $\rightarrow$ buzzard

can be represented as a trophic pyramid.

Figure 14.3 Examples of trophic levels (T_1–T_5) in a food chain

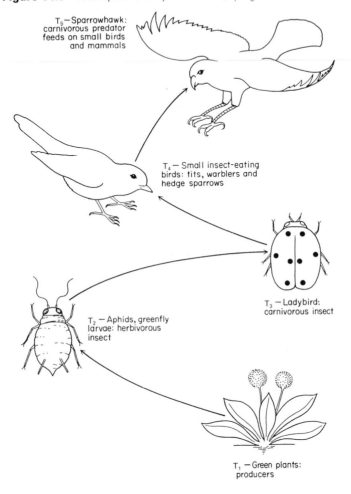

T_5—Sparrowhawk: carnivorous predator feeds on small birds and mammals

T_4— Small insect-eating birds: tits, warblers and hedge sparrows

T_3 —Ladybird: carnivorous insect

T_2 — Aphids, greenfly larvae: herbivorous insect

T_1 —Green plants: producers

Figure 14.6 on page 353 shows how biotic or trophic pyramids are represented by a series of horizontal bars. Each horizontal bar represents a trophic level.

The **pyramid of numbers** diagrammatically shows the numbers of organisms at each trophic level. A great number of producers support a *decreasing* number of consumers. If a pyramid of numbers is constructed for a population of caterpillars feeding on an oak tree, the pyramid will be inverted. The base of the pyramid will be a narrow bar representing *one* oak tree while the bar above it will be wide representing the *large* number of caterpillars.

The **pyramid of biomass** diagrammatically shows the total weight (mass) in kilograms, kg, of living organisms in each trophic level. The weight (mass) *decreases* in each trophic level. If a pyramid of biomass is constructed for a population of caterpillars

Figure 14.4 A food web made up of interrelated food chains

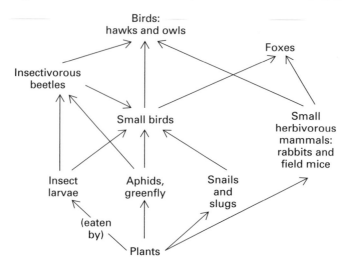

Figure 14.5 Trophic pyramids show in a quantitative manner the numbers, energy or mass (biomass) involved in a food chain

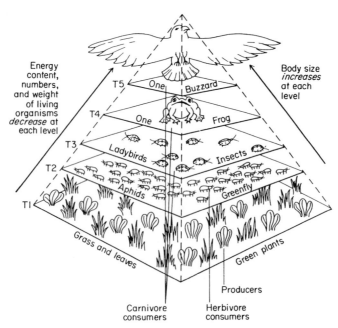

feeding on an oak tree, the pyramid is not inverted. The bar at the base of the pyramid will be very wide, representing the *large* biomass of the tree. The bar above it will narrow, representing the *much smaller* biomass of the caterpillars.

Figure 14.6 Biotic or trophic pyramids

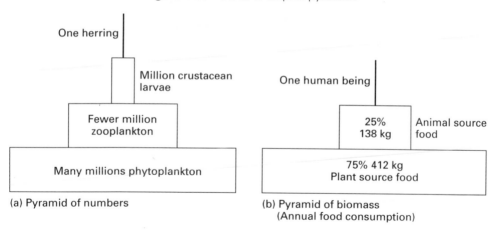

One herring

Million crustacean larvae

Fewer million zooplankton

Many millions phytoplankton

(a) Pyramid of numbers

One human being

| 25% 138 kg | Animal source food |

75% 412 kg Plant source food

(b) Pyramid of biomass (Annual food consumption)

14.5 Energy flow in an ecosystem

Figure 14.7 Energy flow through the biosphere

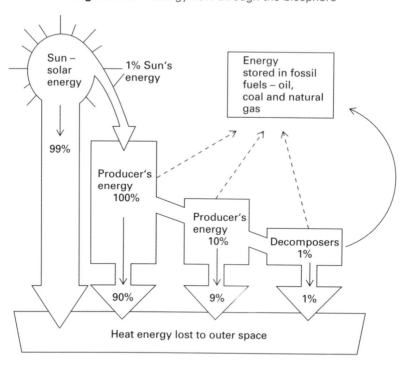

Sun – solar energy

1% Sun's energy

Energy stored in fossil fuels – oil, coal and natural gas

99%

Producer's energy 100%

Producer's energy 10%

Decomposers 1%

90%

9%

1%

Heat energy lost to outer space

Energy flow occurs from the sun through Earth-dwelling organisms into outer space. During the flow, energy is changed into different forms: **light**, **chemical**, **mechanical** or **electrical** energy. During these changes, **heat** energy is always formed and is the form in which energy leaves the natural ecosystem.

solar energy	→	trapped by green plants	→	stored or changed in living organisms	→	released by respiration	→	escapes as heat into outer space

Note that the abiotic environment of rocks, earth and water absorbs almost 99% of solar radiant energy. This is later lost into space by reflection or radiation.

Fossil fuels

Fossil fuels were formed over 300 million years ago during the **carboniferous** period. They contain **stored energy** trapped by green plants and certain minute plankton organisms. The energy in fossil fuels was locked away for millions of years until humans mined and exploited them. This energy source is now rapidly dwindling due to its widespread use by the ever increasing human population.

The Greenhouse Effect

This is a natural process in which some of the heat radiated from the Earth is absorbed in the atmosphere's carbon dioxide and water vapour, which re-emits the heat back downwards to Earth. This atmospheric blanket of gases protects the Earth from violent extremes of temperature by day and night, and is therefore an important factor for supporting and protecting life on Earth.

Pyramid of energy

The pyramid of energy (Figure 14.8) diagrammatically shows the total energy value in megajoules (MJ) in each trophic level of a food chain. The energy loss at each trophic level is clearly shown in the diagram – approximately 90% is lost at each interchange.

Figure 14.8 A pyramid of energy

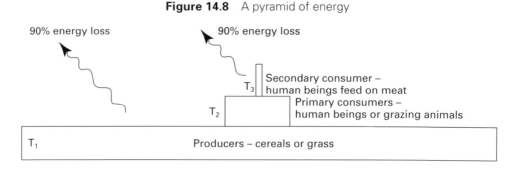

Short food chains, e.g. cereals → humans, lose less energy and are more efficient than long food chains, e.g. grass → sheep → humans.

Question

18 Which human diet, a vegetarian diet or a diet including meat, is the more energy efficient? Explain your answer.

14.6 Cycling of materials

Nutrient element cycling

Nutrient **chemical elements** present in the abiotic environment include carbon, hydrogen, oxygen, phosphorus, sulphur, nitrogen, calcium and iron. These are absorbed and changed chemically into organic **compounds** by green plants, using sunlight energy. After the death of the living organisms the elements return to the abiotic environment. Further use of these chemical elements by other living organisms results in their **cycling** between the abiotic and biotic environment.

This interchange of chemical materials occurs at the same time as energy interchange in all living organisms. The same chemical elements have been reused or recycled in the bodies of different living organisms ever since life began on Earth.

The same amount of chemical material is present on Earth today as when the Earth was first made. (Except for the material sent into outer space in rockets, space probes and satellites.)

The carbon cycle

Living organisms cycle the chemical element carbon through the life support activities of aerobic respiration, anaerobic respiration, fermentation and photosynthesis. Fire or the **combustion** of plant and animal material also produces carbon dioxide.

Figure 14.9 The carbon cycle

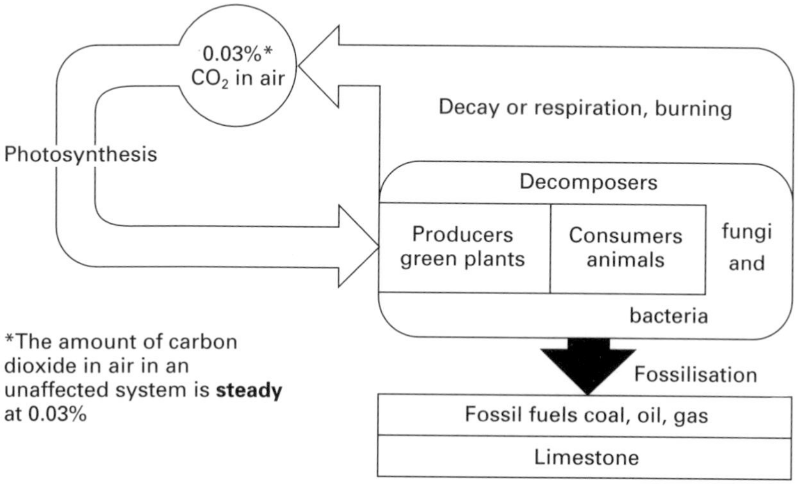

*The amount of carbon dioxide in air in an unaffected system is **steady** at 0.03%

Note that during fossilisation the carbon in fossil fuels and in the fossils in limestone rocks is 'locked away' in a natural carbon cycle and is unaffected by human influences. Large amounts of carbon dioxide dissolve in the water of aquatic ecosystems.

Nitrogen cycle

The nitrogen cycle is the continuous circulation of nitrogen and its compounds in the ecosystem by living organisms. The air nitrogen content is approximately 80%. It is maintained at a fairly constant level through denitrification, volcanic sources and through nitrogen fixation.

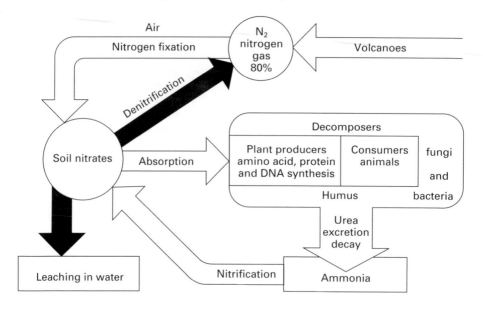

Figure 14.10 The nitrogen cycle

Producers take up nitrates by **root absorption** and use them to make amino acids, protein and genetic material (DNA). Excretion and decomposition convert proteins to ammonia, which is released into the soil.

- **Nitrification** is the process of converting proteins and ammonia into **nitrates** by nitrifying bacteria.

$$\text{Proteins} \rightarrow \text{Ammonia} \xrightarrow[\text{bacteria}]{\text{nitrifying}} \text{Nitrates}$$

- **Nitrogen fixation** is the process of converting gaseous nitrogen in air or in soil air by **nitrogen fixing bacteria** in root nodules or pea or bean plants, and **lightning** in which the electrical discharge during thunderstorms leads to the formation of oxides of nitrogen.
- **Denitrification** is the process of changing soil nitrates into **nitrogen gas** by denitrifying bacteria in water-logged, acid soils.
- **Leaching** is the washing away of soluble nitrates by rain or flood water.
- **Decay** is the process of **putrefaction** and **decomposition** of the remains of producers and consumers through the activity of decomposers (fungi and bacteria) to produce nitrogen-rich **humus**.

Questions

19 Look at Figure 14.9 and explain what could happen to an atom of carbon after it has formed part of a sugar molecule in photosynthesis.
20 Look at Figure 14.10 and explain what may happen to a molecule of nitrogen gas after it leaves the mouth of a volcano.
21 How do the processes of nitrification, nitrogen fixation and denitrification compare?

The water cycle

Figure 14.11 The water cycle

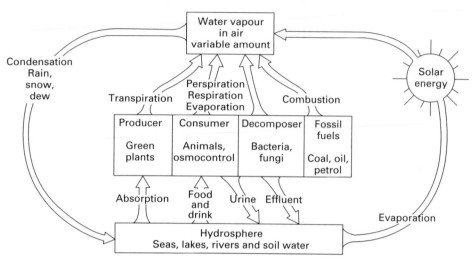

The water cycle is the continuous circulating of water within an ecosystem. It is under the influence of solar energy. This acts as the main circulator by causing evaporation of water in to the atmosphere. **Note:** Water is recycled as a compound, not as separate elements.

- **Evaporation** is the change of physical state of liquid water into gaseous water (water vapour) due to the heat energy of solar radiation.
- **Condensation** is the conversion of gaseous water into either liquid rain or solid snow.
- **Transpiration** by green plants, particularly trees, increases the air water content.
- **Respiration** produces metabolic water.
- **Combustion** produces water vapour, particularly as a product of forest fires.
- **Intake of water** is by root absorption in green plants and by drinking water in animals.
- The **hydrosphere** is where water collects in lakes, rivers and oceans. It also includes soil water.

Questions

22 How is the water vapour content of the air (a) increased, (b) decreased?
23 In what ways are living things involved in the water cycle?

Photosynthesis and respiration balance

About 20% of the air is oxygen. It is the by-product of photosynthesis in green plants. About 0.03% of the air is carbon dioxide. It is the product of respiration of producers, consumers, decomposers and is also produced by the combustion of producers such as trees in forest fires.

The balance of oxygen at 20% and carbon dioxide at 0.03% is maintained because there are more producers making oxygen compared to organisms making carbon dioxide. The producers use carbon dioxide for photosynthesis and the main source of carbon dioxide comes from consumers and decomposers, respiration and from occasional forest fires. This has always been the natural balance until modern humans affected the ecosystem (see Chapter 15).

14.7 Population growth in ecosystems

Limiting factors

The **limiting factor** is that factor or component of the environment which is in short supply. This factor will affect the growth of the population, photosynthesis, respiration or enzyme action. In other words it affects the process of life.

Population

A population is a group of organisms of the same species. The total number of a population can be counted by simple means. The composition of a population varies – some individuals are young, others are old and some may be male or female depending on whether reproduction is sexual or asexual.

Plant population growth

The growth of green plant populations is subject mainly to the abiotic limiting factors which affect photosynthesis, mineral ions, carbon dioxide, water, warmth, space or air containing oxygen and soil for rooting. The biotic factors affecting green plant population growth will include all the herbivore consumers such as sheep, cattle, insects and disease-causing fungal parasitic organisms.

Animal population growth

The growth of animal populations is subject to certain limiting factors:
- The **abiotic** environment is responsible for climate and space to live.
- **Biotic** limiting factors include predators – animals that hunt and kill other animals, the food supply from producers or other consumers, disease-causing organisms, microorganisms and parasites affecting the animal or its food.

Note: Compare with individual organism growth described in Chapter 11.

Population density

Population density is a measure of the number of individuals in a certain area. For example, the number of a certain species within an area of one square metre (m^2) is a measure of population density and indicates the space available for living in.

Population density is determined using a square **quadrat** frame made from wood or metal with all sides measuring one metre. It is placed randomly on the ground and the species within it are identified and their numbers counted as described in Chapter 1. The counts of the different species are recorded in tables and then transferred to bar charts, pie diagrams or maps.

Growth graph of a population

The growth of a population is the result of **reproduction**. The simplest method of reproduction is asexual reproduction. This is seen in bacteria where one individual divides to produce two. Further divisions result in an the increase in numbers from $2 \rightarrow 4 \rightarrow 8 \rightarrow 16 \rightarrow 32 \rightarrow 64 \rightarrow 128$. This is called **exponential growth**.

The graph for population growth of bacteria inside a small laboratory ecosystem, a culture plate, is shown in Figure 14.12. The population growth graph has an 'S' shaped or sigmoid shaped curve.
- The first part of the curve 'A' shows a rapid increase in population numbers, because there are no limiting factors operating. There is a sufficient amount of the abiotic factors, namely light, warmth, nutrients and water, for all the population.

Figure 14.12 Population graph for a population of organisms growing in a restricted area, for example bacteria growing on an agar plate

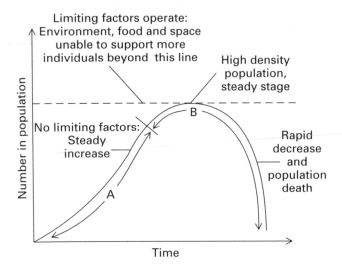

- The second part of the curve 'B' shows no growth in the population followed by a rapid decrease in numbers. This is because limiting factors are operating and components of the abiotic environment are now in short supply with insufficient water, nutrients and space for the bacteria to live in. Death and decay occurs which produces toxic waste and harmful pH conditions which destroy more of the population.

Eutrophication is a natural process involving the excessive enrichment of water in aquatic ecosystems. This environmental factor causes a huge population increase amongst algae which makes the water turn green. This greenness of the water is called an **algal bloom**.

Oxygen levels in the water are reduced as the algae are decomposed by microorganisms and oxygen becomes the limiting factor which causes the death of all organisms in the food chain. The water becomes unpleasant in appearance and smell.

Question

24 Name a limiting factor and describe how it may affect the population of a particular organism.

14.8 Host–parasite interaction

The host–parasite or parasitic symbiont is a major ecological interaction responsible for diseases in human beings and other organisms. Parasitic microorganisms include bacteria, viruses, moulds and protoctists (see Chapter 2) feeding on body fluids or tissues of organisms (see Figure 6.14). Some are beneficial and others are disease-causing pathogens as summarised in Table 14.5 and on page 368.

Table 14.5 Summary of microorganisms

	Bacteria	Viruses	Fungi (moulds)	Protoctists
STRUCTURE	Unicellular; DNA strand; no definite nucleus; 100–2000 μm	Non-cellular; particles or virions; DNA or RNA strand; 10–500 μm	Unicellular thread-like hyphae; definite nucleus	Unicelluar with definite nucleus; over 500 μm (Figure 6.14)
NUTRITION	Mainly saprophytes, parasites and a few symbionts. Feed mainly on protein	Parasites on living cells, which produce the protein needs of the viruses	No chlorophyll; mainly saprophytes or parasites feeding on carbohydrate	Free-living in water or are parasites feeding on various organisms phagocytically
REPRODUCTION	Mainly asexual	By nucleic acid replication in living host nuclei. Cause cell damage to host	Asexual mainly by spores	Asexual mainly; and some sexual
DESTRUCTION	Heat (dry and moist); chemicals (disinfectants); radiation (X and gamma); antibiotics	Heat; chemicals; radiation; certain anti-viral drugs	Heat (dry); chemicals (fungicides); antibiotics (mainly bacteriostatic, some are bactericidal)	Various anti-protoctistal drugs
HARMFUL	Pathogens, causing such diseases as tetanus, STDs, whooping cough, tuberculosis and cholera, and food poisoning; denitrifiers	Pathogens, causing such diseases as AIDS, HIV, polio, common cold, influenza and smallpox	Pathogens, causing such diseases as ringworm. Food, timber, textiles – pests	Pathogens, causing such diseases as amoebic dysentery, malaria
BENEFICIAL	Antibiotic and vitamin B group formers. Soil nitrogen fixers, sewage protein decomposers, and food fermenters	None	Antibiotic and protein producers. Soil humus formers, sewage cellulose decomposers, and food fermenters	Sewage detritus feeders

14.9 Investigating a natural habitat

Plant and animal communities can be studied in a variety of natural habitats, for example a tree, hedgerow, freshwater pond or field. **Fieldwork** requires equipment for collecting samples of organisms for purposes of identification of the species in a habitat. These are then recorded in species lists.

Collecting methods

Short vegetation and litter

For short vegetation and litter as on a lawn, mown field or woodland, a **pooter** (see Figure 1.8) is used to suck up small delicate living organisms after locating them with a magnifier or hand lens.

Long vegetation

For long vegetation as in waist high pasture, a **sweep net** is used and the pooter is then used to pick up the organisms from the net.

Tree and shrubs

A white sheet is placed on the ground beneath the tree or shrub. Alternatively an inverted umbrella can be used under a tree. The branches are batted sharply three or four times with a walking stick, thus dislodging the organisms onto the sheet or into the umbrella. Sweep nets can be used amongst large tree branches.

Ground level organisms

These can be collected with a **pitfall trap** which is a jar sunk into the ground with its top rim level with the ground surface (see Figure 14.13). A cover is made with a flat stone or a tile supported by smaller stones.

Figure 14.13 A pitfall trap for collecting small ground animals

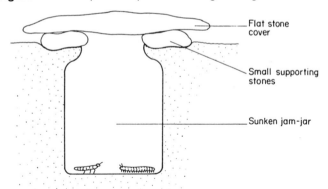

Flat stone cover

Small supporting stones

Sunken jam-jar

Soil organisms

Such organisms, present in turf or leaf mould, are collected by placing the soil or mould sample on a supporting grid of a **Tullgren funnel** (see Figure 14.14) which is above a beaker of water or ethanol solution. The latter kills the organisms.

Figure 14.14 A Tullgren funnel for collecting small soil animals

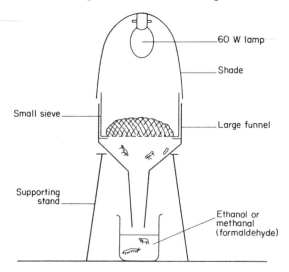

A reading lamp is placed at the funnel open end, to heat the sample for 6 to 8 hours. As the sample dries it also drives the organisms downwards to fall into the water or preserving solution.

Aquatic organisms

These are collected with fine mesh **nets**, or small trawl nets and by means of fine **sieves** on sandy beaches.

Survey and census methods

The following methods are used to obtain information concerning the frequency of organisms in a particular habitat. A record sheet called a **frequency sheet** will indicate the numbers of a particular species. From this a **histogram** can be constructed to show the graphical frequency of organisms in the habitat.

Quadrats are square frames of wood, plastic or wire between 10 cm and 1 m square. A quadrat is laid on the area of habitat to be investigated and all the organisms within the quadrat are collected, identified and recorded on a species or frequency record sheet.

Transects are made along a line across the habitat by connecting a string line between two vertical poles. Every organism touching the string is identified, counted and recorded on a frequency list. Line transects can be made on sea shores and across a hedge or field. Another method is to place a quadrat at stations 5 or 10 m apart along the line and count the species within the quadrat.

Question

25 A hedge is made up of some large shrubs. On one side is regularly mown grass and on the other side is grass that has not been mown for a number of years. Make a plan of how you would investigate the plant and animal life in this habitat.

Investigating soil

The following are some of the main constituents of a soil.

Mineral components

The solid mineral components (see Figure 14.15) of soil are sand, mineral salts, silt, clay and gravel. They are all obtained by weathering of rocks through the action of wind and water erosion, temperature changes and particularly freezing of water in rock crevices which splits the rock.

Figure 14.15 Experiment to separate the solid components of soil by physical analysis

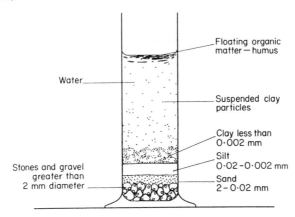

The physical condition of a soil partly depends on its mineral components. Table 14.6 compares sandy and clay soil.

Table 14.6 Comparison of sandy and clay soils

Feature	Sandy soil	Clay soil
Particle size	Large – average 1000 μm	Small – average 25 μm
Air space (unless waterlogged)	Large – abundant air	Small – limited air
Water	Poor – salts leached (improved with humus)	Good – salts retained by strong attraction
Drainage	Rapid – soil dries out, and blows away	Slow – soil wet, sticky and cakes dry
Cultivation	Light – easy to dig	Heavy – hard to dig
Improvement	1 Add clay 2 Add humus to aggregate particles 3 Surface mulch prevents evaporation	1 Add sand 2 Add lime to flocculate or aggregate particles to form crumb 3 Winter digging for aeration

Experiment to estimate the solid mineral components of a soil

Procedure

1 Feel a fresh soil sample for grittiness (sand), stickiness (clay) and silkiness (loam).
2 Shake a sample (30 g) with about 200 cm^3 water in a 250 cm^3 measuring cylinder and leave to settle by sedimentation (see Figure 14.15).
3 Estimate the width of each layer. From the lowest the layers are gravel, sand, silt and clay.

Soil water

Soil water is a dilute solution containing various inorganic mineral salts, forming a film around the mineral soil particles. The film of water moves upwards through the soil air spaces by capillarity and evaporation into the air. Some of the water, called hydroscopic water, is in chemical combination with the soil particles and is not available for plant use.

Experiment to estimate the total water in a soil

Procedure

1 Place a weighed soil sample to dry to constant weight in an oven at 110°C.
2 Re-weigh and determine the loss in weight due to evaporation of the total soil water.
3 The percentage loss can be calculated from

$$\frac{\text{loss in weight}}{\text{original wet weight}} \times 100 = \text{percentage of water in soil}$$

The total soil water varies between 10 and 35%.

Soil air

Soil air is found in the spaces between the soil particles. It consists of less oxygen and more carbon dioxide than atmospheric air. Water-logged bog and marsh soils have no soil air. Soil air is needed for aerobic respiration of green plant roots, soil organisms and certain aerobic microorganisms.

Experiment to estimate the volume of air in a soil

Figure 14.16 describes the method for estimating the air content of a soil sample. The volume of water needed to restore the water in the large measuring cylinder to its original level, will be equal to the volume of air in the soil. The percentage of air in the soil sample is calculated as follows:

$$\text{percentage of air} = \frac{\text{volume of air in soil}}{\text{volume of soil sample}} \times 100$$

Figure 14.16 Experiment to estimate the volume of air in a soil sample

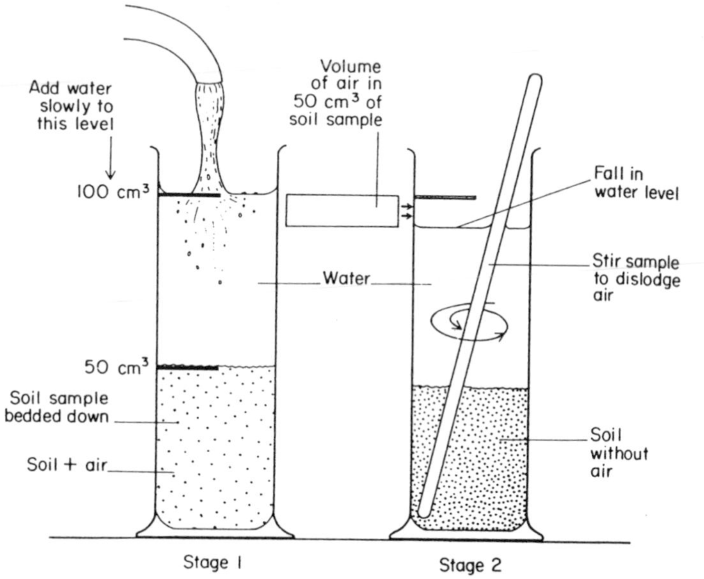

Soil acidity or alkalinity

Every soil will have a certain pH value. Alkaline conditions are associated with chalky soils. Acid conditions are found in poorly drained, peaty and marshy soils. Certain plants, e.g. conifers and rhododendrons, prefer acid soils and some animals, e.g. snails, prefer an alkaline, chalky soil.

Experiment to estimate soil pH

Procedure

1 Mix a sample of soil with distilled water and allow it to settle.
2 Add a piece of Universal Indicator paper and note the colour change in comparison to the pH value colour chart. A pH value of between pH 7 and 1 is acidic whilst values between 7 and 14 are alkaline.

Organic content

The organic content of a soil is due to **humus**, which is the end product of decay of plant and animal material. It is brought about by soil microorganisms, fungi and bacteria.

Humus is an important source of nutrients in the form of nitrates and ammonia for protein synthesis in green plant nutrition. It is also a food for soil organisms such as earthworms and insect larvae.

Humus is a colloidal organic compound which has water absorbing properties. It retains water by hygroscopic action. The adhesive properties of humus improve sandy soils and also cause clay soil particles to cling to the humus fibres.

Figure 14.17 Experiment to estimate the organic or humus content of
a soil sample

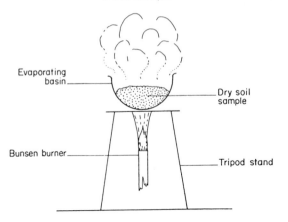

Evaporating basin

Dry soil sample

Bunsen burner

Tripod stand

Procedure

1 Obtain an oven-dried sample of soil.
2 Weigh a sample into an evaporating basin.
3 Heat strongly over a Bunsen burner in order to oxidise all carbon-containing humus and other material.
4 Allow to cool and determine the loss in weight. Calculate the organic content from

$$\frac{\text{loss in weight}}{\substack{\text{weight of original} \\ \text{dry sample}}} \times 100 = \text{percentage of organic matter in the soil}$$

Soil drainage and permeability

The water-holding capacity and drainage of a sandy, clay or peaty loam soil differs due to the mineral content, air space and organic content.

__ Experiment to compare soil drainage _____

Procedure

1 Place equal volumes of samples of different kinds of soils in filter funnels containing a small plug of glass wool. Support the funnels above measuring cylinders.
2 Pour an equal amount of water into each funnel of soil and carefully note the volume of water draining into the measuring cylinder. The amount which drains in a given time is an indication of the soil drainage and water-holding capacity.

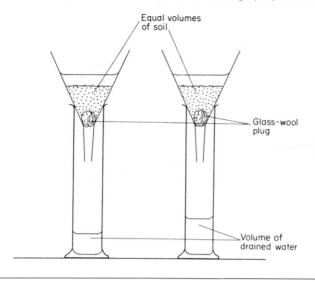

Figure 14.18 Experiment to compare the drainage properties of soils

Equal volumes
of soil

Glass-wool
plug

Volume of
drained water

Question

26 The plant growth in one part of a field is much thicker than in another part of the field. What soil tests would you perform as part of an investigation into the difference in plant growth? Explain the reasons for your tests.

Summary

- An ecosystem is a self-supporting unit which is composed of many different species of organisms and the non-living component of the habitat. (▶ 342)
- The abiotic factors in an ecosystem are the climate and the soil. (▶ 343)
- The biotic components of an ecosystem are the individuals in the different species which make up the populations and communities. (▶ 346)
- The ecological interactions taking place in an ecosystem are competition, predation, symbiosis and succession. (▶ 346)
- A food chain is made up of a series of stages called trophic levels. Food chains link together to form food webs. (▶ 350)
- A biotic or trophic pyramid is used to show the number, biomass or energy in each trophic level of a food chain. (▶ 351)
- Solar energy is trapped by green plants and stored or changed in living organisms before being released by respiration and escaping as heat into space. (▶ 353)
- Carbon and nitrogen are chemicals which are recycled in an ecosystem. (▶ 355)
- Water passes between living things and their abiotic environment in the water cycle. (▶ 357)
- There was a natural balance between photosynthesis and respiration which maintained the composition of the atmosphere before the activities of modern humans began. (▶ 357)

- Population growth is affected by a wide range of limiting factors from the abiotic and biotic components of an ecosystem. (▶ 358)
- Microorganisms are an important component of an ecosystem and affect other living organisms in a variety of ways from causing disease to recycling essential minerals. (▶ 360)
- A habitat can be investigated by using simple apparatus and techniques. (▶ 361)

Figure 14.19   Summary of different methods of nutrition

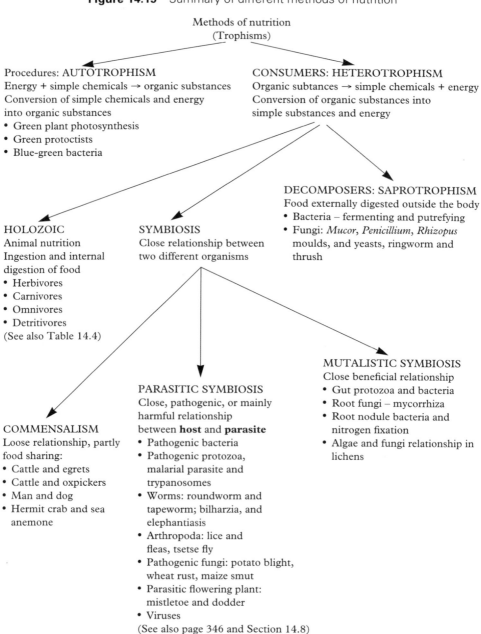

Methods of nutrition
(Trophisms)

Procedures: AUTOTROPHISM
Energy + simple chemicals → organic substances
Conversion of simple chemicals and energy
into organic substances
- Green plant photosynthesis
- Green protoctists
- Blue-green bacteria

CONSUMERS: HETEROTROPHISM
Organic subtances → simple chemicals + energy
Conversion of organic substances into
simple substances and energy

DECOMPOSERS: SAPROTROPHISM
Food externally digested outside the body
- Bacteria – fermenting and putrefying
- Fungi: *Mucor, Penicillium, Rhizopus* moulds, and yeasts, ringworm and thrush

HOLOZOIC
Animal nutrition
Ingestion and internal
digestion of food
- Herbivores
- Carnivores
- Omnivores
- Detritivores
(See also Table 14.4)

SYMBIOSIS
Close relationship between
two different organisms

MUTALISTIC SYMBIOSIS
Close beneficial relationship
- Gut protozoa and bacteria
- Root fungi – mycorrhiza
- Root nodule bacteria and nitrogen fixation
- Algae and fungi relationship in lichens

COMMENSALISM
Loose relationship, partly
food sharing:
- Cattle and egrets
- Cattle and oxpickers
- Man and dog
- Hermit crab and sea anemone

PARASITIC SYMBIOSIS
Close, pathogenic, or mainly
harmful relationship
between **host** and **parasite**
- Pathogenic bacteria
- Pathogenic protozoa, malarial parasite and trypanosomes
- Worms: roundworm and tapeworm; bilharzia, and elephantiasis
- Arthropoda: lice and fleas, tsetse fly
- Pathogenic fungi: potato blight, wheat rust, maize smut
- Parasitic flowering plant: mistletoe and dodder
- Viruses
(See also page 346 and Section 14.8)

The effects of modern humans on the global ecosystem

After you have read this chapter you should be able to:
- compare the impact on the Earth of early and modern humans
- compare modern humans to other animals in the world ecosystems
- understand the factors which have led to an increase in the human population
- distinguish between sustained and unsustained population growth
- interpret population pyramids
- describe how the Earth's resources have been exploited or abused and identify ways in which they can be conserved and managed
- describe how the air, water and land has been polluted and recommend ways in which pollution can be reduced
- describe how biotechnology contributes to human life
- understand what is meant by biological pest control
- understand the advantages and disadvantages of chemical pest control

During the last 10 000 years, **one species** amongst the two million different species has emerged to dominate every other living and non-living thing on Earth. This species is _Homo sapiens_ – the **modern human species**. It has altered the face of the Earth in a way no other living organisms could. Previously all living organisms were controlled by the external environment. Today modern humans have the power to control the external environment but more often they destroy the environment as they use whatever they need from it.

Previously, early humans were in **homeostasis** as they lived together with other organisms in the environment. Now modern humans have altered Earth's homeostasis by increasing their own numbers and causing pollution of the the environment with their wastes.

15.1 The characteristics of modern humans

Modern humans appeared about 10 000 years ago. They were distinguishable from earlier humans as being **intelligent** people capable of **reasoning**, using their memories and having **manual** and **mental** skills which they used in the following ways:
- They developed agricultural skills through artificial selection of organisms and their cultivation.

- They evolved a settled way of life and built protective shelters.
- They developed industrial skills involving the extraction of materials.
- They accumulated scientific knowledge which they used in health care.

Modern humans developed agriculture, industry and culture differently in different parts of the world. The countries of the world can be divided into two groups:

- **Developing countries:** In these countries simple agricultural methods are used and a small surplus is produced. They are not fully industrialised. Most of these countries are found in Africa, Asia and South America. They support 75% of the world's human population.
- **Developed countries:** In these countries advanced agricultural methods are used which produce a large surplus. The countries are fully industrialised. Most of these countries are found in North America, Europe and the western rim of the Pacific including Japan, Australia and New Zealand. They support 25% of the world's human population.

Questions

1 What is meant by early humans being in homeostasis? (see also Chapter 7)
2 How did (a) agricultural skills, (b) a settled way of life, (c) industrial skills and (d) scientific knowledge affect the size of the human population? Explain your answer to each part of the question.
3 What are the key features which distinguish a developing country from a developed country?

Table 15.1 compares developed and developing countries in more detail.

Table 15.1   A comparison of developed and developing countries

	Developed countries	Developing countries
Examples	UK, USA, USSR	Latin America, India, Africa
Natural resources	Industry and resources fully developed	Resources not fully developed; mainly agricultural
Growth rate	Low – 0.6%	High – 2.1%
People aged under 15 years	Low proportion of population – 22%	High proportion of population – 37%
Life expectancy	High (average 73)	Lower (average 60)
Diet	Varied and balanced Low dietary fibre Animal protein 60% diet Plant protein 40% diet Hard saturated lipids Overnutrition	Monotonous and unbalanced High dietary fibre Animal protein 20% diet Plant protein 80% diet Soft unsaturated lipids Undernutrition
Income level	High income	Low income; 50% of world population live in dire poverty
Literacy	High standard of literacy and education	Low standard of literacy and education; 35% of the world is illiterate
Main cause of death	Heart disease, cancer and stroke	Infectious disease, diseases in infants, lung disease

15.2 Modern humans and ecosystems

The natural ecosystem in which early humans lived is compared with the artificial ecosystems made by modern humans in Table 15.2.

Table 15.2 Natural and artificial ecosystem compared

Environmental factors	Natural Ecosystems: primitive human beings controlled by nature	Artificial Ecosystems: made by and controlled by modern human beings
Food	Limited supply, causing undernutrition, starvation and death	Ample supply due to modern agricultural methods, with surplus production
Shelter	Cold, damp, or heat and drought – extremes cause illness and death	Heat-controlled and air-conditioned shelter, with optimum conditions, support good health
Pathogenic diseases	Limited protection by **natural immunity** (see Section 5.10)	Extensive protection by **artificial immunity**, drugs and antibiotics, and modern medicine, surgery and hygiene
Predators	Limited protection by fire, sticks and stones	Total protection by biocides and modern weapons (Biocides kill all living organisms)

Modern humans have the following ecosystem characteristics:
- They are the dominant species amongst all living things because no other species can compete against humans.
- They make a habitat with an artificial ecosystem which is made up mainly of homes in towns and cities.
- They are controllers of certain environmental factors.
- The roles of modern humans in the ecosystems are
  (i) producers of food mountains in developed countries and synthesisers of chemical compounds;
  (ii) consumers of other living organisms;
  (iii) decomposers of materials which has led them to being the most destructive organisms ever.
- They disturb natural mineral cycles by extraction of minerals, and waste energy in the energy flow process.
- They pollute the abiotic environment and habitats with harmful wastes which are the products of industrial activities.
- They are themselves an environmental factor because they are able to exert environmental pressure and affect the global ecosystem.

 Questions
---

4 In what ways do modern humans resemble other animals in an ecosystem?
5 In what ways are humans different from other animals in an ecosystem?

15.3 World human population growth

Figure 15.1 Population graph for humans from earliest time (figures prior to 1825 are estimated)

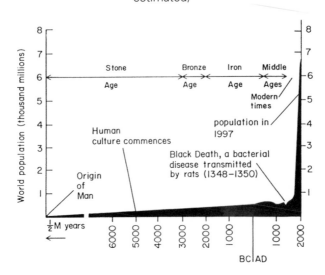

The human population of the word is approaching 6000 million. Figure 15.1 shows the population growth for the world human population in the last half million years including the emergence of modern humans. The rapid and steep increase in world human population during the last 150 years is due to the removal of **limiting factors**. Look at Table 15.2 to see how the limiting factors of food, shelter, pathogenic disease and predators have been removed by modern humans.

The infant mortality rate

The infant mortality rate or **IMR** is the annual number of deaths to infants under the age of one year per 1000 live births.
- World infant mortality rate = 77 per 1000.
- Developed countries IMR = 15 per 1000.
- Developing countries IMR = 86 per 1000.
The IMR is an **indicator** of health conditions, food supply, hygiene and prenatal care in a country. This shows that improvements in health conditions are still required for over 75% of the world population.

Projected population growth

Figure 15.2 shows the massive increase in world human population since 1800.
- Population growth rate = Birth rate – Death rate = %.
- World growth rate = 1.7%
- Developing countries' rate = 2.1%
- Developed countries' rate = 0.6%
Future population growth can be forecast by extrapolation or extending the graph as shown. There are two ways in which this can be done:

Figure 15.2 Projected population growth

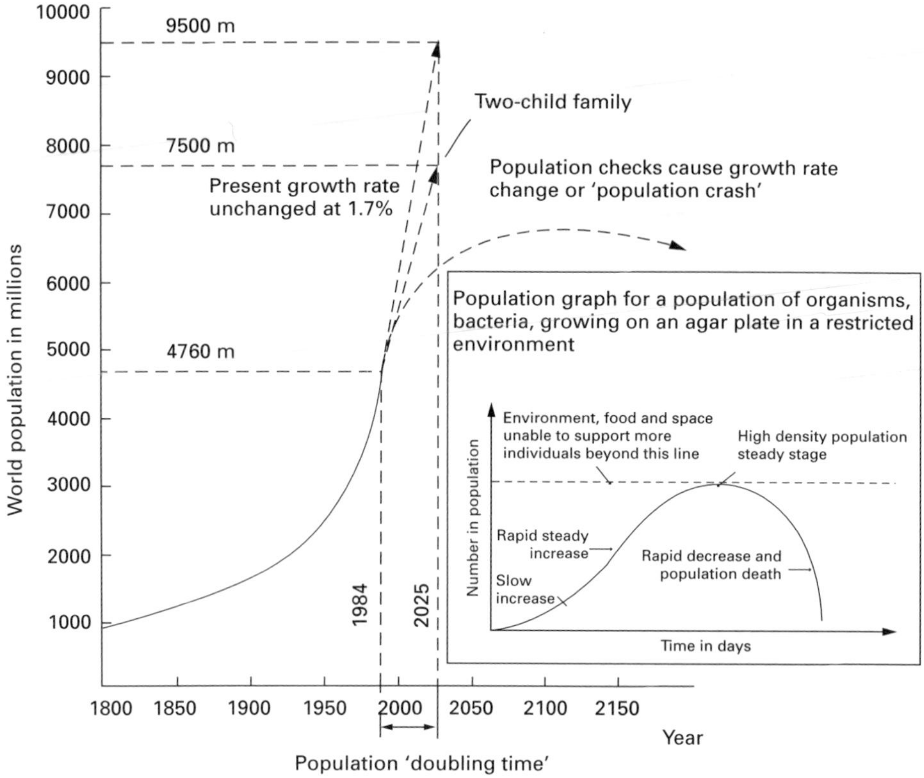

Sustained population growth

This will occur if there is an absence of limiting factors or a continuing food supply, space to live and no harmful components in the environment. At present, with a continuing world growth rate of 1.7%, the world population will reach between 8000 and 9500 million by the year 2020.

Unsustained population growth

This will occur if there are limiting factors present such as a decreasing food supply, less space to live and if harmful components are present in the environment such as pathogens and pollution. This would be a similar situation to the bacteria grown in laboratory conditions on a culture medium (see Figure 14.13). The final result would be a decrease in population with the following consequences:

• food shortage and starvation;
• overcrowding;
• pollution of air and water;
• disease due to resistance of pathogens to antibiotics;
• killing of humans in competition for food, shelter and space.

Questions

6 How has control of environmental factors led to an increase in the human population?

7 How would you expect the IMR in a developing country to change as it becomes more developed? Explain your answer.
8 How would you expect the growth rate to change in a developing country as it becomes more developed? Explain your answer.
9 Look through newspapers over the next few days for evidence of unsustained population growth by using the bulleted items on page 373 to help you. Describe any evidence you find and assign it to one of the five categories. If you do find evidence, what steps can be taken to resolve the problems that you describe? Keep your work to use in question 31.

Age and sex population pyramids

The numbers of people of a particular age and a particular sex vary within a population. These variations can be displayed as **pyramids**.

Expanding populations

These will have a high percentage of the population under the age of 15 years, with a very low percentage of people over the age of 64 years.
- **Africa** has an expanding population with over 45% of the population under 15 years and 3% over 64 years.
- **Developing** countries have an expanding population with 37% under 15 years and 4% over 64 years.
- The **world population** is an expanding population with 33% under 15 years of age and 6% over 64 years.

Stable populations

These have a low percentage population under 15 years of age and a higher percentage over 64 years.
- The **United Kingdom** has a stable population with 20% under 15 years and 15% over 64 years. Most of the population are over 20 years of age.
- **Developed** countries have stable populations with 23% of the population under 15 years and 12% over 64 years. Most of the population are over 20 years.

Figure 15.3 Population by age and sex in developing and developed countries

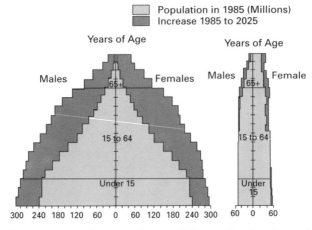

Developing countries: expanding population   Developed countries: stable population

The pyramids in Figure 15.3 show the numbers of people of different ages and different sexes in developing and developed countries.

The features of the population pyramids show that:

- **expanding populations** have a wide base to the triangle. This shows a young population. In Kenya 52% of the population are under 15 years.
- **stable populations** have a narrow base supporting an ageing population. In West Germany 15% of the population are under 15 years of age.

The developing countries have a younger population that the developed countries.

Questions

10 How is a population pyramid made?

11 As a developing country becomes more developed, how would you expect the population pyramid to change? Explain your answer.

Human population control

Humans can control their population growth rate by means of various methods of birth control which are described in Chapter 10. If every family had only two children each, the projected population in the year 2025 could be closer to 7500 million as shown in Figure 15.2.

15.4 Exploitation of the global ecosystem by modern humans

Ecosytem exploitation over the past 5000 years:		**Ecosystem conservation** in the future:
• *Destruction* of the biotic environment • *Pollution* of the abiotic environment • *Abuse* of the ecosytem processes	must be replaced by	• *Preservation* of the remaining biotic envitonment • *Cleansing* and *care* of the abiotic environment • *Wiser use* of the ecosystem processes

The chart shows how the **exploitation** of the global ecosystem in the past must be replaced by **conservation** of the ecosystem in the future. This is *essential* for the survival of life on Earth. The following sections summarise how modern humans have exploited the global ecosystem and how they can conserve or manage the global ecosystem for future generations of life on Earth.

The biotic environment and modern humans

Modern humans first exploited the natural abundance of living organisms in the global ecosystem. Once these organisms are destroyed they can never be replaced and are a **non-renewable resource**.

Table 15.3 Summary of exploitation and possible conservation methods

EXPLOITATION or ABUSE	CONSERVATION or MANAGEMENT
Deforestation: removal of woodlands for land clearance for fuel, grazing, timber products, paper or road and town making	**Reafforestation:** replacement of a variety of tree species, broad-leaved and narrow-leaved
Natural woodland ecosystem covered Britain 2000 years ago; this was a **climax community**. Nowadays single species populations of pine and spruce mar the landscape.	Paper made from trees should be **reused** or **recycled**, to allow new communities of trees to establish themselves
Overkill (humans as predators – see Chapter 14) **Overhunting** and **overfishing** of easily caught species led to **extinction** of Dodo, Great Auk species. Other species are **endangered**, for example herring and whale species.	Fisheries, closed fishing areas and game **management** **Protection** of endangered species in zoos, botanical gardens, nature and wildlife parks **Laws** to protect and preserve wildlife, and allow fisheries stocks to increase
Pesticides Pests are those organisms which are not useful or do not provide pleasure for modern man Chemical pesticides are toxic, **non-biodegradable** organochlorine compounds destroying all organisms without selection	More careful use of chemical pesticides or replacement by **biological** control – the use of a predator organism to eat or weaken the specific pest organism. It is non-toxic, and harmless to other species (see Section 15.6)
Antibiotics: destroy, or slow down the growth of a range of microorganisms which later develop **resistance** against the antibiotic. Only useful for individually treated humans	**Artificial acquired immunity** preferable using **vaccines** or sera against specific organisms. This can be used to protect whole communities of humans (see Section 5.10)

Questions

12 Why should reafforestion programmes be concerned with replacing a *variety* of trees?

13 What commercial advantages are there in growing one kind of tree in a forest?

Feeding processes and modern humans

Modern humans exploited the ability of:
- the soil earth to produce crops. This has led to modern agricultural methods.
- the Earth to produce minerals and fossils. This has led to industrialisation and modern technology.

These agricultural and industrial activities affected the following processes in the global ecosystem: feeding, mineral cycling, energy flow and growth.

Feeding processes affected by modern humans

Table 15.4 Exploitation and conservation related to feeding

EXPLOITATION or ABUSE	CONSERVATION or MANAGEMENT
Monoculture: the rearing of the same *single species* population or crop on the same land area each year, e.g.: **Plants** – cereals, legumes, potatoes, American pine and spruce trees **Animals** – sheep, goats, cattle and indoor batteries of hens and pigs Monoculture maximises soil productivity and harms the soil fertility, attracts pests and leads to soil **erosion**	**Crop diversity** and maintenance of variety in agricultural communities **Crop rotation**: rearing of a *succession* of different crops on the same land area. This improves soil fertility and prevents erosion **Landscaping** of farms into natural ecosystems on marginal land or fallow ground, to preserve and develop natural habitats

Mineral cycling processes and modern humans

The carbon cycle

Note how the amount of CO_2 in the air has increased from 0.03% to 0.04%, an increase of 33% in the 'greenhouse' gases which contribute to global warming.

Figure 15.4 The affected carbon cycle (compare with Figure 14.9)

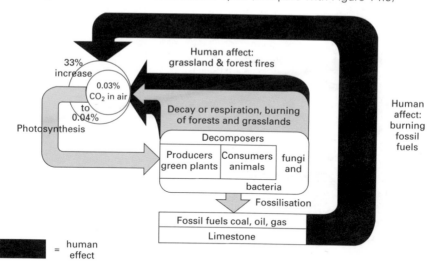

___ **Question** ___

14 Compare Figure 14.9 with Figure 15.4 and look at Table 15.5 overpage, then describe how human activities have affected the carbon cycle.

Table 15.5 Carbon cycle exploitation and conservation

EXPLOITATION or ABUSE	CONSERVATION or MANAGEMENT
Burning of **fossil fuels** coal, gas and oil, once used can never be replaced	**Fossil fuel** usage must be *controlled*, before it is all used up
Tropical forest burning for land clearance	Alternative sources must be found (see Energy flow, page 353)
Grass and straw burning	Reduce the CO_2 levels in air by *increasing*
All these *increase* the carbon dioxide level in air (from 0.03% to over 0.04%, i.e. 33% increase)	the biomass of green plants to *increase* photosynthesis by planting *more* trees.
Non-biodegradable carbon-containing chemical compounds cannot enter the carbon cycle; these substances include many plastics, certain detergents, and the organochlorine pesticides. These substances accumulate on the Earth	Replace the non-biodegradable materials now in use for packaging, etc. with **biodegradable** materials, easily decomposed by saprotrophic organisms Most synthetic detergents are now biodegradable **Recycle** certain plastics
Sewage waste in aquatic ecosystems spreads pathogenic organisms	Biological treatment of sewage in efficient sewage plants using decomposer organisms Recover sludge humus and methane gas

The nitrogen cycle

Figure 15.5 The affected nitrogen cycle (compare with Figure 14.10)

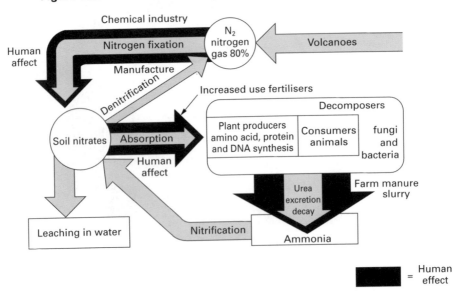

___ Question ___

15 Compare Figure 14.10 with Figure 15.5 and look at Table 15.6, then describe how human activities have affected the nitrogen cycle.

Table 15.6 Nitrogen cycle exploitation and conservation

EXPLOITATION or ABUSE	CONSERVATION or MANAGEMENT
Over-use and production of the following: **Artificial fertilisers:** fertilisers chemically manufactured by nitrogen fixation and extracted by modern humans Highly concentrated forms of nitrates and ammonium salts Affects soil texture and is very soluble; easily washed out or **leached** Less energy-consuming to spread Farmyard Slurry dumping in rivers and lakes	Reduce usage of artificial fertilisers and promote the following: **Organic fertilisers:** natural plant and animal origin materials, compost, straw manure, sewage sludge Dilute form of nitrates and ammonium salts Affects soil by improving texture, and increases **humus** which **holds** water and minerals Nitrogen-fixing bacteria in root nodules of legumes and other species are important nitrogen source for soils Non-leaching artificial fertilisers can be developed to replace the easily leached nitrate fertilisers

The water cycle

Figure 15.6 The affected water cycle (compare with Figure 14.11)

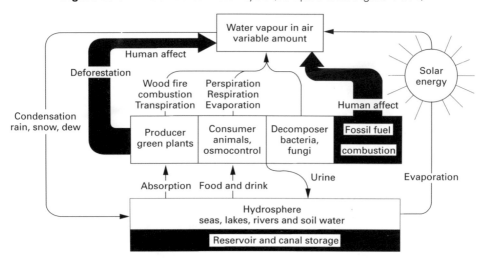

___ Questions ___

16 Compare Figure 14.11 with Figure 15.6 and look at Table 15.7 over page then describe how human activities have affected the water cycle.
17 What steps should be taken to manage the carbon, nitrogen and water cycles more effectively?

Table 15.7 Water cycle exploitation and conservation

EXPLOITATION or ABUSE	CONSERVATION or MANAGEMENT
Deforestation: the clearing of woodland and burning of tropical rain forests, removes the transpiration route for water cycling	**Reafforestation:** the planting of mixed species woodlands, and halting of tropical forest burning and clearance Recycling of paper to save trees
Water storage in reservoirs, canals or underground caverns, removes water from the evaporation route	**Water distillation** from sea water using solar energy needs further development in certain parts of the world
Fossil fuel burning produces water as a product of combustion	Alternative energy sources using solar, tidal, or atomic or wind power. Also biofuels from sewage, refuse, and alcohol fermentation (see Figure 15.7)
Human sewage waste and storm water is allowed to run into rivers and oceans, untreated	Sewage carriage requires large amounts of water; this water can be **recovered** by efficient treatment in sewage farms and treatment plants, and **reused** (recycled)

Figure 15.7 Energy flow and the future

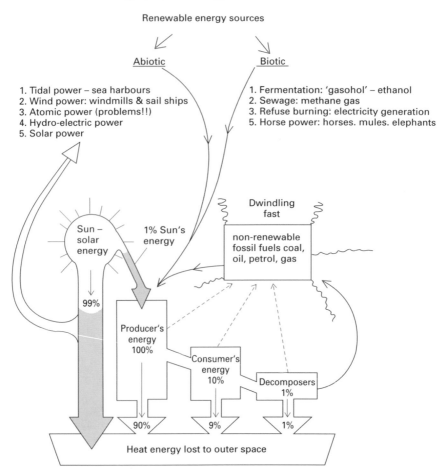

Renewable energy sources

Abiotic

Biotic

1. Tidal power – sea harbours
2. Wind power: windmills & sail ships
3. Atomic power (problems!!)
4. Hydro-electric power
5. Solar power

1. Fermentation: 'gasohol' – ethanol
2. Sewage: methane gas
3. Refuse burning: electricity generation
5. Horse power: horses. mules. elephants

Dwindling fast

Sun – solar energy

1% Sun's energy

non-renewable fossil fuels coal, oil, petrol, gas

99%

Producer's energy 100%

Consumer's energy 10%

Decomposers 1%

90%   9%   1%

Heat energy lost to outer space

Energy flow and the effect of modern humans

Figure 15.7 shows diagrammatically how non-renewable fossil fuels, now being used up at a tremendous rate, should be replaced by **renewable alternative** energy sources, either using the abiotic environment for tidal, wind, atomic, hydro-electric or solar power, or using biotic environment sources of fermentation, 'Gasohol', sewage methane gas (biogas), refuse and rubbish combustion and animal or 'horse' power from horses or other animals such as elephants and oxen.

Table 15.8   Energy exploitation and conservation

EXPLOITATION or ABUSE	CONSERVATION or MANAGEMENT
Fossil fuel energy sources, coal, oil, petrol and natural gas have been used as main or 90% energy sources for technology and industrialisation. They are **non-renewable** energy sources Fossil fuels are valuable sources for organic compounds for drugs, and other materials useful to modern man	**Alternative energy sources** • **Solar energy:** 99% flows away into space unused by modern man. • **Wind energy** and **tidal** energy, insufficient usage • **Atomic energy** used in 10% of UK power stations • **Biofuels** from sewage, or ethanol from fermentation processes not a developed source • **Refuse** burning not a developed source • **Energy saving** programmes and reduced energy consumption in industry, motor vehicles, and homes • **Animal power**, horses, donkeys, elephants, camels and oxen

---

Questions

18 Why will energy sources have to change in the future?
19 What should be important features of the replacement energy sources? Explain your answer.
20 Compare Figure 14.7 wih Figure 15.7, then describe how energy flow is likely to change in the future.

Eutrophication growth process and the effect of modern humans

Table 15.9   Eutrophication exploitation and conservation

EXPLOITATION or ABUSE	CONSERVATION or MANAGEMENT
Eutrophication: Human activities have speeded up this natural population growth process in fresh water and sea water, by adding mineral and organic nutrients in larger quantities than nature would provide, as excreta, phosphates from washing powders, and fertilisers which leach from heavily fertilised land as nitrates and phosphates	This natural growth process in aquatic ecosystems can be conserved through: • abolition or controlled dumping of sewage at sea • nitrate fertiliser agricultural usage reduction • phosphate removal from washing powders, and their replacement with substitutes harmless to the environment. Removal of phosphates in sewage treatment

21 What is eutrophication and why does it occur?

Pollution of habitats

Pollution is the alteration of any part of the environment by modern humans which endangers the life of species in the environment. **Pollutants** are the waste by-products and energy from agricultural, industrial, scientific and other activities by modern humans.

- **Chemical pollutants:** nitrogen oxides, sulphur dioxide, carbon dioxide, smoke, tar, dust and grit, heavy metals such as lead and mercury, hydrocarbons and mineral oils and organic pesticides. Dioxins and PCBs are highly toxic pollutants.
- **Energy pollutants:** waste heat, noise and atomic radiation.

Air pollution

Air pollution is mainly due to the discharge of pollutants into the air from fossil fuel combustions or from industrial machinery.

Figure 15.8   Motor vehicle exhaust gases contribute to air pollution (WHO)

Table 15.10 Air pollution

ABUSE: Pollutant source and effect	MANAGEMENT or CONTROL
Acid rain is due to: Sulphur dioxide from combustion of oil, coal gas, coal and smokeless fuels; nitrogen oxides from fossil fuels and petrol combustion. **Affects:** • humans – lung disease • conifers – leaf fall and photosynthesis • fish eggs – killed in acid water • corrosion of metals and buildings	**Clean-air** laws: • Use sulphur-free fuel or natural gas • Use alternative clean energy sources, e.g. animal power – horses on farms, electricity generated from wind, water or atomic fuel for power stations • Extractors and scrubbers fitted in exhausts or chimneys to remove the acid gases
Lead is a component of certain petrols used in petrol engines **Affects:** • bone in humans • nervous systems in children • absorbed by roadside plants	• Lead-free petrol is now widely available • Engines can be adapted for lead-free petrol • Diesel oil does not contain lead additives
Noise is an energy form of pollutant from heavy industry, machines, engines, weapons, road traffic and discos **Affects:** • nervous disorders • hearing damage and total deafness	• **Noise-abatement** laws in factories and for engines • **Noise insulation**, double glazing of homes • **Ear muffs** and plugs for machinery users
Carbon dioxide levels increasing from 0.03% to 0.04% or 33% increase in 50 years from fossil fuel combustion. **Affects:** • climate causing **global warming** in 'greenhouse' effect*	**Reforestation** – plant more trees **Increase photosynthesis** – plant more green plants Use **clean** alternative energy sources: • animal – horse power • electric power from water, wind, tidal or atomic energy

*__Other greenhouse gases__ include CFCs from aerosols, methane from marsh gas and natural gas, ozone and nitrogen oxides. These gases are increasing in amount in the atmosphere, are able to absorb and re-emit radiant heat from the Earth, and are believed to be factors in global warming.

Water pollution

Water pollution is due to the discharge of wastes or effluents into rivers, lakes and oceans, or accidental oil spillage from tankers.

Table 15.11 Water pollution

ABUSE: Pollutant source and effect	MANAGEMENT or CONTROL
Heavy metal – mercury Chemical work effluent. It is concentrated in food chains **Affects:** • **nervous systems** in fish and other marine animals • **humans** who eat affected fish and shellfish • **human offspring** malformed; mercury crosses the placental barrier	**Effluent purification** at source before discharge; **financial** penalties and fines

Table 15.11 Continued

ABUSE: Pollutant source and effect	MANAGEMENT or CONTROL
Sewage, human and farmyard slurry, and silage effluents **Affects:** • humans by transmitting pathogenic diseases • eutrophication	Efficient **sewage treatment** by biological methods of decomposition; breakdown of human parasite life cycles
Nitrates from leaching of artificial nitrate fertilisers draining into rivers **Effects:** • harmful to babies • eutrophication	**Strict control** of water composition for domestic use; higher analytical standards of water purity for drinking; reduction in use of nitrate fertilisers
Heat, as hot water effluent from power station cooling process, reduces oxygen content	Efficient cooling systems

Terrestrial pollution

Land pollution is caused mainly by soil treatments, agricultural processes and refuse dumping.

Table 15.12 Terrestrial pollution

ABUSE: Pollutant source and effect	MANAGEMENT or CONTROL
Pesticides, persistent non-biodegradable organochlorine pesticides remain in soil **Affects:** • accumulate in fatty tissues of organisms in food chains • concentrated in the predators, e.g. eagles, whose eggs break easily, killing the young	**Total ban** of certain pesticides, e.g. DDT **Alternatives:** • biological control instead of chemical control • use non-toxic chemical pesticides, e.g. derris dust and pyrethrum powder
Refuse in the form of non-biodegradable plastic materials, which accumulate in soil **Affects:** • incomplete burning of these substances produces harmful and toxic dioxins	• Efficient refuse disposal by complete combustion in suitable furnaces of non-biodegradable plastic materials • Encourage aerobic decomposer organisms for biodegradation
Artificial fertilisers (see nitrogen cycle as affected by modern man)	• Use organic fertilisers; farmyard manure and sewage waste

---

Questions

22 How does air pollution affect (a) plants, (b) animals and (c) humans?
23 In what ways can (a) industry, (b) electricity companies (c) foresters and (d) ordinary people reduce air pollution?
24 How could a person be affected by water polluted by mercury without drinking it?
25 What is the advantage of biodegradable plastic over non-biodegradable plastic?

15.5 Recycling

Recycling is an important process of recovering useful re-usable materials, instead of wastefully dumping them. Recycling of materials and metals needs to be more widespread. **Sterile sewage sludge** can be used as an animal protein feed supplement. Farmyard **manure slurry** can also be used to rear earthworms or fly maggots which are 70% protein. These can then be used for feeding poultry, pigs and calves.

Paper can be recycled with comparatively less energy than is used in its production from wood pulp. **Glass** recycling is a process using less energy than glass manufacture. **Metals** used in motor vehicles are recovered at present on the following percentage recovery basis: iron and steel, 90%; lead, 100%; aluminium, copper and zinc, less than 1% recovered.

Of the total consumption of metals in the United Kingdom, the following percentages are recovered from scrap: lead, 30%; aluminium 38%; copper, 20%; zinc 2.5% and tin 39%. This indicates that a greater effort could be made in the collection and recycling of scrap metals.

Certain **plastics** are recyclable. Radioactive wastes, the by-product of nuclear power generation, are hazardous materials requiring special methods of disposal, by burial in sealed containers for over 500 years!

Question

26 Why is recycling good for the global ecosystem?

15.6 Organisms working for humans

Vehicle power

Human beings soon harnessed the energy available to them from slaves, horses, elephants and other mammals. They used this energy to draw agricultural and other kinds of vehicles and to power machinery.

Biotechnology

Biotechnology is the use of certain microorganisms on a large scale to produce large amounts of certain biological products which are useful to human beings. The following are some of the main biological processes.

Biotechnology and food

Food production

Certain moulds, bacteria and single-celled algae are used to make **single-cell protein** (**SCP**) which is also called microbial protein or **mycoprotein**. The protein is used as **textured vegetable protein** (**TVP**) to provide a meat protein substitute. The protein-producing microorganisms are grown on sewage, agricultural and other industrial waste.

Food processing

This involves fermentation or anaerobic respiration for making bread, cheese, butter, yoghurt, vinegar, beer, wines and spirits using certain yeasts and bacteria.

Genetic engineering

Genetic engineering is the biotechnological production of valuable medicines such as interferon, insulin, rennin and human growth hormone in *Escherichia* bacteria. This process involves the transfer of a gene, or DNA segment, from a chromosome of one species, called the **donor**, into the chromosome of another different species called the **host**. In this way the donor gives the host cell a new character or trait or the ability to code for making a special functional protein (see page 303 and Figure 12.6).

The gene for insulin synthesis is transferred from a mammal chromosome and recombined in the chromosome of a harmless human gut bacteria *Escherichia coli*, which is then grown in a sterile culture on a large scale. The transferred gene in the bacteria produces insulin in large amounts. This is due to the favourable growth conditions. This process replaces the slower process of extracting insulin from pig and ox glands.

Antibiotic production

Antibiotics are produced by means of certain bacteria and moulds.

Extracellular enzymes

Proteases, lipases and carbohydrases are extracellular enzymes that are extracted from microorganisms. The enzymes are used in laundry washing powders as **biodetergents**.

Electricity can also be made by enzyme action in biofuel cells.

Vaccines and antibodies

These are produced by inoculating horses or sheep and are used in community immunisation programmes against disease throughout the world (see Section 5.10).

Question

27 In what ways can biotechnology contribute to people's lives?

Biological pest control

Biological pest control is the destruction or weakening of one species by another in a predator–prey or parasite–host interrelationship. Table 15.13 shows how the numbers of five pests can be controlled by other organisms.

Table 15.13 Biological pest control

Target pest	Harmful effect of pest	Controlling agent	Mechanism of control
Mosquito larvae	Adult mosquito female transmits malaria disease	Freshwater fish	Fish feed upon larvae
Aphids – greenfly	Aphids suck sap out of plants affecting plant growth	Ladybirds, hoverfly, flowerfly larvae	Larvae feed on the greenfly
Greenhouse whitefly	Young feed on plant sap	Parasitic wasp	Wasp feeds on the whitefly
Prickly pear cactus plant	Weed makes land difficult to cultivate	Cochineal insect	Insects tunnel through and devour the plants
Rabbits	Devour grazing plants for cattle, etc.	Virus – myxomatosis	Virus infects and kills rabbits

Features of biological control are:

- biological control does not usually eliminate the pest completely;
- the control takes many years to establish itself;
- the method is partly specific in controlling one species;
- initial research is lengthy and expensive;
- it is not toxic to humans and other organisms;
- it does not accumulate with harmful effect in a food chain as occurs in chemical pest control.

Question

28 What are the advantages and disadvantages in setting up a biological pest control programme?

Chemical pest control

Chemicals collectively called **pesticides** will kill living organisms. Of the total pesticides used, 36% are herbicides, killing weeds and 10% are fungicides.

Insecticides

Insecticides which rapidly decompose include the natural plant products derris dust and pyrethrum powder. Long-acting insecticides include the organochlorine compounds DDT, BHC and Dieldrin. These have a very persistent effect and are not decomposed. They remain in the ecosystem indefinitely. Organophosphorus compounds, Parathion and Malathion, are short-acting and non-persistent but are highly toxic to human beings.

Features of organochlorine and phosphorus chemical pesticides are:
- they are rapid acting and some are instantaneous killers;
- they are easy to apply;
- they destroy all pests without selection, killing the harmful and harmless and the pest's natural enemies or predators;
- some pests have developed **heritable resistance** to pesticides. For example, DDT-resistant mosquitoes and houseflies and 'super' rats which are resistant to rodenticides;
- the ecosystem balance is upset by the destruction of ecological niches as a species is wiped out;
- pollution of food, water and air can occur;
- pesticides can be toxic to humans and some are suspected of causing cancer. The harmful effects of organochlorine pesticides are due to dioxins as impurities.

Pesticides in food chains

The persistent organochlorine pesticides which do not decompose in the soil have the property of accumulating in the fatty tissues of animals. Persistent pesticides are sprayed on crops from the air by aircraft. The pesticides are carried world-wide by spray drift and pollute fresh water and sea water.

Persistent pesticides will therefore accumulate along the food chain, reaching the highest concentrations in those predator organisms such as ospreys, herons, buzzards, golden eagles and cormorants, at the top of the **pyramids of biomass**. The accumulating persistent pesticides weaken the egg shells and kill the young birds before hatching.

29 Why is a pesticide that decomposes rapidly better than one that does not decompose?
30 What are the (a) advantages, (b) disadvantages in using chemicals to control pests?

15.17 Biology and the future

At the same time as modern humans have been damaging the world ecosystem they have also been gaining vast amounts of knowledge through scientific investigations. In this chapter, examples of how this knowledge can be used to repair the damage have been given. In the future, greater application of this knowledge will have to be made to secure the survival of all life on Earth.

Questions

31 Look in newspapers over the next few days for evidence of how scientific knowledge is being applied to improve conditions on the planet. Look at the issues which are raised and how problems are being solved. Assess their impact on the world's ecosystem.
32 Compare the work you have done in question 30 with the work you did for question 9. What issues are being dealt with now and which problems must modern humans still begin to solve?

Summary

- Modern humans have made a greater impact on the Earth than early humans. (▶ 369)
- Human beings are part of the global ecosystem. They encourage the growth of producers and act as consumers and decomposers. (▶ 371)
- Several factors have contributed to the size of the world human population. (▶ 372)
- There are two kinds of population growth: sustained and unsustained population growth. (▶ 373)
- The ages and sexes of people in a population can be presented as population pyramids. They can be used to identify expanding and stable populations. (▶ 374)
- The way the modern human population obtains its food has destroyed ecosystems and led to soil degradation. (▶ 376)
- Modern human activities have affected the carbon, nitrogen and water cycles. (▶ 377)
- Modern humans affect the flow of energy through the global ecosystem. (▶ 380)
- Sewage, nitrate fertiliser and phosphates in washing powders cause eutrophication. (▶ 381)
- Sulphur dioxide, nitrogen oxides, lead, carbon dioxide and CFCs are major air pollutants which can be controlled and managed. (▶ 382)
- Heavy metals, sewage, nitrates and heat are major water pollutants that can be controlled and managed. (▶ 383)
- Pesticides, refuse and artificial fertilisers are major land pollutants that can be controlled and managed. (▶ 384)

- Energy and stocks of materials can be conserved by recycling. (▶ 385)
- Biotechnology provides a wide range of useful substances. (▶ 385)
- In biological pest control the pest species becomes the prey in a prey–predator relationship or the host in a host–parasite relationship. (▶ 386).
- Chemical pest control is quick and easy to apply but has many disadvantages such as causing pollution, destroying the balance in ecosystems and threatening the health of humans. (▶ 387)

Glossary

An explanatory list of prefixes and suffixes which form biological terms

The names for biological functions, structures and interrelationships are derived from Latin (L) or Greek (Gk) words. A prefix is a group of letters at the *beginning* of a compound term or word; a suffix is a group of letters at the *end* of the word. As the majority come from Latin, only the Greek origin is shown as (Gk). For example, the muscle of the upper forearm is called the **triceps**; tri- (L) is a prefix meaning three, and -ceps (L) is a suffix meaning head, or the compound term means 'three-headed' muscle.

Prefixes

a- and	
an- (Gk)	'without', e.g. anaemia – without blood
adip-	'fat', e.g. adipose
aero-	'air', e.g. aerobic
aliment-	'food', e.g. alimentary
allelo-	'one another', e.g. alleles
amylo- (Gk)	'starch', e.g. amylase
angio- (Gk)	'container', e.g. angiosperm
annel-	'ring', e.g. annelida
anti- (Gk)	'against', e.g. antiseptic – against septic infection
arterio- (Gk)	'artery', e.g. arterial
arthro- (Gk)	'joint', e.g. arthropoda – jointed limbs
auri-	'ear', e.g. auricle
auto- (Gk)	'self', e.g. autotrophic, self-feeding
bi- (Gk)	'two', e.g. binomial – two names
bio- (Gk)	'life', e.g. biology
bronchi-	'wind-pipe', e.g. bronchitis – inflammation of windpipe
cani-	'dog', e.g. canine tooth
capill-	'hair', e.g. capillary vessel – hairlike tube
cardio- (Gk)	'heart', e.g. cardiac veins
carni-	'flesh', e.g. carnivore – flesh-eater
centi-	'a hundred', e.g. centipede – with 100 feet
cephalo-	'head', e.g. cephalothorax – combined head and thorax of spiders and crustaceans
chloro- (Gk)	'green', e.g. chlorophyll – green leaf
chromo- (Gk)	'coloured', e.g. chromosome – coloured body

cuti-	'skin', e.g. cuticle
cyto-	'hollow vessel' or cell, as in cytoplasm
dent-	'tooth', e.g. dentine
derm- (Gk)	'skin', e.g. dermis
dia- (Gk)	'across', e.g. diaphragm
di- (Gk) and **diplo-**	'two/twice', e.g. diploid
ecto-	'outside', e.g. ectoskeleton
endo-	'inside', e.g. endoskeleton
epi- (Gk)	'upon', e.g. epiglottis
eryth- (Gk)	'red', e.g. erythrocyte – red blood cell
gamet-	'marry', e.g. sexual gametes
gastr-	'stomach', e.g. gastric juice
glyco- (Gk) and **gluco-**	'sweet', e.g. glucose and glycerine
haemo- (Gk)	'blood', e.g. haemoglobin
haplo- (Gk)	'one/single', e.g. haploid
hepato- (Gk)	'liver', e.g. hepatic artery
herb-	'grass/herb', e.g. herbivore
hetero- (Gk)	'different', e.g. heterotrophic and heterozygous
homo- (Gk)	'same', e.g. homozygous
hyper- (Gk)	'above' or 'over', e.g. hyperthermia – overheated
hypo- (Gk)	'under' or 'below', e.g. hypothermia – underheated, and hypodermic
inter-	'between', e.g. intercellular
intra-	'inside', e.g. intracellular
karyo-	'cell nucleus', e.g. karyotype
kilo- (Gk)	'one thousand', e.g. kilogram
lacto-	'milk', e.g. lactic acid
leuco- (Gk)	'white', e.g. leucocytes – white blood cells
lipo- (Gk)	'fat', e.g. lipids
lympho-	'watery', e.g. lymph
meio- (Gk)	'lessening' or 'halving', e.g. meiosis
menstru-	'monthly', e.g. menstruation
micro-	'small', e.g. microorganisms
milli-	'one thousandth', e.g. milligram
mito- (Gk)	'thread', e.g. mitochondrion and mitosis
mono- (Gk)	'one', e.g. monocotyledon
multi-	'many', e.g. multicellular
nephro- (Gk)	'kidney', e.g. nephron
neucleo- and **nucle-**	'a little nut', e.g. nucleus
oestro-	'female', e.g. oestrogens
osmo- (Gk)	'push', e.g. osmosis
per-	'through', e.g. permeable
photo- (Gk)	'light', e.g. photosynthesis
pleuro- (Gk)	'rib chest', e.g. pleura
proto- (Gk)	'first', e.g. proteins – major nutrients
ren-	'kidney', e.g. renal vessels
rhizo- (Gk)	'root', e.g. mycorrhiza and rhizome
sapro- (Gk)	'decaying' or 'putrid', e.g. saprotrophic – eats dead matter
spermo- (Gk)	'seed', e.g. spermatozoa, or 'animal seed'
thermo- (Gk)	'heat', e.g. thermometer
thrombo- (Gk)	'blood clot', e.g. thrombosis
trans-	'across', e.g. translocation

uni-	'one', e.g. unicellular
ur-	'urine', e.g. urogenital, and urea
vacu-	'empty', e.g. vacuole
vas-	'vessel', e.g. vas deferens and vascular
vent-	'air movement', e.g. ventilation
xylo- (Gk)	'wood', e.g. xylem – wood vessel
zoo- (Gk)	'animal', e.g. zoology
zygo- (Gk)	'paired together', e.g. zygote

Suffixes

-aemia (Gk)	'blood', e.g. anaemia – no blood
-angium (Gk)	'container', e.g. sporangium
-ase (Gk)	used for enzyme names
-ceps	'head', e.g. biceps muscle
-cide	'killing', e.g. bactericide
-duct	'pipe', e.g. oviduct
-ennial	'yearly', e.g. biennial – two yearly
-lysis (Gk)	'setting free', e.g. catalysis and hydrolysis
-phyll (Gk)	'leaf', e.g. chlorophyll – green leaf
-phyta (Gk)	'plant', e.g. spermatophyta – seed plants
-rhoea (Gk)	'flow', e.g. diarrhoea – flow of faeces
-some (Gk)	'body', e.g. autosome – body cell, and chromosome – coloured body
-spire	'breathe', e.g. inspire, respire
-trophic (Gk)	'feeding', e.g. autotrophic – self-feeding
-tropic (Gk)	'turning direction', e.g. phototropic movement
-vore	'eating', e.g. herbivore – plant-eating
-zyme (Gk)	'fermentation' or 'leavening', e.g. enzyme

General index

Page numbers in **bold** type refer to definitions or principal descriptions.

Generic name index

Acer (sycamores and maples) 260
Aesculus (horse chestnut) 250
Agaricus (mushrooms) 51
Alopecurus (foxtail grass) 55
Amoeba 49, 50, 146, 157, 158, 169, 170, 220, 229, 246, 278, 279, 288
Anopheles (mosquito) 92
Aphis (greenfly) 92
Archaeopteryx (bird-like fossil) 329

Biston (moths) 336–7

Canis (dogs) 64, 65, 104, 199, 323
Cheiranthus (wallflower) 54
Chlamydomonas (green algae) 49, 50
Chorthippus (grasshoppers) 59, 285
Clupea (herring) 94
Columba (pigeon) 63
Cuscuta (dodder) 28

Drosophila (fruit flies) 278, 297, 302
Dryopteris (ferns) 53

Elodea (Canadian pondweed) 74
Epeira (spiders) 60
Equus (horses) 329

Felis (cats) 223
Funaria (moss) 52

Gallus (fowls) 227
Gasterosteus (stickleback) 61
Glossina (Tse-tse fly) 59

Helianthus (sunflowers) 284
Helix (snails) 56, 57
Hosta (variegated plant) 73
Hydra 157, 158, 247, 261–2

Ilex (holly) 78

Julus (millipede) 58

Lacerta (lizards) 62
Lumbricus (earthworm) 56, 57, 157, 221, 235, 262

Mirabilis (Marvel of Peru, four o'clock flowers) 312, 314, 315
Mucor (fungus) 248–9
Musca (housefly) 92, 93

Oryctolagus (rabbit) 64, 105, 199
Ovis (sheep and goats) 104

Pandalus (shrimp) 58
Paramecium 229
Pellia (liverwort) 52
Penicillium (mould) 51

Periplaneta (cockroaches) 284, 285
Pieris (butterflies) 92, 93, 286
Pinus sylvestris (conifer) 53
Pisum (peas) 305, 308–9, 310–11
Poa (meadow grass) 255
Prunus (cherries and plums) 260

Quercus (oaks) 260

Rana (frogs) 62, 263, 287
Ranunculus (buttercups) 114
Rhizobium (fungus) 248–9
Rosa (roses) 260

Saccharomyces (yeasts) 140, 247
Sagittaria (arrowheads) 322
Sambucus (elders) 289
Scyliorhinus (dogfish) 60, 61
Solanum (potatoes) 252, 322
Spirogyra (green algae) 49, 50

Taenia (tapeworms) 55, 56, 93, 139, 245, 247, 262
Trichophyton (skin, hair and nail fungi) 106
Tulipa (tulips) 251

Vicia (beans) 283, 284, 290

Zea (maize) 17, 18, 283